158
Structure and Bonding

Aims and Scope

The series *Structure and Bonding* publishes critical reviews on topics of research concerned with chemical structure and bonding. The scope of the series spans the entire Periodic Table and addresses structure and bonding issues associated with all of the elements. It also focuses attention on new and developing areas of modern structural and theoretical chemistry such as nanostructures, molecular electronics, designed molecular solids, surfaces, metal clusters and supramolecular structures. Physical and spectroscopic techniques used to determine, examine and model structures fall within the purview of *Structure and Bonding* to the extent that the focus is on the scientific results obtained and not on specialist information concerning the techniques themselves. Issues associated with the development of bonding models and generalizations that illuminate the reactivity pathways and rates of chemical processes are also relevant

The individual volumes in the series are thematic. The goal of each volume is to give the reader, whether at a university or in industry, a comprehensive overview of an area where new insights are emerging that are of interest to a larger scientific audience. Thus each review within the volume critically surveys one aspect of that topic and places it within the context of the volume as a whole. The most significant developments of the last 5 to 10 years should be presented using selected examples to illustrate the principles discussed. A description of the physical basis of the experimental techniques that have been used to provide the primary data may also be appropriate, if it has not been covered in detail elsewhere. The coverage need not be exhaustive in data, but should rather be conceptual, concentrating on the new principles being developed that will allow the reader, who is not a specialist in the area covered, to understand the data presented. Discussion of possible future research directions in the area is welcomed.

Review articles for the individual volumes are invited by the volume editors.

In references *Structure and Bonding* is abbreviated *Struct Bond* and is cited as a journal.

More information about this series at
http://www.springer.com/series/430

I. David Brown • Kenneth R. Poeppelmeier
Editors

Bond Valences

With contributions by

S. Adams • B.R. Bickmore • I.D. Brown • J.A. Enterkin •
F.C. Hawthorne • M.W. Lufaso • K.R. Poeppelmeier •
R.P. Rao • M. Schindler • P.M. Woodward

Springer

Editors
I. David Brown
Brockhouse Institute for Materials Research
McMaster University
Hamilton
Ontario
Canada

Kenneth R. Poeppelmeier
Department of Chemistry
Northwestern University
Evanston
Illinois
USA

Additional material to this book can be download from http://extras.springer.com

ISSN 0081-5993 ISSN 1616-8550 (electronic)
ISBN 978-3-642-54967-0 ISBN 978-3-642-54968-7 (eBook)
DOI 10.1007/978-3-642-54968-7
Springer Heidelberg New York Dordrecht London

Library of Congress Control Number: 2014941937

Printed on acid-free paper

Springer is part of Springer Science+Business Media (www.springer.com)

Preface

Bond valence theory has developed slowly over the last 100 years as a method of analyzing and validating the structures of inorganic materials. The various rules of the model have hitherto been strictly empirical, having grown out of the ionic model and the bonding rules proposed by Linus Pauling in the early part of the last century. The remarkable success of the model, however, suggests that it must reflect some underlying theory. Such a theory is developed in the first part of this volume. Classical electrostatics applied to a simple, but physically correct, picture of the atom provides a chemical bond definition that can be used to derive not only the theorems of the bond valence model but also the rules of the traditional ionic and covalent ball-and-stick models.

Recent developments have seen applications of bond valence theory extended from the simple validation of crystal structure determinations to the prediction of the structure and bonding geometry of complex materials, to the exploration of structure-related properties such as ionic conduction, and to the study of the chemistry of surfaces and interfaces. These topics are the subject of the contributions to the second part of this volume.

It is our hope in producing this volume that supplying a theoretical base for the bond valence model, and illustrating some of its more recent applications, will inspire a greater interest in, and appreciation of, the underlying concepts and applications of this powerful bonding model.

Ontario, Canada — I. David Brown
Illinois, USA — Kenneth R. Poeppelmeier

Contents

Struct Bond (2014) 158: 1–10
DOI: 10.1007/430_2013_134

Published online: 7 November 2013

Historical Introduction

I. David Brown

Abstract The bond valence theory grew out of Linus Pauling's electrostatic valence principle, but its development was slow until crystal structure determination became sufficiently accurate to make clear how the valence of a bond correlates with its length. Armed with this quantitative link with experiment, the theory has subsequently found many uses in analysing, modelling and predicting the structures of complex crystals, surfaces and liquids. Its theorems show how the physical properties of complex materials can be understood as the consequence of their chemical structure. The theory is increasingly finding new uses in solid-state chemistry and condensed matter physics.

Keywords Bond valence · History of bond valence · Pauling's electrostatic valence principle

Contents

I.D. Brown (✉)
Brockhouse Institute for Materials Research, McMaster University, Hamilton, ON, Canada L8S4M1
e-mail: idbrown@mcmaster.ca

1 The Origin and Development of the Bond Valence Theory

In 1929 Pauling [1] published a seminal paper entitled "The Principles Determining the Structures of Complex Crystals." He pointed out the weaknesses of the ionic model that had recently been proposed by Born and Landé [2], specifically that with the resources then at hand, it would be impossible to use the ionic model to obtain any useful understanding of the complex mineral structures that were being determined at the time. What he proposed was a series of five heuristic principles that were not to be regarded as rigorous laws, but rather signposts that could provide insight into the complex interactions that were responsible for the structures of crystals.

His first principle stated that the lengths of bonds can be determined by adding the ionic radii of the two terminal ions and that the coordination numbers of the cations can be determined from the ratio of these radii. More important was the second principle: the electrostatic valence principle, which he stated as:

> In a stable coordination structure the electric charge of each anion tends to compensate the strength of the electrostatic valence bonds reaching it from the cations at the centers of the polyhedron of which it forms a corner, that is, for each ion:

$$\zeta = \sum_i \{Z_i/\nu_i\} = \sum_i \{s_i\} \tag{1}$$

In these equations ζ is the valence of the anion, Z_i is the valence, and ν_i the coordination number of the ith cation, and s_i is the strength of the bond between the anion and cation i. This equation defines the Pauling bond strength as the valence of the cation divided by the number of bonds the cation forms.

The chemical community of the time was more focused on the structures of molecules than the structures of complex minerals, and Pauling's rules played only a minor role in the chemical curriculum, but they were adopted by mineralogists who found them a valuable tool for proposing and checking the validity of new mineral structures.

Pauling did not expect these principles to give quantitative predictions, since he recognized that the atomic radii, first proposed by Bragg [3] in 1920 to account for the lengths of bonds, could not be universally applied to ionic, covalent and metallic compounds. Furthermore, the distances between a the same pair of bonded atoms were not always equal to the sum of these radii, but depended in part on the coordination number of the atom, being larger when the coordination number was larger. Further, the electrostatic valence principle (1) was rarely obeyed exactly; deviations of up to 40% could sometimes be found between the anion valence and the sum of the Pauling bond strengths. Baur [4] drew attention to the quantitative correlation between the length of a bond and the degree of the apparent over- or underbonding around the anion. He showed that if the sum of the Pauling bonds strengths was larger than the anion valence, the measured bonds were longer than the sum of the radii, but if the sum was smaller, the bonds were shorter.

In 1947 Pauling [5] himself proposed that the bond length, R, was correlated with what he called the bond number, *n* (the number of electrons that formed the bond), as given by Eq. (2):

$$R_1 - R = 0.300\log(n) \tag{2}$$

where R_1 is the length of a bond with a bond number of 1.0. Curiously, this paper was concerned with the structures of metals which are not normally described using the ionic model, but the idea of a correlation between the length and strength (valence) of a bond (see [78] Section 6.1) was picked up in 1951 by Byström and Wilhelmi [6] to describe the variation in the bond lengths in $(NH_4)_2Cr_2O_7$ and V_2O_5. Later Zachariasen [7, 8] and Zachariasen and Plettinger [9] applied similar relations to uranates and borates, Evans [10] to vanadates and Kihlborg [11] to molybdates. In 1969 Clark et al. [12] expressed the length of the bond as a power series with three constants which they fitted to the cations found in the pyroxenes they were studying.

With the increased accuracy of the X-ray diffraction studies of crystals in the 1960s, it became possible to see how the variations in bond length correlated with other properties. In 1964 Slater [13] updated Bragg's atomic radii, and in 1969 Shannon and Prewitt [14] published a set of ionic radii that depended on coordination numbers as well as formal oxidation states. This table was updated by Shannon [15] in 1976. In 1991 O'Keeffe and Brese [16] proposed a set of atomic radii that could, with corrections for electronegativity, be applied to metals and homoionic as well as heteroionic bonds. Meanwhile in 1970 Donnay and Allmann [17] had used a bond valence–bond length correlation to locate hydrogen atoms in minerals. Even with the improved X-ray diffraction methods, it was not possible at that time to reliably detect the single electron associated with the hydrogen atom and crystal structure reports routinely left the hydrogen atoms out of the list of atomic coordinates, but Donnay and Allmann showed how, by calculating the experimental bond valence sums from the known bond lengths, the locations of the missing hydrogen atoms were readily revealed by the presence of underbonded oxygen atoms. They used a more complex function to describe the correlation. For the short bonds this followed the power law of (3)

$$S = S_0(< R > /R)^n \tag{3}$$

where S is the bond valence, S_0 the valence of the a bond with the average length $<R>$ and n is a constant fitted to each coordination sphere. The valence of longer bonds was given by a tangent to this curve to ensure that S extrapolated to zero at a distance R_{max} determined from the ionic radii. In 1973 Brown and Shannon [18] used a simplified form (4) of the Donnay and Allman correlation to cover the whole range of bond lengths:

$$S = (R/R_0)^{-n} \tag{4}$$

Here S_0 has been set to 1.0 valence unit (vu) making R_0 the notional length of a bond of unit valence. They point out that even though the bond valence never becomes zero in this formulation, it is sufficiently small at distances of around 3 Å that it matters little whether one includes longer distances or not. They fitted the bond valence parameters, R_0 and n, to give bond valences sums equal to the valence (formal oxidation state) of each cation and found that the bond valence parameters for a given cation-anion pair were transferable between the different compounds in which these bonds were found. They determined these parameters for bonds between oxygen and each of the cations in periods 2 to 4 of the periodic table. Three years later Brown and Wu [19] extended this list to the rest of the periodic table and in 1982 Slupecki and Brown [20] added parameters for bonds to sulfur. Brown and Shannon pointed out many of the uses to which bond valences (called "bond strengths" in their paper) could be put: calculated from the bond lengths they could reliably reproduce the formal atomic valences of both the cations and anions, even when the coordination environment was distorted. They could identify the oxidation states of the individual vanadium atoms in vanadium oxides and the proportions of different cations on mixed occupancy sites; they could also be used to determine the radii of ions in any coordination state, known or unknown. The agreement between the sum of the bond valences and the atomic valence could be used to confirm the correctness of a crystal structure or to locate the undetected hydrogen atoms. Brown subsequently applied the model to analyze the structural chemistry of lone pair cations [21], hydrogen bonds [22, 23], and acetates and trifluoroacetate ions [24], and O'Keeffe and collaborators applied it to alkali metal oxides [25] and nitrides [26].

About the same time, Mackay and Finney [27] suggested that the bond valences in a network of chemical bonds might be predicted using equations similar to the Kirchhoff equations used to solve electrical circuits see [78] Section 6.1. Brown [28] later demonstrated that predictions of bond distance could be made with an accuracy of better than 0.1 Å for most bonds using the valence sum rule together with an iterative method of equalizing the valences of the bonds, and later he [29] and O'Keeffe [30] both confirmed Mackay and Finney's conjecture by showing that a set of Kirchhoff-like network equation gave good bond length predictions. The Kirchhoff node rule corresponds to the valence sum rule [78 equation 14a] and the Kirchhoff loop rule corresponds to the equal valence rule [78 equation 14b] that insures that the valence of an atom is distributed as uniformly as possible among the bonds it forms. A decade later Preiser et al. [31] showed that each bond in the ionic model is formally an electric capacitor, allowing Brown [32, p. 19] to derive the network equations directly by treating the bond network as a capacitive electric circuit and assuming that when a structure is in equilibrium all the bonds have the same capacitance.

In 1979 in a paper entitled "Non-Existent Silicates" Dent-Glasser [33] deliberately did not ask the obvious question "why do certain silicate exist?", but instead asked the less obvious question "why do some silicates not exist?". Among the missing silicates were alkali metal orthosilicates and framework silicates containing transition metals. Dent–Glasser's question led Brown [34] in 1981 to formulate the valence matching rule [78 section 4]. He assumed that the typical valences of the bonds formed by an atom were measures of their bonding strength, and since every bond has an atom at each end, stable bonds would only be formed if both atoms had similar bonding strengths. This rule was implicit in Pauling's [1] electrostatic valence rule as well as in the "trivial topological constraint" pointed out by O'Keeffe and Hyde [35] in 1982, that the total number of bonds formed by the cations must equal the total number of bonds formed by the anions. Trivial as this constraint may appear, it is a powerful restriction that is closely related to the valence matching rule. Valence matching was quickly picked up by Hawthorne [36, 37] who applied it in an extensive series of papers in which bond valence concepts are used to explore the systematics of various classes of mineral structures as described in more detail by Hawthorne and Schindler in [79]. In particular Hawthorne et al. [38] showed how, by applying this principle at the level of individual atoms, one could predict the kinds of short-range order that might be expected in disordered systems.

Interest in bond valences picked up in 1985 after Brown and Altermatt [39] published an extensive list of bond valence parameters for use in calculating the valences of bonds to oxygen. Unlike the earlier work, they adopted Pauling's exponential correlation in the form of Eq. (5) partly because it is mathematically more versatile, but primarily because they found they could use the same b parameter (0.37 Å) for all the bonds they examined:

$$S = \exp((R_0 - R)/b) \tag{5}$$

Here S is the bond valence, R the bond length and R_0 and b are the fitted bond valence parameters. In 1991 Brese and O'Keeffe [40], assuming the same constant value for b, found that substituting a different anion, such as sulfur, for oxygen, changed the values of R_0 by a fixed amount that was independent of the cation. This allowed them to propose bond valence parameters for the bonds between a wide range of cations and anions. In two later papers, O'Keeffe and Brese [16, 41] proposed a method for generating R_0 from the properties of the two terminal atoms, but the parameters given by this procedure have proven not to be sufficiently accurate for routine structure analysis. On the other hand, the parameters listed by both Brown and Altermatt [39], as well as those listed by Brese and O'Keeffe [40], are now widely used, chiefly for validating newly determined crystal structures by checking how well the valence sum rule is obeyed. Over the last 20 years some 50 further papers have reported new or improved bond valence parameters, most of them having b set to 0.37 Å so that R_0 can be uniquely calculated for each coordination environment. Alternatives to Eq. (5) for describing the bond valence-bond length correlation have been proposed by Ziolkowski [42], Naskar et al. [43], Valach [44], and Mohri [45], but these have not found widespread favor. In a seminal paper in 2001, Adams [46] showed that increasing the coordination number around an atom, by increasing the bond cut-off distance, causes an increase in the value of b and a decrease in that of R_0, the values stabilizng only after the cut-off length reaches around 5 Å. He also showed that b should not be treated as a universal constant as he found it increases from 0.37 Å to around 0.5 Å as the difference between the softness of the two terminal atoms increases. Subsequent work by other authors [47–53] confirmed that the correct value of b for cations with lone pairs of electrons is close to 0.5 Å.

In 1989 O'Keeffe [30] drew attention to a number of structures in which the bond valence sums differed significantly from the expected atomic valence, an effect he attributed to the steric strain caused by nonbonded contacts [78 section 8]. Around the same time O'Keeffe and Hanson [54] and Brown [55] showed that such steric strains in the high-temperature superconductor $Ba_2YCu_3O_7$ were partially relieved by the transfer of electrons from a copper reservoir layer to the superconducting CuO_2 layer, explaining in part why the superconductivity varied with composition. Brown [56] followed this with a 1992 study on the chemical effects of lattice strain in crystals. If shifting valence from one bond to another cannot relieve the stress, its presence leads to violations of the valence sum rule. In 1995 Garcia-Muñoz and Rodriguez-Carvajál [57] showed that the instability caused by steric stress can be quantified using the global instability index, G, which is the root of the mean square deviation between the atomic valence and bond valence sum [78 Eq. 21]. They showed that in a series of isostructural rare earth compounds, those for which G would exceed 0.2 vu were sufficiently unstable to force them to adopt a different structure type. Since then G has proved a remarkably robust measure of structural instability and has been widely used particularly for compounds in which steric stress leads to interesting and sometimes technologically useful physical properties. In 2001 Lufaso and Woodward [58] wrote the program SPUDS to predict the crystal structures of perovkites by finding the structure with the smallest value of G as described in [80].

As early as 1978 Waltersson [59] had created the first valence map [78 section 6.2] to find the location of a lithium atom in Li_2WO_4 where the presence of the heavy tungsten atom masked the weak signal from lithium. Brown [29] later showed that such maps could be used to trace out ionic diffusion paths, an idea that has been exploited by Adams [60] and Adams and Swenson [61–65] in a series of papers exploring diffusion, not only in crystals but also in simulated amorphous structures. A program to calculate valence maps, VALMAP, has been written by Gonzáles-Platas et al. [66].

During the 1980s and 1990s the idea slowly developed that the bond valence could be treated as a vector, having a magnitude equal to the bond valence and a direction parallel to the bond (Chap. 2.6.2). It arose from the idea that if the bonds are uniformly distributed around an atom, the sum of the valence vectors should be close to zero. However, it was not stated as a rule until Harvey et al. [67] formulated it in 2006. Zachara [68] explored the concept in a slightly different form the following year. The vector valence sum of an atom is closely related to the slope of the valence map at the location of the atom, a zero sum indicating a minimum in the valence map. A recent study by Bickmore et al. [69] shows that in cases where the bonds are not uniformly distributed, e.g., when the central atom has a stereoactive lone pair [78 section 7.1], the valence vector sum provides a convenient measure of the deviation of the atomic environment from centrosymmetric symmetry.

Recently bond valence theory has been applied in several other areas. Structure simulations are normally carried out using Monte Carlo methods in which the positions of atoms in an array are changed in such a way as to minimize a cost function, typically representing the potential energy of the system. Since the simulation depends on the parameters chosen for the cost function, it is useful to check that the valences of the resulting bonds obey the valence sum rule, so that, if necessary, the parameters of the cost function can be adjusted [70, 71]. The reverse Monte Carlo method is used to simulate amorphous structures using a cost function based on the observed X-ray scattering pattern. Not all the solutions produced by this process are chemically plausible, so chemical restrictions must be included in the calculation. Bond valence restraints, which are ideal for this purpose, were used by Norberg et al. [72] in their reverse Monte Carlo program RMCProfile. So far no one has demonstrated that structures can be simulated using only the rules of bond valence theory though this should be possible. Such a simulation would have the advantage of using fewer computer resources while giving a complementary picture of the structure.

Studies of the surface between a solid and an aqueous solution have traditionally been based on a macroscopic picture in which the surface is assumed to be formed of alternating charged layers, but the introduction of bond valence into these studies has caused a slow evolution towards a microscopic model in which the surface is treated as composed of atoms with residual valence [73–75]. The misunderstandings and difficulties encountered in this transformation make an interesting case study [76]. The topic of surfaces is taken up in [79, 80, 81] of this volume.

In spite of the success of the bond valence theory, its justification has always been empirical; none of the theories of chemical bonding developed over the past century has suggested that there is any theoretical basis for the concept of a chemical bond, much less has it been possible to derive the empirical theorems that the bond valence appears to obey. This started to change in 1999 with the paper of Preiser et al. [31] who showed, following up on a much earlier idea of Bragg [77], that in the ionic model, the bond valence could be identified with the electrostatic flux linking the cations to the anions. This work showed how, by exploiting the electric field rather than the electric potential, the chemical bond could be defined and its properties derived, preparing the way for the full theoretical development of the model presented here in [78].

In recent years bond valence theory has been put to many other uses, some routine and others exploring new applications of the theory as described in other chapters of this volume. After a long period of slow development dating back almost a 100 years, the bond valence theory is beginning to find its place among the varied models used to describe chemical structure. Later chapters in this volume describe some of the exciting developments currently underway and point to the unexplored potential that will form the substance of some future history of the bond valence theory.

References

1. Pauling L (1929) J Am Chem Soc 51:1010–1026
2. Born M, Landé A (1918) Sitzungber Preuss Akad Wissen Berlin 45:1048–1068
3. Bragg WL (1920) Phil Mag Six Ser 40:169–189
4. Baur WH (1970) Trans Am Crystallogr Ass 6:129–155
5. Pauling L (1947) J Am Chem Soc 69:542–552
6. Byström A, Wilhelmi AK (1951) Acta Chem Scand 5:1003–1010
7. Zachariasen WH (1954) Acta Cryst 7:795–799
8. Zachariasen WH (1963) Acta Cryst 16:385–389
9. Zachariasen WH, Plettinger HA (1959) Acta Cryst 12:526–530
10. Evans HT Jr (1960) Z Kristallogr 114:257–277
11. Kihlborg L (1964) Arkiv Kemi 21:471–495
12. Clark JR, Appleman D, Papike J (1969) Miner Soc Am Spec Pap 2:31–50
13. Slater JC (1964) J Chem Phys 41:3199–3204
14. Shannon RD, Prewitt CT (1969) Acta Cryst B25:925–946
15. Shannon RD (1976) Acta Cryst A32:751–767
16. O'Keeffe M, Brese NE (1991) J Am Chem Soc 113:3226–3229
17. Donnay G, Allmann R (1970) Am Min 55:1003–1013
18. Brown ID, Shannon RD (1973) Acta Cryst A29:266–282
19. Brown ID, Wu KK (1976) Acta Cryst B32:1957–1959
20. Slupecki O, Brown ID (1982) Acta Cryst B38:1078–1079
21. Brown ID (1974) J Solid State Chem 11:214–233
22. Brown ID (1976) Acta Cryst A32:24–31
23. Brown ID (1976) Acta Cryst A32:786–792
24. Brown ID (1980) J Chem Soc Dalton Trans 1118–1123
25. McGuire NK, O'Keefe M (1984) J Solid State Chem 54:49–53
26. Brese NE, O'Keeffe M (1992) Struct Bond 79:307–378
27. Mackay AL, Finney JL (1973) J Appl Cryst 6:284–289
28. Brown ID (1977) Acta Cryst B33:1305–1310
29. Brown ID (1987) Phys Chem Miner 15:30–34
30. O'Keeffe M (1989) Struct Bond 71:161–190
31. Preiser C, Loesel J, Brown ID, Kunz M, Skowron A (1999) Acta Cryst B55:698–711
32. Brown ID (2002) The chemical bond in inorganic chemistry; the bond valence model. Oxford University Press, Oxford
33. Dent-Glasser L (1979) Z Kristallogr 149:291–325
34. Brown ID (1981) The bond valence method: an empirical approach to chemical structure and bonding. In: O'Keeffe M, Navrosky A (eds) Structure and bonding in crystals, vol II. Academic Press, New York, pp 1–30
35. O'Keeffe M, Hyde BG (1982) J Solid State Chem 44:24–31
36. Hawthorne FC (1985) Am Min 70:455–573
37. Hawthorne FC (1992) Z Kristallog 201:183–206
38. Hawthorne FC, Della Ventura G, Oberti R, Robert J-L, Iezzi G (2005) Canad Miner 43:1895–1920
39. Brown ID, Altermatt D (1985) Acta Cryst B41:244–247
40. Brese NE, O'Keeffe M (1991) Acta Cryst B47:192–197
41. O'Keeffe M, Brese NE (1992) Acta Cryst B48:152–154
42. Ziolkowski J (1985) J Solid State Chem 57:267–290
43. Naskar JP, Hati S, Datta D (1997) Acta Cryst B53:885–894
44. Valach F (1996) Polyhedron 116:699–706
45. Mohri F (2000) Acta Cryst B56:626–638
46. Adams S (2001) Acta Cryst B57:278–287
47. Krivovichev SV, Brown ID (2001) Z Kristallogr 216:245–247

48. Locock AJ, Burns PC (2004) Z Kristallogr 219:259–266
49. Hu S-Z (2007) Acta Phys-Chem Sin 23:786–789
50. Brown ID (2009) Acta Cryst B65:684–693
51. Mills SJ, Christy AG, Chen ECC, Raudsepp M (2009) Z Kristallogr 224:423–431
52. Sidey V (2010) Acta Cryst B66:307–314
53. Krivovichev SV (2012) Z Kristallogr 227:575–579
54. O'Keeffe M, Hanson S (1988) J Am Chem Soc 110:1506–1510
55. Brown ID (1989) J Solid State Chem 82:122–131
56. Brown ID (1992) Acta Cryst B48:553–572
57. Garcia-Muñoz JL, Rodriguez-Carvajál J (1995) J Solid State Chem 115:324–331
58. Lufaso MW, Woodward PM (2001) Acta Cryst B57:725–738
59. Waltersson K (1978) Acta Cryst A34:901–905
60. Adams S (2000) Solid State Ion 136–137:1351–1361
61. Adams S, Swenson J (2000) Phys Rev Lett 84:4144–4147
62. Adams S, Swenson J (2000) Phys Rev B63:054201
63. Adams S, Swenson J (2001) Phys Rev B64:023204
64. Adams S, Swenson J (2002) J Solid State Ionics 154(5):151–159
65. Adams S, Swenson J (2005) J Phys Condensed Matter 17:S87–S101
66. Gonzáles-Platas J, Gonzáles-Silgo C, Ruiz-Péres C (1999) J Appl Cryst 32:341–344
67. Harvey MA, Baggio S, Baggio R (2006) Acta Cryst B62:1038–1042
68. Zachara J (2007) Inorg Chem 46:9760–9767
69. Bickmore BR, Wander MFC, Edwards J, Maurer J, Shepherd K, Meyer E, Johanson, WJ Frank RA, Andros C, David M (2013) Am. Min. 98:340–349
70. Rossano S, Farges F, Ramos A, Delaye J-M, Brown GE Jr (2002) J Non Cryst Solids 204:167–173
71. Bickmore BR, Rosso KM, Brown ID, Kerisit SJ (2009) J Phys Chem A 113:1847–1857
72. Norberg ST, Tucker MG, Hull S (2009) J Appl Cryst 2:179–184
73. Bickmore BR, Rosso KM, Nagy KL, Cygan RT, Tadanier CJ (2003) Clays Clay Miner 51:359–371
74. Bickmore BR, Tadanier CJ, Rosso KM, Monn WD, Eggett DL (2004) Geochim Cosmochim Acta 68:2025–2042
75. Bickmore BR, Rosso KM, Tadanier CJ, Bylaska EJ, Doud D (2006) Geochim Cosmochim Acta 70:4057–4071
76. Brown ID (2009) Chem Rev 109:6858–6919
77. Bragg WL (1930) Z Kristallogr 74:237–305
78. Brown ID (2013) paper 'Bond Valence Theory' in this S&B volume
79. Hawthorne FC, Schindler M (2013) 'Crystallization and dissociation in aqueous solution: a bond valence approach' in this S&B volume
80. Lufaso MW, Woodward PM (2013) 'Using bond valences to model the structures of ternary and quaternary oxides' in the current S&B volume
81. Bickmore BR (2013) 'Structure and acidity in aqueous solutions and oxide-water interfaces'
82. Poeppelmeier K, Enterkin J (2013) 'Bonding at Oxide Surfaces' both in this S&B volume

Struct Bond (2014) 158: 11–58
DOI: 10.1007/430_2012_89

Published online: 19 April 2013

Bond Valence Theory

I. David Brown

Abstract This chapter shows that a unified concept of a chemical bond can be derived from a theoretical picture of the atom in which the Coulomb forces are described using the electric field rather than the electric potential. The localized chemical bond and its valence arises naturally from this picture, allowing the theorems of electrostatics to be used to describe the formation and properties of any chemical structure composed of localized bonds. There is no distinction made between ionic and covalent bonds. An empirical correlation links the theoretical bond valence to the experimental bond length. The resulting picture of chemical structure predicts where bonds will form, how long they will be, and in what direction they will point. It indicates the conditions for chemical stability, suggesting which reactions a compound might undergo either in solution or at a surface. Electronic anisotropies are handled in an ad hoc manner, in which the VSEPR theory of lone pairs is extended to cases where the lone pairs are inactive or only partially stereoactive. Steric constraints leading to compressed or stretched bonds are quantified by observing the difference between the real and theoretical structures. The potential of the bond valence theory is only beginning to be exploited.

Keywords Bond · Bond valence · Bond valence theory · Bonding geometry · Bonding strength · Lewis acids and base strengths · Lone pairs · Steric effects · Structure prediction

---

I.D. Brown (✉)
Brockhouse Institute for Materials Research, McMaster University, Hamilton, ON, Canada L8S 4M1
e-mail: idbrown@mcmaster.ca

Contents

1 Introduction

The first chapter in this volume [1] describes how the bond valence theory developed over the last hundred years as an empirical research tool for analyzing inorganic crystal structures. The justification for the theory has always lain in its experimental success rather than its theoretical underpinning. This chapter attempts to rectify this lack of a theoretical basis. It starts with a simple physical picture of the atom, the core-and-valence-shell picture, from which the rules of bond valence theory are derived using the concepts of classical physics. Although a full quantum analysis is clearly essential to an understanding of the structure of the atom and its spectrum, equilibrium chemical structures involve only the ground electronic state, which greatly simplifies the theory, allowing a development in terms of classical electrostatics in which quantum effects are introduced as needed on an ad hoc basis. Apart from the obvious desirability of having a physical picture of bond valence, putting the model on a sound theoretical basis has two particular advantages. In the first place, it allows one to appreciate the scope of the theory and to understand better how it might be extended. In the second place, a simple theory of chemical structure based on a proper physical description of bonding is ideal for introducing students to structural chemistry, a topic taken up in the Chap. 9 of this volume [2].

This chapter starts with a description of the core-and-valence-shell picture of the atom, followed by a derivation of the principal theorems of bond valence theory. The later sections of the chapter illustrate the ways in which these theorems are applied over a range of chemical bonds and structures.

2 Core-and-Valence-Shell Picture of the Atom

The concept of a chemical bond as a localized interaction between two neighboring atoms has been a central part of chemistry for the past century and a half, yet our current description of chemical bonds is still empirical; it is a collage of ill-defined and largely incompatible models that are based on assumptions that do not always correspond to physical reality. The ionic and covalent models are mutually incompatible, and both the Lewis and orbital models have serious flaws [3, 4]. They do not conform to modern views of atomic structure, and consequently their predictions sometimes fail. While the bond valence theory belongs to this tradition of localized bond models, it is derived from a realistic, though simplified picture of the atom, one that is compatible with more sophisticated atomic descriptions. It can be used to derive powerful and quantitative theorems about chemical structure. The rules of both the traditional ionic and covalent models can be derived as two special cases of this model (Sects. 5 and 7.2).

What most clearly distinguishes bond models from other models of chemical structure is the centrality of the concept of valence, that is, the number of electrons that an atom uses for bonding. In order to determine the valence, we must be able to count how many valence electrons each atom contains. This is not possible in electron density models such as the quantum theory of atoms in molecules (QTAIM), where the individuality of the electron is lost as soon as it enters the atom [5]. A different picture of the atom is needed, one that tracks the *functions* of the individual electrons, rather than their *locations*.

A second distinguishing feature of bond models is their treatment of the chemical bond as a localized interaction between two neighboring atoms; everything beyond the first neighbor shell is ignored. The long-range Coulomb potential, which is a central component of most other bonding models, extends well beyond the nearest neighbors, which is why the concept of a chemical bond has never been derived from models expressed in terms of the Coulomb potential, whether treated using Newtonian or quantum mechanics. Because the bonds are local, the number of bonds formed by an atom, its coordination number, is also an important quantity in bond models.

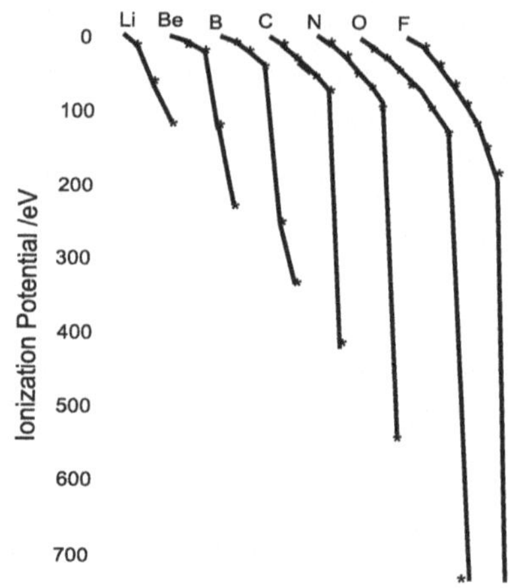

Fig. 1 Ionization potentials of elements from the second period. Successive ionization potentials are displaced to the right

These two distinguishing features of chemical bond models, namely, the atomic valence and the bond as an interaction between neighboring atoms, point to the need for a different way of describing both the structure of the atom and the force responsible for the bond. The emphasis on the valence of an atom requires a model in which the valence electrons can be conceptually separated from the remaining (core) electrons, and the emphasis on localization requires that the long-range Coulomb potential be replaced by the electrostatic field which provides an equivalent, but complementary, description, one in which neighboring charges are linked by electrostatic flux as first suggested by Pauling [6] and Bragg [7]. In any locally neutral distribution of positive and negative charges such as are found in molecules and crystals, the Faraday lines of field, which provide an intuitive picture of this flux, always connect positive charges to their neighboring negative charges, and since the electrostatic flux (represented by the number of lines of field) is equal to the charges (valence electrons) that generate them, the concepts of valence and electric flux fuse into a picture which is both intuitive and quantitative. It is a picture from which the theorems of the bond valence theory naturally follow. This theory is completely equivalent to the traditional energy-based classical theories, but unlike the traditional models which require that all the long-range interactions be treated explicitly, valence-and-flux models are conceptually and computationally simpler because all the interactions are local, and the long-range relaxations are implicit rather than explicit.

The justification for separating the valence electrons from the core is based on the observed ionization potentials: the energies required to successively remove electrons from an atom. The large increase in ionization energy after the first few electrons have been removed allows us to separate the weakly bound electrons ($\lesssim$100 eV), namely the valence electrons, from the core electrons that are too strongly bound to the atomic nucleus to be able to take part in bonding (Fig. 1). The result is the core-and-valence-shell picture of an atom, one in which the valence electrons are conceptually distinguished from the core electrons.

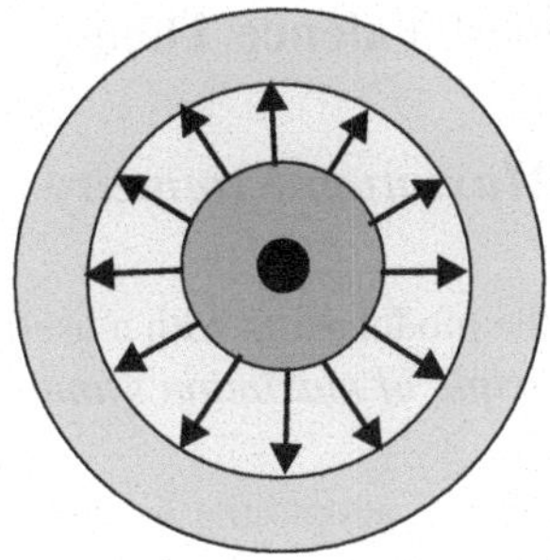

Fig. 2 The core-and-valence-shell picture of an atom. The core is shown as *dark gray*, the valence shell in *light gray*, and the nucleus is shown in *black*. The number of lines of field (electrostatic flux) is proportional to the negative charge on the valence shell as well as the positive charge on the core

The core-and-valence-shell picture of an atom is illustrated in Fig. 2. The electrons are separated into those in the valence shell (light gray) and those that remain strongly bound in the core (dark gray). Because all the electrons are held by a strong central attraction to the nucleus (black), both the core and the valence shell remain essentially spherically symmetric, even when the atom is bonded to other atoms. The justification for this assumption is that the relaxation of the electron density on bond formation is small enough (~1%) to be ignored in this approximation. In the bond valence theory, atoms are always assumed to be electrically neutral unless otherwise stated, meaning that the core and the valence shell are not only spherically symmetric but they also carry equal and opposite charges. The small ionization energy of the electrons in the valence shell implies that this shell lies close to the surface of the atom and carries a negative charge equal to the number of electrons in the valence shell. The atomic core contains both the nucleus and the core electrons and carries a net positive charge of the same magnitude. The relative location of the valence shell and core is not relevant in this picture since the electrostatic flux linking the core to the valence shell depends only on the number of electron charges it links, not on their location. It is convenient to display them as spherical in the picture shown in Fig. 2 as this is similar to the physical distribution of electrons in the atom. The picture would be even more similar to the real atom if the core and valence shell were shown as overlapping each other, but it is more convenient to visualize them separated by a gap, since this provides room to display the lines of electrostatic field that link them.

This model explicitly does not give a picture of the true electron density since the physical location of the electrons is not relevant to the model and, indeed, cannot be derived from the model. Superimposing the valence shell and core does not yield a true physical picture of the atom because the purpose of the model is not to reproduce the true electron density, but rather to keep track of the roles played by the valence shell and core electrons. The model's validity does not depend on its ability to predict the electron density, which it is neither intended, nor is it able to do, but on its ability to predict the bonding structures in crystals and molecules. In this respect it performs at least as well as any other model and in many respects better.

3 Theorems of the Bond Valence Theory

3.1 The Principle of Maximum Symmetry

The formal development of the model starts with a useful heuristic assumption that underlies the model: the ***principle of maximum symmetry***:

> A system in stable equilibrium adopts the highest symmetry consistent with the constraints acting on it.

(1)

The justification for this statement is that a symmetric system is always at an energy extremum with respect to any distortion that lowers its symmetry, and that the system is stable only if this extremum is a minimum. We need no further justification to explain why a compound has high symmetry. If the symmetry is lower than expected, there must be a constraint that causes the symmetry to be lost. Important constraints are those imposed by the rules of chemistry (chemical constraints) and those imposed by requirements of three dimensional space (steric constraints).

There are two atomic properties that are central to the bond valence theory: the **valence of the atom**, that is, the number of electrons the atom uses for bonding and the size of the atom, conveniently represented in bond valence theory by the atom's **coordination number**, that is, the number of bonds that it forms, as discussed in more detail below.[1]

The theory assumes that only the valence-shell electrons are used for bonding because these lie close to the surface of the atom where they overlap with the valence shells of other atoms on bond formation. For elements in periods two and three of the Periodic Table, the concept of a valence shell is well defined because of the large difference in ionization energy between the electrons of the valence shell and those of the core. For elements in higher periods, the gap is less well defined and the concept of a valence shell becomes more problematic, particularly for the transition metals discussed in Sect. 7.4.

3.2 Assumptions and Theorems

We start with a formal statement of the assumptions that underlie the model. Giving these explicitly not only serves to introduce the model but also to define the range of compounds that the model describes.

---

[1] The terms used in this theory are shown in bold type and are defined in the Glossary (Appendix 1 in this volume).

1. All atoms are assumed to be *spherically symmetric*, even when bonded to other atoms. The rational for this assumption is that the relaxation of the electron density on bond formation is limited to a few percent. While this relaxation may be important for understanding the nature of the chemical bond and calculating its energy, it has little effect on the resulting chemical structure. The relaxation of the electron density can be ignored in modeling these structures, and the assumption of spherical symmetry places some useful constraints on the structural properties of the atoms.
2. In bond valence theory, atoms (with one exception) are always treated as *uncharged*. A bond is formed when the valence shells of two neutral atoms overlap. The electrons from the two atoms spin-pair, but they are still counted as being part of their original spherically symmetric valence shells. This assumption simplifies the description of the bond since it avoids using such elusive concepts as electron transfer, ionic character, and atomic charge. The only exception to this assumption is the ionic model which is derived from the bond valence theory in Sect. 5. Even in this case the transfer of electrons from the cation to the anion is a matter of formal bookkeeping; it does not imply any physical movement of the electrons.
3. Since a bond is formed by the pairing of electron density in the region where the valence-shells of two atoms overlap, each atom is assumed to contribute the *same number of valence electrons* to the bond. This number is known as the **bond valence**. The bond valence is therefore also equal to the number of electron pairs that constitute the bond. In general the bond valence is not an integer.
4. It is assumed that there is a large difference in the *ionization energy* between the valence shell electrons and those in the core. This allows us to ignore the core electrons and focus on those in the valence shell. This assumption holds for light main group elements, but is questionable for transition metals (Sect. 7.4) and heavy main group elements. The advantage of this assumption is that it allows us to identify clearly which atoms have well-defined valences and which atoms may not. Being able to identify why the rules may not work in a given compound makes it easier to suggest how the model might be adapted.
5. It is also assumed that all the *excited states* of the atom lie too far above the ground state to be involved in bonding. This again is valid for light main group elements, but like assumption 4 is not always valid for transition metals and heavy main group elements. Again it identifies where the model might not work and why.
6. The valence shell may contain both *bonding and nonbonding (lone pair) electrons*. The electron density of the valence shell always remains spherically symmetric (see assumption 1 above), but the way the electrons in the shell function, being either bonding or nonbonding, may result in different parts of the valence shell functioning in different ways. If the bonding and nonbonding functions are not uniformly distributed, the lone pair is said to be *stereoactive*. The properties of atoms with lone pairs are described in Sect. 7.1.

7. The bond is assumed to be *localized* between the two bonded atoms. This excludes compounds with delocalized bonding such as metals and aromatic compounds. Delocalization occurs when the valence shells of the different atoms meld into an extended band so the valence electrons are no longer constrained to remain on their own atom. Recognizing this assumption may suggest ways of extending the model to delocalized systems.

As stated above, the valence of an atom is defined as the number of electrons it uses for bonding. The valence of a bond is defined as the number of valence electrons that an atom contributes to a particular bond. From these definitions, the *valence sum rule* immediately follows.

> The sum of the valences of all the bonds formed by an atom is equal to the valence of the atom.

(2)

From assumption 3 above, each of the bonded atoms contributes an equal number of electrons to the bond. This is known as the *Equal Contribution Rule*:

> Each of the two terminal atoms contributes the same number of electrons to the bond that links them. This number is therefore equal to the number of electron pairs that form the bond and so is equal to the valence of the bond. It is notrestricted to integers because the electron density does not consist of identifiable individual electrons.

(3)

Since the flux is equal to the charges it links, the flux terminating on the electrons contributed to the bond is also equal to the valence of the bond (see Fig. 2).

It is apparent that the closer two atoms approach each other, the greater the number of electrons that lie in the **overlap region**, hence shorter bonds are expected to have larger bond valences. There is no simple way to calculate this relationship, but it can be determined empirically by comparing the predicted valence of a bond with its measured length. Even though the individual bond valences are not normally known a priori, the valences of the atoms that form the bonds are known. It is possible to determine the bond valence–bond length correlation by insuring that the bond valences determined from observed bond lengths add up to the valences of the atoms at both ends of the bond. These correlations are empirically found to be transferrable between all the bonds of the same type, i.e., all the bonds that have the same two terminal atoms in the same valence state. It is this correlation that links the theory to experiment. The problem of determining these correlations is discussed more fully in this volume by Adams [8].

The existence of a correlation between the theoretically determined bond valences and the experimentally observed bond lengths is what validates the model. Without this link, the abstract theorems of the bond valence theory developed below would have no relevance to the real world. The bond valence theory is

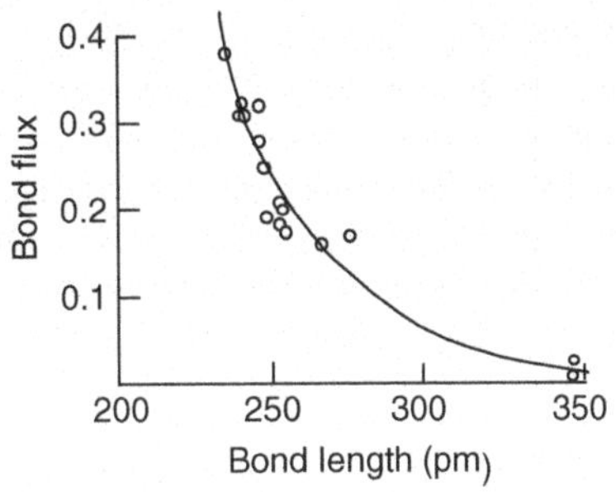

Fig. 3 The correlation between bond valence (bond flux) and bond length for Ca–O bonds. The *small circles* show the bond fluxes calculated for particular compounds (Fig. 3.1 from [11] by permission of the Oxford University Press)

justified by noting that bond valences calculated from the observed bond lengths obey the theorems of the model.

The difficulty in calculating this correlation from quantum theory is not the only reason for determining it empirically. An empirical determination automatically takes into account all the factors that affect the bond length. For example, strong bonds are better able than weak bonds to draw the ligands closer to each other, but in turn the resulting repulsion between the ligands tends to stretch the bonds. With an empirically determined correlation such stretching is automatically taken into account.

Although there is no theory that predicts the algebraic form of this correlation, over a limited range of bond lengths the simple two parameter equation (4) works well [9].

$$S_{ij} = \exp\left(\left(R_0 - R_{ij}\right)/b\right) \tag{4}$$

where R_{ij} is the length and S_{ij} the valence of the bond between atoms i and j, and R_0 and b are the empirically determined **bond valence parameters**. Values for these parameters compiled from various sources can be found at reference [10]; their determination is discussed by Adams [8]. An example of this correlation is shown in Fig. 3.

The form of Eq. (4) gives rise to an important theorem, *the distortion theorem* (5) [12, 13]:

> If some of the bonds formed by an atom are lengthened and others shortened, their bond valence sum will increase if the average bond length is held fixed, or alternatively, the average bond length will increase if the bond valence sum is held fixed.

(5)

If an atom lies at the center of a cavity that is too large, the bonds will all be longer than expected, and the atom will appear to have a valence sum that is too low. By moving the atom away from the center of the cavity some of the bonds become shorter and others longer, but the valence of the shorter bonds increases more than the valence of the longer bonds decreases, leading to an increase in the bond valence sum. In cases where an atom occupies a large cavity, this mechanism is found to increase the bond valence sum to match the atomic valence while keeping the average bond length constant. This is effected either by the atom being displaced from the center of the cavity, or by the ligands moving to ensure that the bonds have different lengths. This theorem is central to the work described by Lufaso and Woodward in [14].

4 Bonding Strength and Valence Matching

The valence sum rule is not sufficient by itself to determine the valences of individual bonds, but the principle of maximum symmetry (1) implies that the bonds formed by an atom will have similar valences. In this case the valences of all the bonds in the coordination sphere are expected to be close to their average which can be calculated using Eq. (6) provided we know the number of bonds the atom forms.

$$S_{\text{average}} = V/N \tag{6}$$

where S is the valence of a bond, V is the valence of the atom, and N the number of bonds the atom forms (its coordination number).

The coordination number depends primarily on the relative sizes of the atom and ligand, but may be affected by other factors. In most cases, the coordination numbers observed for a given **atom type** (an element in a given valence state) tend to cluster around their average value. For example when sodium is bonded to oxygen, all coordination numbers between 3 and 12 are known, but 80% of the sodium atoms have coordination numbers between 4 and 7 with an average of 6.4, giving $S_{\text{average}} = 0.16$ **valence units** (vu).

If the value of N is known or can be inferred from other considerations, this value should be used in Eq. (6). If N is not known, a standard value, N_{O}, the average of the observed coordination numbers with oxygen ligands, can be used as a best guess instead, as shown in Eq. (7) [15]. The resulting value of S_{O} is known as the **bonding strength of the atom**, since, in the absence of any better information, the bonds formed by the atom are expected to have valences close to S_{O}.

$$S_{\text{O}} = V/N_{\text{O}} \tag{7}$$

The bonding strengths of many elements in the Periodic Table are shown in Tables 1–3. The bonding strength, which is characteristic of each atom type, has two important uses. Firstly, the bonding strengths, S_{E}, of the different elements when in their highest valence state, V_{max}, i.e., when all the electrons in the

Table 1 The Periodic Table with electronegativities and cation bonding strengths (in valence units, vu) to oxygen

1	2	3–12	13	14	15	16	17	18
H[b]								
0.8								
0.2								
Li	Be		B	C	N	O	F	Ne
0.20	0.50		0.87	1.35	167	(2.0)	(2.3)	
Na	Mg		Al	Si	P	S	Cl	Ar
0.16	0.33		0.57	1.00	1.25	(1.6)	1.75	
K	Ca	Se–Zn	G$_A$	Ge	As	Se	Br	Kr
0.13	0.27		0.65	0.89	1.13	(1.5)	(1.5)	
Rb	Sr	Y–Cd	In	Sn[a]	Sb[a]	Te[a]	I[a]	Xe[a]
0.12	0.23		0.50	0.68	0.85	1.0	(1.2)	(1.3)
Cs	Ba	La–Hg	Tl[a]	Pb[a]				
0.11	0.20		0.49	0.70				

Bonding strengths are calculated using Eq. (7) (except those in parentheses which are estimated)
[a]These cations are also found in lower oxidation states with one or more lone pairs
[b]Hydrogen has two bonding strengths (see Sect. 8.1.1)

Table 2 Simple anion bonding strengths (S_B)

Col #	15	16	17
Valence	−3	−2	−1
	N	O	F
	−0.75	−0.5	−0.25
	P	S	Cl
			−0.17
		Se	Br
			(−0.10)
		Te	I
			(−0.08)

Anion bonding strengths are based on coordination numbers of 4 for N, O and F, and 6 for Cl

valence shell are bonding, follow the same ordering as their electronegativities as expressed in Eq. (8).

$$S_E = V_{max}/N_O \tag{8}$$

S_E provides a convenient scale of **electronegativity** within the bond valence theory, because even though it differs numerically from the traditional electronegativity scale, it is derived from the basic parameters of bond valence theory, namely the valence and coordination number. It is the scale used in this chapter.

Secondly the definition of bonding strength in Eq. (7) leads to an important rule for predicting structures. The equal contribution rule (3) states that the two bonded atoms should contribute the same valence to the bond, which means that a bond will only be formed if both atoms have the same bonding strength. Some latitude is

Table 3 Complex ion bonding strengths

	Residual valence	S_A or S_B
NH_4^+	1	0.12
ClO_4	−1	−0.1
BF_4	−1	−0.1
NO_3	−1	−0.11
HCO_3	−1	−0.17
SO_4	−2	−0.17
H_2PO_4	−1	−0.18
H_2O	0	−0.2
HPO_4	−2	−0.22
CO_3	−2	−0.22
PO_4	−3	−0.25
BO_3	−3	−0.33
SiO_4	−4	−0.33
OH	−1	−0.4

Anion bonding strengths are based on coordination numbers of 4 for N, O and F

allowed by atoms adopting different coordination numbers, but it is found that stable bonds rarely form between atoms whose bonding strengths in Tables 1, 2, and 3 differ by more than a factor of 2. This is expressed in the *valence matching rule* (9):

$$\text{Stable bonds will normally form only if } 0.5 < S_A/S_B < 2.0 \qquad (9)$$

where S_A and S_B are the bonding strengths of the two bonded atoms, A and B. If the coordination number is known or can be reasonably guessed, then a better estimate of the bonding strength can be made. The bonding strength can therefore be influenced by local conditions, but it tends to lie close to the value given by Eq. (7). There are exceptions to the valence matching rule. As described in Sect. 7.1, the presence of a lone pair in the valence shell of one or both atoms provides a flexibility that removes the upper limit in the expression (9) with interesting consequences for the coordination geometry.[2]

If one knows which atoms are present in a compound, the inequality (9) can be used to determine which atoms will bond to each other, information that can be used to generate the **bond network**, as described in Sect. 11.2.

[2] Coordination numbers depend on the nature of both the bonded atoms and are sometimes different even between the same pair of atom types in different compounds. S_O is taken arbitrarily to be an atomic property, but if the bonded atom is not oxygen, a different value of N might be more appropriate. For example, carbon has a coordination number of three with oxygen, but four with hydrogen. Reducing the coordination number increases the bonding strength and this is the usual mechanism by which the bond valences of two bonded atoms can be made exactly equal.

5 Valence Compounds and the Ionic Model

Most bonds are formed between atoms having different electronegativities. It is convenient to call the atom having the lower electronegativity the **cation** (or A, for Lewis Acid) and the atom having the higher electronegativity the **anion** (or B, for Lewis Base). If both atoms have the same electronegativity, the choice of which is the cation and which the anion is arbitrary; either can be used. Although in traditional models cations are assumed to carry a positive charge and anions a negative charge, in bond valence theory all atoms, including cations and anions, remain electrically neutral unless otherwise stated. The terms "cation" and "anion" are used only as labels to identify which of the two atoms forming a bond has the smaller, and which the larger, electronegativity.

> A **valence compound** is defined as one in which every atom is uniquely labeled as either a cation or an anion, and every bond has a cation at one end and an anion at the other, i.e., there are no bonds between two cations or between two anions.

(10)

A bond network with this property is said to have a **bipartite graph**, and a corollary to this definition is that the bond network of a valence compound contains only even-membered rings since an odd-membered ring necessarily contains a **homoionic bond**.

The valence shell of the cation carries a small charge and is linked to its core by a weak electric field. Consequently it lies far from the nucleus. On the other hand, the valence shell of the anion carries a large charge and is held closer to the nucleus. As a result, the bond overlap between the cation and the anion occurs closer to the anion as shown in Fig. 4. Quantum mechanics places restrictions on the number of electrons that can be accommodated in the overlap region, and while the nature of these restrictions is complex, it is conveniently summarized by the *octet rule* (11) which states that[3]:

---

[3] When the valence shells of two atoms overlap, they split into a low-energy bonding level localized in the overlap region between the atoms and a high-energy antibonding level localized behind the atoms. Because the overlap region is closer to the anion, the bonding level has more of the character of the valence shell of the anion and the antibonding level more the character of the cation. If more electrons are available for bonding than can fit into the anion-like bonding level, they must necessarily occupy the antibonding level, tending to destabilize the bond. The most stable bond is formed when the bonding level is full and the antibonding level is empty, a condition that is expressed by the octet rule. When both the bonding and antibonding levels are full, there is no bonding advantage in overlapping the valence shells, which explains the inertness of the noble gases. The repulsion that prevents the atoms from merging is provided by the electrostatic repulsion that occurs when their cores overlap (the overlap in this case providing no bonding advantage) as well as by the electrostatic repulsion between the two nuclei.

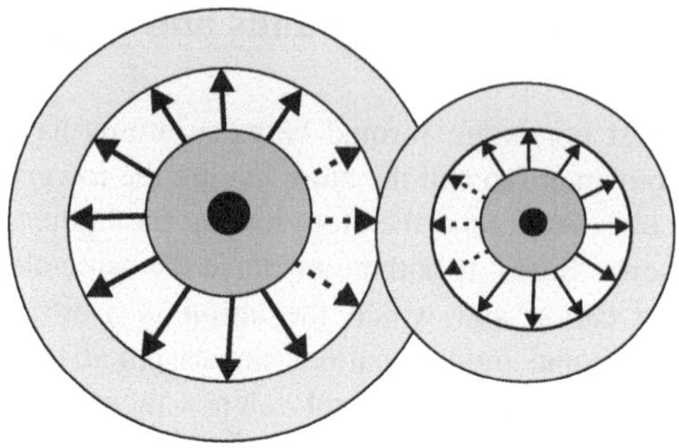

Fig. 4 The overlap region of the bond between a cation (*left*) and an anion (*right*) occurs closer to the anion. The *dotted field lines* are those that contribute to the bond

All the valence electrons of the cations must be accommodated within the valence shells of the anions. (11)

Since for many anions, particularly those in periods two and three of the Periodic Table, the valence shell is capable of holding only eight electrons; the rule (11) is known as the octet rule. The neutral anion normally has a valence shell that is more than half filled, limiting the amount of space available for the electrons of the cations. Since the equal contribution rule (3) limits the anion and cation to contributing the same number of electrons to the bond, the anion is not able to use all its valence shell electrons for bonding. For example, oxygen has six electrons and two vacancies in its valence shell. As it can accommodate only two valence electrons from the cations, it can use only two of its own valence shell electrons for bonding. The valence of oxygen when acting as an anion is only two; the remaining four electrons are nonbonding and form two nonbonding lone pairs. Thus when an atom acts as an anion it has a smaller valence than when acts as a cation, and by convention, anion valences are taken as negative as shown in Table 2 [16]. The role played by the nonbonding electrons is discussed in Sect. 7.1.

For valence compounds, the bond valence theory can be converted to the **ionic model** simply by formally reassigning the cation valence electrons to the valence shells of the anions that already accommodate them. This does not require that these electrons be physically moved from the cation to the anion, since the anion and cation-bonding electrons already occupy the same overlap region, and in any case the core-and-valence-shell model contains no information about the actual location of the electrons. Bond valence theory is only a means of keeping track of the way the electrons function; reassigning the cation valence electrons to the anion is no more than a book-keeping operation. It makes no difference to the physical description of the bond. The advantage of this approach is that it changes the way we view the bond by eliminating the need to speculate on the precise location of the bonding electrons.[4]

[4] The core-and-valence-shell diagrams, used here to illustrate the formation of bonds, are purely schematic. The pairing of electron densities that forms the bond occurs at some place where the electron density of the two atoms overlaps, but its location depends on how the atoms are defined and in any case cannot be identified experimentally.

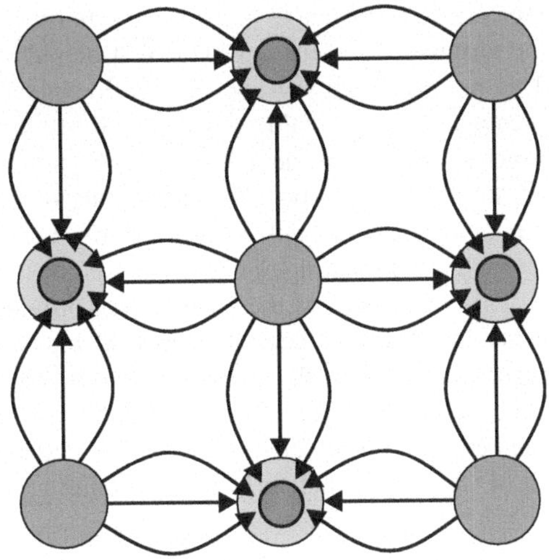

Fig. 5 A schematic picture showing how the lines of electrostatic field define the bonds in a neutral array of positively and negatively charged atoms. The valence electrons are shown in very *light gray* around the (negative) anion

Reassigning the cation electrons to the anion valence shell leaves the cation with the positive charge of its core which is equal to its positive valence, and the anion with a negative charge equal to its negative valence. Further the electrostatic flux that links the core of the cation to its bonding electrons (cf. Fig. 4) is still present, but as the bonding electrons are now all assigned to the anions, this flux links the cores of the positively charged cations to the valence shells of their negatively charged anion neighbors. If both ions are now shrunk to a point, one retrieves the ionic model: an array of point charges linked together by electrostatic flux, each ionic charge equal to the valence of the cation or anion it represents.

This leads to an interesting extension of the core-and-valence-shell picture. Where the valence of a bond was previously defined as the flux linking the cation core to the electrons it contributes to the bond, in the ionic model it is defined by the same flux which now links the cation to the anion. If the positions of the atoms in the array are known from experiment, this flux can be directly calculated. The calculation involves extensive computation, but Preiser et al. [17] have shown that in structures in equilibrium, the correlation between the bond flux and bond length is the same as the correlation that had previously been observed between the bond valence and bond length, showing that the electrostatic flux and bond valence are indeed the same.

It is sometimes assumed that the long range of the Coulomb potential makes it impossible to define a localized bond in the ionic model, but in fact the ionic model is the *only* model that provides a useful and unambiguous definition of a bond. Since the electrostatic flux that links two ions is equal to the valence of the bond that links them, a bond only exists between ions if they are linked by flux. A simple picture of the electrostatic field in the ionic model is provided by the lines of field that connect a cation to its first shell of anion neighbors as shown in Fig. 5.

The paradox of a localized interaction having a long-range influence can be understood by considering what happens if the valence of a particular bond is increased. According to the valence sum rule (2), the valences of the bonds between the terminating atoms and their first neighbors must be correspondingly reduced. This in turn triggers an increase in the valence of the bonds between the first and second neighbors. In this way the bond network relaxes by means of a wave passing from one atom to the next throughout the structure. Just as the collective behavior of an ant colony is driven by the actions of each individual ant responding to its immediate environment, so the collective behavior of a molecule or crystal is driven by each atom reacting to changes in its local environment.

This derivation of the ionic model provides not only a natural definition of a bond, but it also defines the scope of the model. The assumptions on which the derivation is based show that far from being confined to compounds whose bonds have "ionic character," the ionic model can be used for any valence compound. It can be used to describe not just NaCl but also SF_6, CO_2, CH_4, CH_3COOH, and O_2, all of which have networks with bipartite graphs.

6 Geometry of Valence Compounds

6.1 Predicting the Bond Lengths

The ionic model can be used to derive a number of theorems that apply to valence compounds.

The first is the *electroneutrality rule* (12). Because the atoms of the core-and-valence-shell model are all electrically neutral and all the charges have been conserved during the derivation of the ionic model, the array of charged ions in the ionic model must also be electrically neutral.

> In a valence compound, taking the cation valences as positive and the anion valences as negative, the sum of the valences (charges) of a the atoms (ions) is zero.

(12)

More importantly, we can use the ionic model to predict the electrostatic flux or valence of the individual bonds, and from these we can predict their lengths. If the positions of the atoms are already known, the bond flux can, in principle, be calculated using Coulomb's law, but this is computationally intensive, and as it requires a prior knowledge of the structure, the result is not a prediction. Fortunately there is a simpler approach that requires no prior knowledge of the atomic positions. All that is required is a knowledge of the bond network, that is, which atoms are linked by bonds.

In the ionic model, a bond consists of two equal and opposite charges (ions) linked by an electrostatic field. This makes the bond an electric capacitor. Each bond in a bond network can therefore be replaced by a capacitor, converting the

Fig. 6 The bond network shown in Fig. 5 can be represented as a capacitive electrical network

bond network into a capacitive electrical network as shown in Fig. 6 which illustrates the equivalent circuit for the bond network shown in Fig. 5.

The charges, Q_i, at each node in this circuit are the atomic valences, and the distribution of charges (equal to the bond fluxes), F_{ij}, on the plates of the capacitor linking atoms i and j, can be found by solving the set of Kirchhoff equations (13a) and (13b):

$$Q_i = \sum_j F_{ij} \tag{13a}$$

$$0 = \sum_{\text{loop}} P_{ij} \tag{13b}$$

P_{ij} is the potential difference between atoms i and j. If there are n atoms, there are $n - 1$ independent equations of the type (13a) (one equation is redundant since the total charge of the compound is zero), and if there are m bonds, there are $m - n + 1$ independent equations of type (13b). The electric potential is related to the charge on the capacitor by the capacitor equation $F_{ij} = P_{ij}C_{ij}$. All that is required to solve the resulting set of equations is the capacitance, C_{ij}, of each of the m different bonds. The principle of maximum symmetry (1) implies that the values of C_{ij} will all be equal, and in practice this is found to be the case for equilibrium structures. Two types of constraint break this symmetry: anisotropies in the electronic structure of the ion, e.g., lone pairs and Jahn–Teller distortions discussed in Sect. 7, and steric stresses, e.g., atoms in cavities that are too large, such as the hydrogen atom in hydrogen bonds, as discussed in Sect. 8. In this section, we consider only those structures in which these constraints are absent.

Substituting the bond valence, S_{ij}, for the bond flux F_{ij} and the atomic valence, V_i, for the charge, Q_i, and noting that the bond capacitances, being equal, cancel out

Fig. 7 The finite bond network corresponding to the infinite network shown in Fig. 4. A is the cation, B the anion

of the equations, one gets the **network equations** (14a) and (14b) linking the valences, S_{ij}, of the bonds to the valences, V_i, of the atoms.

$$V_i = \sum_j S_{ij} \tag{14a}$$

$$0 = \sum_{\text{loop}} S_{ij} \tag{14b}$$

Since the valence V_i is known for all the atoms, this set of m equations can be solved to give S_{ij} for each of the m bonds. This calculation is implemented in the program BONDVAL [18]. The bond valences calculated from these network equations are known as **ideal bond valencces**. If they are the same as the experimental values, the assumptions of the bond valence theory are validated [17], but if they are different, they indicate that one of the additional constraints described above is present. In principle it should be possible to model these constraints by choosing suitable bond capacitances, but in practice the capacitances themselves often depend on the context in ways that are not always transparent [19].

For molecules, the bond network is finite and generating and solving the network equations (14a) and (14b) is relatively straightforward, but for non-molecular compounds the network is infinite which makes it impossible to solve the network equations unless, as in a crystal, the network is composed of identical copies of a repeating unit. In this case the infinite bond network can be reduced to a finite network by extracting one repeating unit (formula unit) from the crystal and reconnecting the broken bonds. The resulting finite network is similar to that of a molecule, except that some atoms will be connected by more than one bond. The infinite two-dimensional network of Fig. 5 reduces to the finite network, also known as the quotient graph, shown in Fig. 7. In Fig. 5, each cation A forms four bonds to four different anions, B, resulting in the two atoms of the formula unit appearing in the quotient graph (Fig. 7) being linked by four separate bonds. The local environment of each atom in the quotient graph is the same as that of the corresponding atom in the infinite network, and the network equations that solve the quotient graph give the same bond valences as they would for the infinite network. In the example shown, the network equations can be solved by inspection; if the cation has a valence of +1 and the anion a valence of −1, all the bonds have valences of 0.25 vu (valence units).

The network equations (14a) and (14b) predict the valences of the bonds based on knowing only the composition of the compound and the way the atoms are connected. From these predicted bond valences, it is possible, using the bond valence–bond length correlation (4), to predict the lengths of each bond, and provided there are no additional electronic or steric constraints, these distances are found to differ from the observed distances by around 0.02 Å, an accuracy comparable with that achieved by other methods, both classical and quantum. Ways in which the bond network can be predicted are described in Sect. 11.2.

6.2 Predicting the Bond Angles

Predicting the bond angles is not as straightforward as predicting bond lengths, since they are more sensitive to steric effects which can only be taken into account once the bond network has been mapped into three-dimensional space.

If the bonds are all equivalent in the bond graph, they are expected to be uniformly distributed around the atom. The principle of maximum symmetry (1) predicts that three equivalent bonds will be arranged with threefold (triangular) symmetry as in BF_3 and CO_3^{2-}, four equivalent bonds will have tetrahedral symmetry as in CH_4 and SO_4^{2-}, and six will have octahedral symmetry as in SF_6 and NaCl. These are the most frequently found coordination environments. The only high symmetry arrangements of five or seven bonds are planar and unlikely to be found in a three-dimensional structure. Five and seven coordination must have at least two inequivalent bonds and are notably less common than three, four, or six coordination. They are found only when a more symmetric arrangement is not possible.

Nonuniform environments occur if the ligands are different or have different environments in the bond graph so that the bonds are no longer equivalent. The symmetry may also be lost if electronic anisotropies such as stereoactive lone pairs are present (Sect. 7.1). In this section, we assume that lone pairs, if present, are not stereoactive.

Any part of the flux linking the core of an atom to its valence shell belongs to one of the bonds formed by the atom (see Fig. 4), and because the atom is spherically symmetric, the solid angle, Ω, subtended by this flux is a proportionate part of the total solid angle of 4π steradians surrounding the core, leading to Eq. (15).

$$\Omega_{ij} = 4\pi S_{ij}/V_i \tag{15}$$

An approach based on this idea has been shown to work well for calculating bond angles in tetrahedral coordination, as for example around the sulfur atom in sulfates [20], and as shown in Sect. 7 it is a useful approach to exploring the extent to which lone pairs are stereoactive.

If measuring the solid angle occupied by a bond is not convenient, an alternative approach is to use the ***valence vector, S***. This is a vector that represents the electric flux linking the core to the overlap region. Its magnitude is equal to the magnitude of the flux and its direction is taken to be along the line connecting the two atoms as

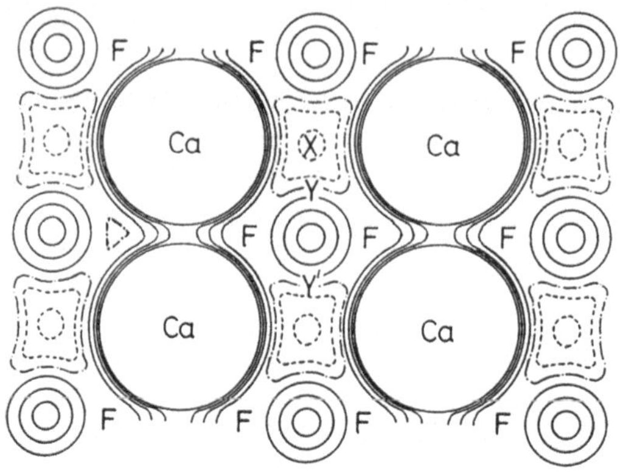

Fig. 8 A valence map of fluorine in a (110) plane in CaF_2. *Dotted lines* are negative; contours greater than 1.8 not shown. X and Y show possible sites for interstitial F (Fig. 11.8 from [11] by permission of the Oxford University Press)

an approximation to the net direction of the flux lines in the bond. Since the total flux is distributed uniformly around the core, the sum of the bond valence vectors should be zero in the absence of any perturbing influence, a condition known as the *valence vector sum rule* (16), which provides a way of expressing one of the constraints acting on the bond angles. This approach has been explored by Harvey et al. [21] and Zachara [22].

$$0 = \sum_j \mathbf{S_{ij}} \tag{16}$$

If this sum is not zero, it gives a measure of how far the bonding electrons in the valence shell deviate from spherical symmetry [23].

Another technique that is useful if the positions of most of the atoms are known, but the location of others is not, is the ***valence map*** [24]. Figure 8 shows the map for fluorine in CaF_2. A different representation of a valence map can be seen in Fig. 10 in Chap. 4 [14]. It is calculated by moving the fluorine atom systematically over the positions on a grid spanning the positions of the calcium atoms in the structure. At each point its bond valence sum is calculated from the distance to its neighbors. The resulting grid of bond valence sums represents a map that has high values when the fluorine atom is placed too close to the position of a calcium atom and a minimum when it is placed at the center of a cavity. In Fig. 8, the sites normally occupied by fluorine are labeled F, and possible sites for interstitial fluorine are marked X and Y.

The valence vector sum can also be treated as a field since it can be calculated for an atom placed at any point in the structure, not just the known site of an atom. Mathematically it represents the slope of the valence map, $U(\mathbf{r})$, at the position $\mathbf{r}$:

$$\sum \mathbf{S}(\mathbf{r}) = b.\nabla U(\mathbf{r}) \tag{17}$$

where b is the bond valence parameter (assumed here to be the same for all the bonds). The minimum in the valence map is the point where the valence vector sum is zero, and if the correct location of the atom is at the center of one of the cavities, the valence sum at the minimum should be the same as the valence of the atom itself. If the cavity is too small for the atom, the minimum will be larger than this, if the cavity is too large, the minimum will be smaller. If the atom does not lie at the center of the cavity, e.g., as a result of application of the distortion theorem (5), it is expected to lie on the contour whose value is equal to the valence of the atom. The position of the atom on this contour depends on which neighbor has the most deficient valence sum.

The valence map can be used for locating weakly scattering and weakly bonding atoms such as lithium in cases where it has not been detected by X-ray diffraction, but more commonly it is used for tracing possible diffusion paths in crystalline and amorphous solids as discussed in [25].

6.3 Non-valence Compounds and Homoionic Bonds

The network equations (14a) and (14b) can only be used if the graph of the bond network is bipartite, that is, if every bond has a cation at one end and an anion at the other. In inorganic compounds, and particularly in organic compounds, this condition is not always satisfied. Although this restricts the application of the bond valence theory, the core-and-valence-shell picture of the atom is still valid, as is the description of the chemical bonds this picture gives.

There are a variety of ways in which a non-bipartite bond graph can be converted to a bipartite graph, although some information is usually lost along the way.

1. The two cations, or two anions, that form the bond can be considered as a single pseudo-atom. This makes the bond graph bipartite. This solution works well for cations like Hg_2^{2+} which are traditionally described in this way. In the mercurous cation, the Hg–Hg bond is formed by an electron pair and has a valence of 1.0 vu. If the Hg_2^{2+} cation is in an asymmetric environment, the valence of the external bonds formed by the individual Hg atoms may not be the same, in which case the two mercury atoms may contribute different numbers of electrons to the Hg–Hg bond. Although this violates the equal contribution rule (3), the valence of the bond is correctly given by the average of the contributions of the two mercury atoms. Hg–Hg bonds are known in a number of mercury complexes, and not all of these are electron pair bonds, but as expected, the length of the bond is found to correlate with its valence (the number of electron pairs that form it) in the same way as any other bond [26].
2. An alternative approach to the mercurous cation is to treat the bonding electron pair as a pseudo-anion, E^{2-}. Inserting such an anion into the bond again makes the graph of the bond bipartite and does not require that the two Hg–E bonds have the same valence. Since it is not possible to locate the pseudo-anion, the individual valences of the Hg–E bonds cannot be found from their bond lengths,

but the sum of these lengths does correspond to the charge of the pseudo-anion and correlates with the Hg–Hg bond length [26].

3. In some cases an atom may act as an anion towards some of its ligands and as a cation to others. For example, in the NO_2^- ion, the oxygen atoms are the anions with a valence of −2 vu and the nitrogen is a cation with a valence of +3 vu. But the nitrogen is also a Lewis base and can coordinate to a transition metal to form a bond in which the nitrogen acts as the anion. In this case, the valence sum at the nitrogen is calculated by adding the bond valences having regard to their sign. For example, if the nitrite group bonds to copper, the Cu–N bond might have a valence of −0.4 vu, increasing the valence of the N–O bonds from 1.5 to 1.7 vu to give a sum of +3.0 around nitrogen (see Fig. 6 in [2]):

$$V_{\mathrm{N}} = 2S_{\mathrm{NO}} + S_{\mathrm{CuN}} = 2 \times 1.7 - 0.4 = 3.0 \text{ vu}$$

With the N–O bonds having a valence of 1.7 vu, the residual valence on each of the oxygen atoms is reduced from −0.5 vu in the free nitrite ion to −0.3 vu and the residual valence of the nitrite group as a whole is then

$$2 \times (-0.3) + (-0.4) = -1.0 \text{ as expected.}$$

This is equivalent to splitting the nitrogen atom into two, an N^{2-} anion bonded to copper and an N^{5+} cation bonded to oxygen linked by an N–N bond of 1.8 vu. The N–N bond is entirely fictitious; its valence depends on the valence of the Cu–N bond but in any case is not susceptible to measurement. As in the other cases, splitting the nitrogen atom into two results in the creation of a bipartite bond graph which can be solved using the network equations (14a) and (14b).

7 Electronic Constraints

Electronic constraints arise from changes in the electronic structure of the atom itself. There are several ways in which these occur. Some are initiated by the atom's environment, some occur spontaneously as a result of degenerate electronic states as predicted by the Jahn–Teller theorem which states that the environment of an atom will distort if such a distortion removes the degeneracy. Electronic constraints manifest themselves as a loss of symmetry either in the bond lengths or in the bond angles. Changes in the bond lengths cause the equal valence rule (14b) to be violated, but valence sum rule (14a) continues to be obeyed. Changes in the bond angles generally do not violate the network equations (14a) and (14b), but do affect the vector valence sum (16) which is no longer expected to be zero [23].

The most common electronic constraint involves the rearrangement of nonbonding "lone pair" electrons that are found around strongly bonding main group elements as discussed in Sect. 7.1. Other constraints are found in the transition metals, leading to the characteristic dipolar distortions found around octahedrally coordinated d^0 atoms, or in the so-called Jahn–Teller quadrupolar distortion found around Cu^{2+} and Mn^{3+} described briefly in Sect. 7.2.

7.1 Role of Lone Pairs

All anions, and main group cations in lower oxidation states, contain lone pairs [16]. The treatment here focuses on anions, but the same principles apply to cations. In some compounds, such as MgO, the lone pairs on the oxygen anion have no effect on the bonding geometry; the cations are distributed uniformly around the anion as expected from the principle of maximum symmetry (1). But in other compounds, the lone pairs provide a constraint that destroys this symmetry; for example, the bonds formed by the oxygen atoms in the sulfate ion, SO_4^{2-}, are not uniformly distributed; there is one strong bond to sulfur and a number of weaker bonds to other cations. The different types of asymmetric bonding induced by the lone pairs in such compounds are described by the well-established valence shell electron pair repulsion (VSEPR) model [27]. The rules of this model can be carried over with little change into the bond valence theory with one difference: the underlying picture of the VSEPR model, in which the electron pairs in the valence shell are assumed to repel each other, is replaced by the more flexible core-and-valence-shell picture described in Sect. 2 [16].

The behavior of lone pairs in bond valence theory is best illustrated by an example. When oxygen acts as an anion, its valence is −2 and it commonly forms between two and six bonds, so a typical coordination number can be taken as four. This gives oxygen an anion bonding strength, S_B, of −0.5 vu (Table 2). According to the valence matching rule (9), oxygen should form stable compounds with any cation having a bonding strength between 0.25 and 1.00 vu. Table 4 lists the binary compounds that oxygen forms with the elements of periods two and three of the Periodic Table. For each compound listed, Table 4 gives the bonding strength, S_A, of the cation (taken from Table 1) and the ratio of the cation to anion bonding strengths, S_A/S_B. Those compounds that have S_A/S_B between 0.5 and 2.0 satisfy the valence matching rule (9) and are shown in normal type. As expected, these are all stable crystalline solids. Those compounds with ratios less than 0.5, shown in *italics*, are unstable and can be formed only with difficulty as they readily pick up water and CO_2 from the atmosphere to form basic carbonates which provide a better valence match (see Tables 1, 2, and 3). Those shown in bold type have a ratio greater than 2.0 and, according to the valence matching rule, they ought not to exist. The reason they occur is attributable to the presence of the lone pairs on the oxygen atom.

According to the assumptions of the model, the electron density in the valence shell remains spherically symmetric even when the atom forms bonds. Although the electron density of the valence shell is always distributed with spherical symmetry, there is no requirement that the way this electron density functions is also uniformly

Table 4 Oxides of atoms of rows two and three of the Periodic Table

	S_A (vu)	S_A/S_B	State	S_2	Comment
Li_2O	*0.2*	*0.4*	*Solid*	*0.2*	*Unstable, picks up water*
BeO	0.5	1	Solid	0.5	Stable
B_2O_3	0.87	1.7	Solid	0.2	Stable
CO_2	**1.35**	**2.7**	**Gas**	**−0.22**	**Stable but dissolves in water to give CO_3^{2-}**
N_2O_5	**1.67**	**3.3**	**Gas**	**−0.11**	**Reactive, forms NO_3^-**
Na_2O	*0.16*	*0.33*	*Solid*	*0.16*	*Unstable, picks up water*
MgO	0.33	0.67	Solid	0.33	Stable
Al_2O_3	0.57	1.14	Solid	0.43	Corundum, stable
SiO_2	1	2	Solid	−0.33	Quartz, stable, forms minerals with SiO_4^{4-}
P_2O_5	**1.25**	**2.5**	**Solid**	**−0.25**	**Reacts with water to form H_3PO_4**
SO_3	**1.5**	**3**	**Solid**	**−0.17**	**Reacts with water to form H_2SO_4**
Cl_2O_7	**1.75**	**3.5**	**Liquid**	**−0.08**	**Reacts with water to form $HClO_4$**

Items in *italics* and **bold** do not obey the valence matching rule (9)
S_A is the bonding strength of the cation, S_B is the bonding strength of the anion (= 0.5), and S_2 is the bonding strength of the secondary bonds, all in valence units (values in parentheses are calculated for the corresponding complex ion). Values in **bold type** refer to the corresponding complex ion.

distributed. Some parts of the valence shell may be bonding and other parts nonbonding.

In compounds with $S_A/S_B < 1.0$, the bonds are arranged around the anions symmetrically, as expected from the principle of maximum symmetry. The coordination is tetrahedral around oxygen in BeO and octahedral in MgO, in all cases the symmetry is as high as the constraints of three-dimensional space allow. Both the bonding and the lone pair electron density are uniformly distributed around the valence shell; each portion of the valence shell contains some electron density that is bonding some that is nonbonding (lone pair).

For compounds with $1.0 < S_A/S_B < 2.0$, the valence matching rule is still satisfied. The compounds are stable, but the lone pairs start to show some stereoactivity as can be seen in the structures of silica and alumina. Silica adopts various structures whose Si–O–Si angles vary from 130° to 180°. In corundum, Al_2O_3, two of the four bonds formed by oxygen are short (primary) and two are long (secondary) with the valence of the primary bonds being equal to the bonding strength of Al^{3+}, 0.57 vu (1.83 Å), and the valence of the secondary bonds being equal to 0.43 vu (1.93 Å). This incidentally allows pairs of AlO_6 octahedra to adopt the unusual feature of sharing a face, which also happens to be a topological requirement of mapping its bond network into three-dimensional space.

If the ratio S_A/S_B is greater than 2.0, the oxygen is required to form a stronger bond to the cation than its normal bonding strength allows. If there were no lone pairs present, extra electrons would have to be brought in from other parts of the valence shell, thus destroying the spherical symmetry of its electron density. This would cost more energy than could be recovered by forming the bond, but because the oxygen anion already has nonbonding electron density in the overlap region, it can match the bonding strength of the cation by converting lone pair electron density in the overlap region to bonding while converting an equivalent amount of bonding electron density elsewhere into lone pairs. In this way, it can reassign the function of the electron density while retaining the spherical symmetry of the valence shell.

It is only necessary to reassign enough electron density to meet the bonding requirements of the cation, resulting in an asymmetry in which the valence of the primary bonds on one side of the oxygen anion is equal to the bonding strength, S_A, of the cation, leaving a smaller residual valence to form weaker secondary bonds of valence S_2 on the other side [28]. The relative strengths of the primary and secondary bonds are shown in Fig. 5 of [2]. For example, the oxygen atoms of the PO_4^{3-} ion form one strong bond (1.25 vu) to phosphorus and several weaker bonds of about 0.25 vu to other cations. In extreme cases all the bonding electron density is found on one side of the atom, leaving only lone pair electron density on the other. In CO_2 the separation between the valence electrons and the lone pairs is so complete that the oxygen atoms in carbon dioxide molecules are unable to form any further bonds. Carbon dioxide remains a gas under ambient conditions.

To achieve a complete separation between the valence electrons and the lone pairs as in CO_2, oxygen must form one fully saturated bond of 2.0 vu, but fluorine forms fully saturated bonds at the much small valence of 1.0 vu. Many fluorides, such as CF_4, contain saturated bonds, and since a saturated bond prevents fluorine from forming a second bond, fluorides of cations with a bonding strength of 1.0 vu or higher usually form isolated molecules with an unreactive surface. On the other hand, oxygen, because of its higher valence, rarely forms fully saturated bonds because few cations can form bonds with valences greater than 2.0 vu. Oxide molecules such as SO_3 have saturated bonds, but since such molecules, unlike CF_4, have space for additional ligands in their coordination sphere, they tend to link into chains or to react with water to form SO_4^{2-} with enough bonding electrons residing on the surface of the complex to bond to weakly bonding cations such as sodium.

The core and valence shell model can be used to examine the effects of lone pairs on the bonding geometry. Because every atom has a spherically symmetric electron density, the electric field linking the core to the valence shell is also spherically symmetric. If lone pairs are present in the valence shell, some of the electrostatic flux (valence) will link to lone pair electrons and some to bonding electrons. Although the total flux is distributed symmetrically, its function as either bonding flux or lone pair flux need not be.

As shown in Sect. 6.2, the solid angle subtended by a bond is proportional to the valence of the bond as given in Eq. (15). The area occupied by the bonding overlap region is related to the solid angle, $4\pi S_{ij}/V_i$, by (18)

$$\text{area of the overlap region} = R_i^2\Omega_i = 4\pi R_i^2 S_{ij}/V_i \tag{18}$$

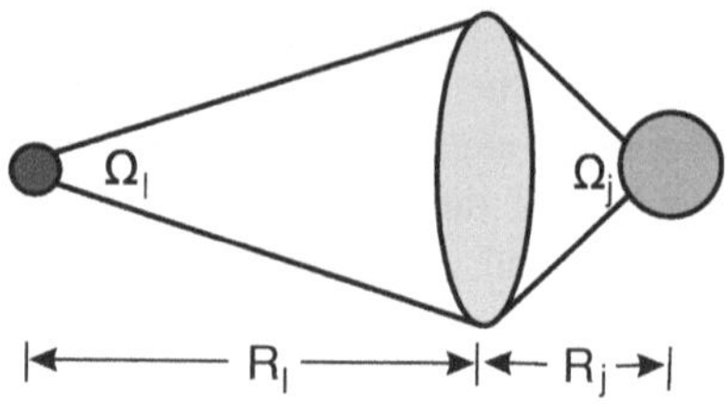

Fig. 9 The overlap region (*light gray*) subtends a larger angle at the anion (*medium gray*) than at the cation (*dark gray*)

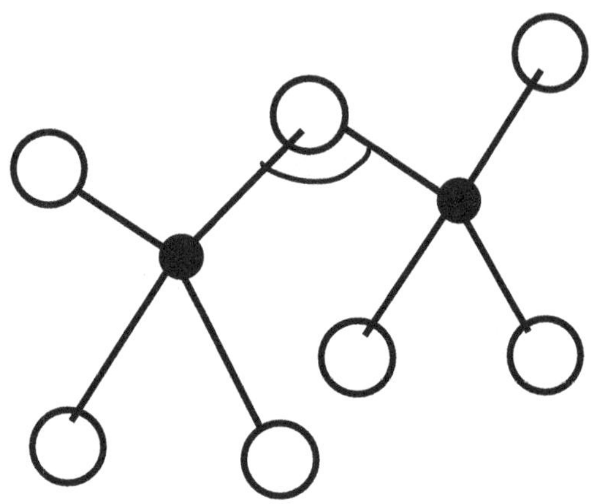

Fig. 10 An M_2O_7 complex. The angle at the bridging oxygen is marked

where R_i is the distance from the overlap region to atom i. Since the overlap region is common to both bonded atoms, it defines the solid angle, Ω_j, of the bond at the other atom j as given by Eq. (19) and shown in Fig. 9.

$$\Omega_j R_j^2 = 4\pi R_i^2 S_{ij}/V_i \tag{19}$$

The relative sizes of the two solid angles thus depend on the location of the overlap region given by R_i/R_j. The overlap region is not defined by bond valence theory, though it is expected to lie closer to the anion as can be seen in Fig. 4. We might therefore expect the solid angle subtended by the flux at the anion to be larger than the solid angle at the cation.

The presence of lone pairs gives rise to two possible extreme geometries. In the first, the lone pairs are not stereoactive and the bonding electrons and lone pairs are both distributed with spherical symmetry. The bonding flux and nonbonding flux are also uniformly distributed, and each part of the valence shell, i.e., each overlap region, contains both bonding and nonbonding electrons. Consider, for example, an M_2O_7 complex composed of two MO_4 tetrahedra sharing a common bridging oxygen shown in Fig. 10. The valence of the M–$O_{bridging}$ bonds is 1.0 vu, and since this saturates the oxygen atom, if the two lone pairs of the bridging oxygen are uniformly distributed around the valence shell, the two M–$O_{bridging}$ bonds are collinear, giving an M–O–M angle of 180° with the overlap regions of each bond occupying one half of the sphere.

Table 5 Predicted and observed M–O–M angles in M_2O_7 complexes

	Solid angle of the bond as a fraction of a sphere		M–O–M (degrees)	
	M	O	Predicted	Observed
$Si_2O_7^{6-}$	0.25	0.45	168	140–180
$P_2O_7^{4-}$	0.20	0.36	143	122–156
$S_2O_7^{2-}$	0.17	0.30	127	114–121
Cl_2O_7	0.14	0.26	115	115

In the other extreme geometry, when the lone pairs of the oxygen atom are fully stereoactive, the bonding and nonbonding electrons are completely separated, with the four non-bonding electrons occupying two-thirds of the valence shell and the two bonding electrons occupying the remaining one-third. Each bond occupies just one-sixth of the total sphere. In this case, the lone pairs are said to be stereoactive. The most symmetric geometry of six electrons around the oxygen atom is an octahedron. If the two M–O bonds are *cis*, as expected for the dipolar valence shell of oxygen, the M–O–M angle is 90°.

In both of these extreme geometries, the fluxes of the M–$O_{bridging}$ bonds are 1.0 vu, but the distribution of the bonding flux and the solid angle it subtends at the nucleus depends on the degree to which the lone pairs are stereoactive. If they are completely inactive, each of the two bonds subtend a solid angle of $(1/2) \times 4\pi$ steradians resulting in an M–O–M angle of 180°, but if they are fully active, each bond subtends an angle of only $(1/6) \times 4\pi$ steradians and the angle between the bonds is only 90°.

The M_2O_7 complexes with M = Si, P, S, and Cl all have geometries that are intermediate between these extremes, with the M–O–M angles decreasing as the valence of M increases from 4 to 7. All of these complexes except Cl_2O_7 are anions, and they all have the same geometry apart from the lengths of the M–$O_{terminal}$ bonds and the M–O–M angle. Setting the valence of the bridging M–O bond to 1.0 vu in Eq. (19), the solid angle subtended by its overlap region at the bridging oxygen equals $4\pi(R_M/R_O)^2/V_M$, where R_O is the distance between the overlap region and the bridging oxygen. The factor $(R_M/R_O)^2$ is not known a priori, but the physically reasonable value of 1.8 accounts for the decrease in the M–O–M angle as one progresses from Si to Cl as shown in Table 5. One can understand why this angle changes even though there is no change in the valence of the bridging M–O bond. As the valences of the terminal M–O bonds increase, these bonds occupy a larger portion of the solid angle around M, causing the M–$O_{bridging}$ bond to focus into a smaller solid angle. This leaves a larger solid angle at the bridging oxygen for the lone pairs, making them more stereoactive.

Bickmore et al. [23] have adopted an alternative approach to the description of stereoactivity by showing that the valence vector sum (Eq. 16) can be used as a measure of the stereoactivity of the lone pair, noting the correlation between vector valence sum and the valence of the shortest primary M–O bond, which as shown above is equal to the bonding strength of the cation.

In cases where the lone pairs are fully stereoactive, the VSEPR and bond valence theories make similar predictions for the bonding geometry because both models assign the same localized regions of the valence shell to the lone pairs and the bonding electrons, but they differ in the way they make this assignment. The VSEPR model is useful for predicting the overall geometry when the lone pairs are stereoactive because it is simpler, but bond valence theory is more quantitative and can predict the extent to which the lone pairs will be stereoactive. The full power of the VSEPR model can therefore be incorporated into bond valence theory.

7.2 Covalent Models

The short primary bonds that are responsible for the stereoactivity of the lone pairs on anions are normally described as covalent, but bond valence theory makes no distinction between ionic and covalent bonds, since the bond valence is a variable that runs continuously across the whole spectrum of bond types. The ionic–covalent distinction does, however, reflect a marked difference between the structural chemistry of those compounds that obey the valence matching rule (9) in which the bonding is generally weak (less than about 0.8–1.0 vu), and those where the presence of anion lone pairs permits the formation of much stronger bonds by making the lone pairs stereoactive as described in Sect. 7.1. The bonds that obey the valence matching rule are those traditionally described as ionic, while bonds formed by atoms with stereoactive lone pairs are those traditionally described as covalent. Even though the bonds form a continuous series in which such a distinction is not necessary, the term covalent can be usefully applied to the primary (short) bonds formed by anions with stereoactive lone pairs.

The term “covalent” was originally applied to the bonds in the ball and stick picture of structure used in organic chemistry. It is therefore interesting to note under what conditions the ball and stick picture can be derived from bond valence theory. The concept of a bond arose in the mid nineteenth century from the study of the chemistry of carbon compounds. The original model was simple: it envisioned atoms as having a fixed number of hooks each of which could form a bond by linking to a similar hook on other atoms. This ball and stick picture, in which the balls represent the atoms and the sticks the bonds, can be derived from the bond valence theory if all the atoms have coordination numbers that are equal to their valence. Examples are silicon ($V = N = 4$), molybdenum ($V = N = 6$), and the two elements that form hydrocarbons: carbon ($V = N = 4$) and hydrogen ($V = N = 1$).[5]

In hydrocarbons the valence of the C–C, C–H, and H–H bonds is always 1.0 vu and since carbon and hydrogen both have the same electronegativity, these bonds are electron pair bonds and are equivalent to each other.

[5] The coordination number of carbon with oxygen is three (cf., Table 1), which is also the coordination number found in elemental carbon (graphite). The coordination number of four is found only in compounds in which carbon is bonded to hydrogen or a halogen, or in compounds such as diamond that are formed under pressure. See Sect. 8.1.1 for a fuller discussion of the coordination number of hydrogen.

The consequence is that for hydrocarbons, bond valence theory reduces to the simple ball and stick model in which all the bonds are formed by a single electron pair and the coordination numbers are fixed at four for carbon and one for hydrogen. The principle of maximum symmetry (Eq. 1) requires that carbon's four bonds be directed to the corners of a tetrahedron. Carbon and hydrogen atoms arc linked to each other by identical electron pair bonds to form extended networks in which hydrogen, having a coordination number of only one, always terminates the network, thus ensuring that all hydrocarbons are molecules. Double and triple bonds are formed by linking two carbon atoms by a tetrahedral edge or face, respectively. Bond valence theory does not use π bonds to describe these compounds but both descriptions are consistent with quantum mechanics.

This simple model is often extended to nitrogen ($V = N = 3$), oxygen ($V = N = 2$), and halogens ($V = N = 1$), but these elements have a higher electronegativity than carbon and hydrogen. They are anionic with stereoactive lone pairs, but with sufficient residual valence to provide Lewis base sites that can act as receptors of secondary bonds such as hydrogen bonds (see Sect. 8.1.1). Although it is common to consider that the base function on oxygen is supplied by the lone pairs, lone pairs are, by definition, nonbonding. Oxygen and nitrogen act as the receptor of hydrogen bonds as the result of the ability of their bonding electrons to rearrange themselves within the valence shell in response to their environment, diverting some of the valence previously used to form the C–O or C–N bonds into accepting hydrogen bonds. When this happens, the valences of the C–C and C–H bonds will no longer be exactly 1.0 vu, and the bonds will no longer be pure electron pair bonds. The simple ball and stick model no longer strictly applies. One can treat the hydrogen bonding as a perturbation, but it is a perturbation that can significantly change the properties of the molecule. Unlike the rules of the simple ball and stick model, the rules of bond valence theory require the hydrogen bonds to be explicitly taken into account, as can be seen from the examples given in Sect. 8.1.1. Bond valence theory can predict, or at least provide limitations on, the structure and properties of many organic molecules, specifically those that have localized bonds and are valence compounds.

7.3 *Complex Ions*

Cations are Lewis acids, and anions are Lewis bases, and their respective bonding strengths are also their Lewis acid and Lewis base strengths. So far this discussion has focused on simple ions, that is, ions that consist of a single atom, in which the Lewis acid and base strengths can be calculated using Eq. (7). Complex ions are those composed of more than one atom. They differ from simple ions in that the different atoms in the complex may each have their own Lewis acid or base strength.

Each atom in a complex ion is necessarily bonded to one or more atoms within the complex to form a network of strong internal (covalent) bonds. For example, the sulfur in the sulfate anion, SO_4^{2-}, is a hexavalent cation, S^{6+}, linked by bonds of 1.5 vu to the four oxygen anions. In the ammonium cation, NH_4^+, nitrogen is a trivalent anion, N^{3-}, linked by bonds of 0.75 vu to the four hydrogen cations. When the bonding requirements of the internal bonds are satisfied, the individual atoms forming the complex may have a residual valence that is available to form bonds external to the complex, often considered to be ionic bonds. In the sulfate ion, the oxygen uses only −1.5 vu internally, leaving −0.5 vu for forming external secondary bonds. In the ammonium ion, the hydrogen uses +0.75 vu to form the internal bonds leaving +0.25 vu for forming external bonds. The unused valence of the peripheral atoms is known as their **residual valence**. Assuming that all the oxygen atoms in the sulfate ion are equivalent and they each form four bonds, one to sulfur and three external to the complex, each oxygen has a bonding strength of $-0.5/3 = -0.17$ vu. This is then the **Lewis base strength** of the oxygen atoms and, therefore, the Lewis base strength of the sulfate ion as a whole. Similarly, each hydrogen atom in the ammonium ion normally forms two extra bonds, so its **Lewis acid strength** is $+0.25/2 = +0.12$ vu, similar to that of rubidium. Because the acceptor hydrogen bond normally has a valence of 0.2 vu, the ammonium ions can also bond to anions with larger bonding strengths. Since most of the atoms on the surface of a complex form more than one external bond, the Lewis acid or base strength is normally smaller than the residual valence. Complexes may be anions or cations, or neutral molecules, and any complex may contain some atoms that function as Lewis acids and others that function as Lewis bases, but in all cases the sum of the residual valences over all the atoms in a complex, taking into account their sign, is equal to the total residual valence, or formal charge, of the complex.

Neutral molecules, such as water or ammonia, if they have any chemical activity, must contain both Lewis acid and Lewis base functions, with the condition that their residual positive valence must be numerically equal to their residual negative valence. A consequence is that the Lewis acid and base functions must work together; in neutral molecules the total valence of the external bonds formed by the Lewis acid functions must equal the total valence of the external bonds formed by the Lewis base functions. It is helpful to distinguish the Lewis acid and Lewis base functions by arrows on the bonds directed from the acid to the base.

Complex ions and molecules have more flexibility in adapting to their environment than simple ions since they can redistribute their residual valence among the different atoms that form the surface of the complex ion by changing the valences of the internal bonds. Both the bonding strengths displayed by such complexes, and their internal geometries, may vary depending on what counterions are present in the compound.

7.4 Transition Metals

The valence shells of transition metals are less well defined than those of the light main group elements because the valence s–p shell of one period overlaps with the d shell of the period that lies above it in the Periodic Table. Only the s–p shell is involved in the bonding of atoms of groups 1–3 because for these atoms the d shell lies at a higher energy and is not involved in bonding. Only the s–p shell is involved in bonding in atoms of groups 12–17 because for these atoms the d shell lies below the s–p shell; it is full and part of the core. In the remaining groups 4–11, the s–p and d shells overlap and all their electrons take part in bonding. Since the s, p, and d electrons lose their identity in the valence shell, they cannot be distinguished from each other and there is no gap in the ionization energies to define the boundary between the valence shell and the core. One consequence is that atoms in these groups can adopt a variety of different valences. Once the valence is known, the bond valence theory can be applied in the same way as it is applied to main group atoms. In some cases, there are electronic constraints that can be introduced into the bond valence theory using simple ad hoc models. These cases are discussed individually in Sects. 7.4.1–7.4.3 below. Bickmore et al. [23] have shown how the valence vector sum can be used to measure the size of the non-centrosymmetric distortions displayed by transition metals.

7.4.1 Distortions Around Transition Metals with d^0 Configurations

The early transition metals in high-valence states show a strong aversion to being in the centrosymmetric environment of an octahedral field. They either avoid six coordination or if they adopt it, their octahedral coordination is distorted. Since the tetrahedron does not have an inversion center, it is unaffected by this distortion.

The distortion takes the form of an off-center displacement of the transition metal from the center of its octahedron, sometimes known as the Second-Order Jahn–Teller (SOJT) effect. It results from a dipole distortion that can be described using the spherical harmonics associated with the p shell. The distortion becomes stronger as one moves across the Periodic Table. Scandium(III) shows no tendency to distort its octahedral environment. Titanium(IV) usually adopts a regular octahedral coordination, but the titanium atom is easily driven off-center, for example, by occupying a cavity that is too large as in $BaTiO_3$ (Sect. 8). Although this displacement is predicted by the distortion theorem (5), it is stabilized by the d^0 electronic effect. If the ligands are topologically equivalent, as they are in $BaTiO_3$, the direction of displacement can be changed by application of an external electric field. The d^0 distortion is best observed in vanadium in its +4 (d^1) and +5 (d^0) valence states. In these states, vanadium can adopt either octahedral or tetrahedral coordination, but octahedral coordination is sufficiently distorted that vanadium compounds typically adopt bond networks that support the distortion. The strongly bonded VO^{3+} group [$s(V\text{–}O) \approx 1.80$ vu] found as part of the octahedral environment of vanadium is so striking that it is sometimes called the vanadyl cation. Its strong bond is usually trans to a very weak bond which in some cases is entirely absent. In chromium(VI) the effect is so strong that Cr^{6+} is never found in an octahedral environment, preferentially forming the tetrahedral CrO_4^{2-} anion even though chromium is large enough to accommodate six ligands.[6]

7.4.2 Jahn–Teller Distortions

The Jahn–Teller theorem states that if an atom finds itself in a degenerate ground state it will distort if such a distortion can remove the degeneracy. The theorem is very general and applies to most electronically driven distortion, but the name has become attached to a specific tetragonal distortion of the octahedral environment of certain transition metals in particular oxidation states, namely copper(II) and manganese(III). The distortion causes the two axial bonds in an octahedral environment to be longer than the four equatorial bonds. Both atoms have a degeneracy in their ground state which can be understood using a one-electron (or one hole) picture which provides a simple quantum description of the effect. Removing the degeneracy results in a quadrupolar distortion of the core electron density associated with the spherical harmonics that describe the d shell. A tetragonal distortion that makes the two axial bonds longer than the four equatorial bonds removes the degeneracy predicted for high spin d^4 (Mn^{3+}) and the d^9 (Cu^{2+}) systems.

This distortion is closely related to the observation that nickel(II), palladium(II), and platinum(II) are often found with square four coordination rather than tetrahedral configuration. This can be thought of as a more extreme distortion of the same kind in which the weakly bonded ligands are removed entirely.

7.4.3 Late Transition Metals in Low-Valence States

The late transition metals have nearly full d electron shells capable of back bonding, i.e., the transition metal acts as a σ-bonding cation towards an anionic ligand while at the same time acting as π bonding anion towards the same ligand acting as a cation. In the bond valence theory, these bonds are composed of two oppositely directed bonds with valences S_σ and S_π, respectively. Since S_π is a negative number, the valence of the bond used for calculating the valence sum is the difference in the magnitudes of the two valences:

$$S = |S_\sigma| - |S_\pi| \tag{20}$$

which results in small, possibly negative, values for the apparent valence of the bond. If the transition metal forms all of its bonds in this way, it can achieve a bond

[6] The changes in the distribution of the valence electron do not mean that the atom loses its spherical symmetry. Because the core and valence shell have similar energies, the distortion in the valence shell can be compensated by the distortion of the core.

valence sum of zero. Transition metals in groups 9 and 10 sometimes have apparent atomic valences of zero.

Although the apparent valence of the bond is small and possibly zero, the bond itself can be quite strong, since both partial bonds are contributing to its strength. The total number of electrons forming the bond is given by the sum of the magnitudes of the bond valences:

$$\text{Number of valence electron pairs in the bond} = |S_\sigma| + |S_\pi| \quad (21)$$

In this way strong transition metal–ligand bonds are formed that contribute little to the atomic valence, leading to complexes in which the formal valence may even be zero. Bond lengths in these cases do not provide much information about the bond valences. There are suggestions that the bond-valence parameters in this region may depend on the ability of the ligand to accept π bonds, which in turn may depend on the coordination number of the ligating atom [29, 30]. This is an area where more work is needed.

8 Steric Constraints

Steric constraints are those that arise when a bond network cannot be mapped into three-dimensional space without straining the ideal bond lengths calculated with the network equations (14a) and (14b); some bonds may have to be stretched and others compressed. The result of these strains is that the network equations no longer give good predictions of the observed bond lengths, but the ideal bond lengths predicted by these equations provide a convenient reference which allows both the nature of the strain (either tension or compression), and its magnitude (Eq. 22), to be determined, both being useful in assessing the nature of the mechanical stresses in the bonding system.

$$\text{strain} = (R_{\text{obs}} - R_{\text{ideal}})/R_{\text{ideal}} \quad (22)$$

Here R_{obs} is the measured bond distance and R_{ideal} is the distance predicted by the network equations (14a) and (14b).

If an atom finds itself in a cavity that is too large for its bonds to adopt their ideal length, the bonds must be stretched. According to the distortion theorem (5), the environment of the atom will distort in such a way as to make the bond lengths unequal in order that the bond valence sums becomes equal to the atomic valences. As mentioned in Sect. 7.4.1, this contributes to the distortion around titanium(IV) in $BaTiO_3$. In many cases, such distortions are found in compounds where electronic distortions are also expected, the two effects being mutually supportive.

In all cases of steric constraint, the observed bond distances violate the equal valence rule (14b) and, in some cases, the valence sum rule (14a) as well. A simple measure of the strain in compounds where the bond valence sum rule is violated is the global instability index, G.

$$G = \left(\sum_i \{(V_i - \sum_j S_{ij})^2\}/n \right)^{1/2} \tag{23}$$

where n is the number of atoms, i, in the formula unit. Experimental uncertainties in well-determined structures account for values of G around 0.05 vu. Structures with G greater than this are strained but few stable structures are found with $G > 0.2$ vu. Examples of the use of G can be found in [14].

There are two causes of steric strain: either close contacts between nonbonded atoms (Sect. 8.1) or incommensurations in the natural spacings of different parts of the structure (Sect. 8.2). Each of these is discussed separately.

8.1 Nonbonding Contacts

Steric effects resulting from nonbonding contacts are well-known in organic chemistry where they are invoked to explain why crowded molecules are difficult to prepare, but this kind of strain is also found in inorganic compounds, most notably in the hydrogen bond where it is the origin of the O–H. . .O asymmetry discussed in Sect. 8.1.1.

8.1.1 The Chemistry of Hydrogen and Hydrogen Bonds

Apart from the chemically inert helium, hydrogen is the only element in the first period of the Periodic Table. It differs from all other elements in having no electrons in its core and no possibility of a lone pair in its valence shell. Without a core, the nucleus is unshielded, and unlike the nuclei of other atoms, it can penetrate into the valence shell of any atom that it bonds to as shown in Fig. 11. Since the hydrogen atom continues to have spherical symmetry even when bonded to other atoms [31], its bonding electron is arranged spherically around the nucleus, allowing it to overlap completely with the valence shell of the bonded atom. Without any lone pairs to block parts of its bonding environment, hydrogen cannot use the same mechanism as other anions, such as fluorine, for terminating the bond network (Sect. 7.1). Where it does terminate the network, as in hydrocarbons, a different mechanism is used as discussed below.

When hydrogen bonds to an atom such as oxygen, the whole hydrogen atom, both nucleus and its valence electron, is drawn into the oxygen valence shell. Without a lone pair to block the formation of a second (acceptor) bond, the hydrogen nucleus can attract another anion (oxygen for example) to form a second bond. Following the principle of maximum symmetry (1), we would expect the two

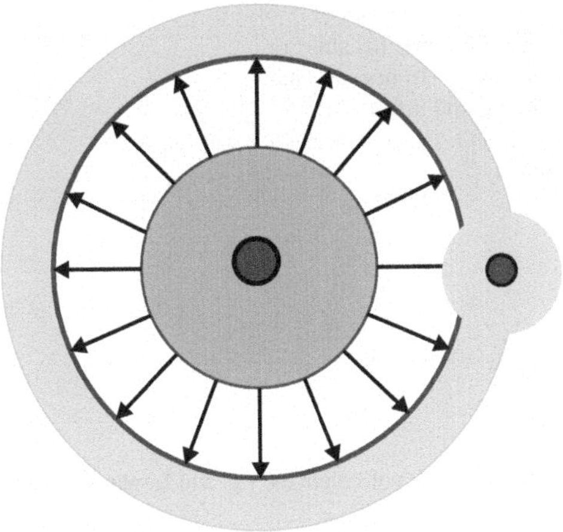

Fig. 11 Without a core, the nucleus of the hydrogen atom can penetrate the valence shell of a bonded atom

bonds to have the same valence, 0.5 vu, and length, 1.10 Å, with the hydrogen atom lying at the midpoint between the two oxygen atoms. However, if the hydrogen atom were to lie within the valence shells of both anions, the valence shells of the two anions would have to overlap, and since both valence shells are already effectively full, this is not possible. The repulsion between the oxygen atoms causes the bond to stretch from the expected O...O distance of 2.20 to 2.42 Å, and according to the distortion theorem (5), the hydrogen atom will move from the point midway between the two oxygen atoms to increase its bond valence sum to 1.0 vu [11, p. 75ff]. Equilibrium is observed when hydrogen forms a strong bond of 0.8 vu to the oxygen atom with the larger base strength (donor) and a weak bond of 0.2 vu to the other oxygen (acceptor). For this reason, hydrogen is unusual in that it has two cation bonding strengths, 0.8 and 0.2 vu which are labeled as donor and acceptor (Table 1). Other constraints within the compound may force the hydrogen to be placed either less or more symmetrically, giving rise to three possible types of hydrogen bond. The normal O–H...O hydrogen bond has one short bond of 0.97 Å (0.8 vu) to the donor oxygen and one longer bond of 1.90 Å (0.2 vu) to the acceptor. As expected from the principle of maximum symmetry, all three atoms are collinear as shown in Fig. 12b. This geometry has the lowest energy and is the configuration most often found. More symmetric hydrogen bonds have a shorter O...O distances and are found only if the energy needed to bring the valence shells of the two oxygen atoms closer can be recovered elsewhere in the structure (Fig. 12a). For example, hydrogen phosphate ions, HPO_4^{2-}, with a bonding strength of −0.22 vu, retain their tetrahedral symmetry better if they form more symmetric hydrogen bonds. A more asymmetric hydrogen bond having a longer O...O distance is found if the acceptor has a bonding strength smaller than −0.2 vu (Fig. 12c). The perchlorate ion, ClO_4^-, with a bonding strength of −0.08 vu forms only weak hydrogen bonds [32].

More symmetric hydrogen bonds are kept linear by the strong repulsion between the oxygen atoms. Longer hydrogen bonds are usually bent since the increased

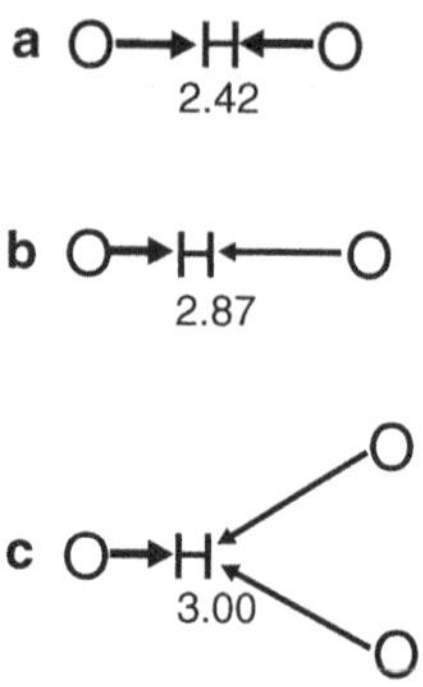

Fig. 12 Three types of hydrogen bond: (**a**) short or symmetric, (**b**) normal, and (**c**) long and bifurcated. Typical O. . .O distances are shown

Table 6 Residual valences, V_R, and Lewis acid and base strengths of the second row hydrides in valence units

	Lewis acid		Lewis base		Boiling point
	V_R	Acid strength	V_R	Base strength	/K
H_4C	0	0	0	0	109
H_3N	0.6	0.2	−0.6	−0.6	240
H_2O	0.4	0.2	−0.4	−0.2	373
HF	0.2	0.2	−0.2	−0.07	293

V_R is the residual valence associated with the Lewis function
Acid strengths are calculated by assuming the hydrogen bonds have a valence of 0.2 vu and the base strengths by assuming the anions have a coordination number of 4

O. . .O distance relieves the stress that forces the oxygen atoms apart. For longer hydrogen bonds, the O–H. . .O angle decreases from 180° with the increase in asymmetry. Weak hydrogen bonds are frequently bifurcated, i.e., they involve more than one acceptor bond (Fig. 12c), so that the valence of the donor O–H bond remains close to 0.8 vu.

It is interesting to compare the properties of the hydrides formed by the anions of the second period: H_4C, H_3N, H_2O, and HF. Their formulas are written to emphasize that in each case hydrogen is the cation, since it has the lower electronegativity. Because these are all neutral molecules, the sum of the valences of any external bonds they form must be zero. Table 6 shows the estimated Lewis acid and base strengths of these compounds.

The electronegativity difference is greatest for HF with the hydrogen atom acting as the Lewis acid and the fluorine as the Lewis base. The residual valence of the Lewis acid function (hydrogen) must equal the residual valence of the Lewis base function (fluorine), both of which are therefore determined by the intrinsic anisotropy of the hydrogen bond formed by HF. The existence of the complex $F–H–F^-$ anion with a symmetrical hydrogen bond shows that the Lewis acid strength of the hydrogen can be as large as 0.25 vu, though for most bonds it is likely smaller as assumed in Table 6.

The next largest electronegativity difference is found in water, H_2O, where it is well attested that the Lewis acid strength of the hydrogen is close to 0.2 vu. The residual bond valence contributed by the two hydrogen atoms is +0.4 vu, which means that the residual valence at oxygen must be −0.4 vu. Assuming oxygen has a coordination number of four, two internal and two external bonds, this leads to water having a Lewis base strength of $-0.4/2 = -0.2$ vu.

Ammonia, H_3N, forms three hydrogen bonds, hence the nitrogen atom with a coordination number of four is expected to form only one external bond, the other three bonds being formed internally to hydrogen. As shown in Table 6, this gives nitrogen the relatively large base strength of 0.6 vu, even though the hydrogen bonds individually have Lewis acid strengths similar to those in water. Methane, H_4C, is unique in this series in that the four hydrogen atoms saturate the coordination number of carbon, leaving no possibility for methane to have a Lewis base function, and without a Lewis base function its hydrogen atoms must have a Lewis acid strength of zero. They are unable to form hydrogen bonds and terminate the bond network.

Methane, without any Lewis acid or base function, is a gas with the lowest boiling point of the group. Water is the molecule whose Lewis acid and base strengths are ideally matched to each other, which is why water is a liquid or solid at ambient temperatures. Ammonia and hydrogen fluoride are poorly matched to themselves since their acid and base strengths differ by a factor of 3. According to the valence matching rule (9), the bonds that H_3N molecules form with themselves should not be stable. The same is true for HF, which is why both are gases and tend to react with compounds that form better hydrogen bonds. Ammonia is the strongest base, and is well-known for its ability to coordinate strongly to cations such as transition metals which typically have bonding strengths around 0.5 vu. In each case where hydrogen forms hydrogen bonds, it is bonded to its donor, N, O, or F, by a donor bond of 0.8 vu which means that the lone pairs of the donor atoms are usually stereoactive for the reasons given in Sect. 7.1. The three donor N–H bonds in ammonia are arranged pyramidally, and the two O–H bonds of water are not collinear.

When hydrogen is found in molecules and complex ions, it is usually chemically active except in pure hydrocarbons and there its lack of activity is not an intrinsic property of the hydrogen atom, but only the result of the coordination of carbon being saturated, leaving it no opportunity to act as a Lewis base. Hydrogen is left with a bonding strength of zero, but in all other compounds it forms hydrogen bonds that play a significant role in the cohesion of molecules and the formation of crystals. Where methyl groups are linked to other atoms having Lewis base functions, as in the acetate ion, $H_3CCO_2^-$, C–H...X bonds can be formed. Even though they are typically very weak, of the order of 0.03 vu, they can have a significant effect on structure and properties because organic compounds usually contain many hydrogen atoms. In the acetate ion C–H...O bonds are responsible for the length of the C–C bond being reduced from the single bond length of 1.54 to 1.51 Å as further described in Sect. 10.1 [11, pp. 108–9].

The ammonium ion, NH_4^+, which is isoelectronic with methane, has a residual valence of +1.0 vu, ensuring that the hydrogen atoms have a residual valence of +0.25 vu with the valence of the N–H bonds reduced to 0.75 vu. In amine groups where one ammonia hydrogen atom is replaced by an organic residue, R, the strength of the hydrogen bonds determines the valence of the N–R bond and hence the bonding strength of any Lewis base functions on R.

8.2 Incommensuration

Steric strain can also arise when two different components of a crystal structure have natural translation distances that are incommensurate with each other. Since a crystal has full translational symmetry, all its components must possess translations that are commensurate in all three dimensions. If the natural translations of the components are not equal, there are two possibilities. In the first case, the interactions within the components are larger than the interactions between components, so that each component adopts its own spacing. The result is an incommensurate crystal with two independent lattice spacings along one or more of the crystal axes, as found in $Hg_{2.68}AsF_6$ in which the AsF_6^- anions form a lattice that contains channels occupied by linear chains of mercury atoms whose bond length is incommensurate with the spacing of the AsF_6^- lattice [33]. On the other hand, if the interaction between the two components is stronger than the interaction within one or both components, then the spacing of each component will change in order that they both adopt the same lattice spacing. This is the situation in $BaTiO_3$ where the three-dimensional TiO_3^{2-} framework contains cavities that are too small for the barium atoms, causing the framework to expand. This requires the Ti–O bond to be stretched and the Ba–O bonds to be compressed, with a consequent deviation of these bond lengths from their ideal values. When this happens, both of the network equations (14a) and (14b) may be violated, but the loop rule (14b), being the weaker constraint, will always be violated.

$BaTiO_3$ is a member of the perovskite series of structures which provide a good illustration of this type of steric strain. With the composition ABX_3, ideal bond lengths can be calculated for both the A–X and B–X bonds, but since the structure is cubic with only one adjustable parameter, the lattice parameter, it is in general impossible to find a lattice parameter that simultaneously matches both distances. Consequently one set of bonds must be stretched and the other compressed. According to the distortion theorem (5), relaxation involves distortion of the environment of the cation with the stretched bonds, leading to structures in lower symmetry space groups. As they describe in [14], Lufaso and Woodward [34] have used bond valences in their program SPuDS to predict the distortion that will be found in a perovskite of a given composition.

9 Properties of Water

9.1 Structure

As shown in Sect. 8.1.1, water molecules have a Lewis acid strength of 0.2 vu through hydrogen and a Lewis base strength of −0.2 vu through oxygen. Consequently water molecules are perfectly valence-matched to each other. It is not surprising that water is a liquid, or on cold days a solid, rather than a gas. Water is undoubtedly the most significant molecule on our planet, and although it has been extensively studied, its properties are still not fully understood.

The oxygen atom of a water molecule can form four hydrogen bonds with other water molecules, two as the hydrogen-bond donor and two as the acceptor. Since there are no other species present to exert additional constraints, the bonds in ice or liquid water are expected to be normal, i.e., they should be linear with an O...O separation of 2.87 Å, and according to the principle of maximum symmetry, the bonds are expected to be arranged tetrahedrally around each oxygen atom. This is the structure of ice, but the packing efficiency of tetrahedral structures is poor and they tend to have low densities. A denser packing can be achieved if the O–H...O bonds are bent which requires the hydrogen bonds to be longer and more asymmetric. This is possible in the liquid where the atoms are not held in a rigid framework. A range of hydrogen bond geometries is expected in liquid water. Most hydrogen bonds are normal and approximately linear, but a small number are significantly bent and more asymmetric. This picture differs from that of a commonly used model of water that assumes each hydrogen atom either forms a normal hydrogen bond or no hydrogen bond at all, with no intermediate state considered. Realistic simulations of the structure of liquid water agree well with the expectations of the bond valence theory [35].

9.2 Solubility

The dissolution of solids in water is a chemical reaction in which the atoms of the solid react with the surrounding water molecules and are taken into solution. Solubility is determined by whether the Lewis acid and base strengths of the cations and anions of the solid are better matched to each other than they are to the molecules of water. If they are better matched to each other, the solid is insoluble, but if they are better matched to water they will dissolve. Applying the valence matching rule (9) to the Lewis acid and base strengths of water (±0.2 vu), one would expect water to form stable bonds with anions and cations having bonding strengths between 0.1 and 0.4 vu. Sodium ($S_{Na} = 0.16$ vu) and chlorine ($S_{Cl} = -0.17$ vu) are well matched to each other, but they are also well matched to water. NaCl readily dissolved in water, but equally readily crystallizes out when the water is removed. Magnesium ($S_{Mg} = 0.33$ vu) and carbonate

($S_B = -0.22$ vu) are reasonably well matched to each other but both also lie within the range of solubility and can be dissolved. $MgCO_3$ is only slightly soluble and easily crystallizes to form the familiar stalactites and stalagmites found in caves. Magnesium and sulfate ($S_{sulfate} = -0.17$ vu) are less well matched to each other and sulfate is well matched to water. $MgSO_4$ dissolves in water with the formation of discrete $Mg(H_2O)_6{}^{2+}$ ions in solution. The Mg–O bonds of 0.33 vu result in the hydrogen atoms of the coordinated water molecules having a bonding strength of 0.33/2 = 0.17 vu. This allows both the sulfate and the magnesium ions to form normal hydrogen bonds with the surrounding water molecules, but recrystallizing this compound from water does not yield crystals of the original $MgSO_4$ but rather crystals of $Mg(H_2O)_6SO_4$ (H_2O) in which all the valences are better matched. Any compound composed of ions with bonding strengths larger than 0.4 vu, such as silica SiO_2 ($S_{Si} = 1.0$ vu, $S_O = -0.5$ vu), is insoluble. A revealing example of the valence matching rule is found in the fluorides of the alkaline earths which are an exception to the usual rule that the properties of elements vary monotonically as one moves down the Periodic Table. Fluorine has a bonding strength of −0.25 vu, somewhat larger than that of water, the alkaline earths have bonding strengths of 0.33 vu (Mg), 0.27 vu (Ca), 0.23 vu (Sr), and 0.20 vu (Ba). The best matches for fluorine are calcium and strontium with the result that the fluorides of these two elements are less soluble than the fluorides of either magnesium or barium.

9.3 Solution Chemistry

Thermodynamic studies of water divide cations into those that increase the entropy during dissolution (structure breaking) and those that decrease the entropy (structure making). As one might expect, there is a close relationship between the molar entropy of solution and the bonding strength of the cation [11, p. 57]. Cations with bonding strengths less than 0.2 vu are structure breaking, those with bonding strengths in the range of 0.3–0.4 vu form complexes with a single coordination sphere of water while those with larger bonding strengths form complexes with a double coordination sphere, the outer sphere being attached to the inner sphere by hydrogen bonds with a larger valence than those normally found between water molecules. As shown in Sect. 9.2, magnesium with a bonding strength of 0.33 vu forms six bonds to water, but the $Mg(H_2O)_6{}^{2+}$ complex forms 12 hydrogen bonds with a bonding strength of 0.17 vu, distributing the two valence units of magnesium over a the 12 hydrogen atoms to give a complex with a bonding strength that is a good match for the solvent water. In this example, the water of the hydration sphere acts like a transformer, changing the bonding strength of magnesium from 0.33 to 0.17 vu to allow it to form normal hydrogen bonds with water. Iron(III) forms six bonds of 0.5 vu to water, resulting in the water molecules forming hydrogen bonds to the second coordination sphere of 0.25 vu. As this valence is larger than that of the hydrogen bonds in the surrounding liquid, the second coordination sphere is

Table 7 Bonding strength of protonated orthosilicates

Complex	V	V_R	N	$S_B = V_R/N$
SiO_4	−4	−4	12	−0.33
$HOSiO_3$	−3	−3.2	10	−0.32
$(HO)_2SiO_2$	−2	−2.4	8	−0.3
$(HO)_3SiO$	−1	−1.6	6	−0.27
$(HO)_4Si$	0	−0.8	4	−0.2

V is the formal valence of the complex, V_R is its residual valence when the formation of hydrogen bonds is taken into account, N is the number of external bonds the oxygen atoms are expected to form, and S_B is the Lewis base strength of the complex ion

strongly attached to give $Fe(H_2O)_6(H_2O)_n^{3+}$ with $n \sim 9$, a number that gives a bonding strength equal to that of water.

Similarly anions, such as perchlorate ($S_{perchlorate} = -0.08$ vu), with a small bonding strengths tend to disrupt the structure of liquid water, while those such as phosphate ($S_B = -0.25$ vu) with larger bonding strengths tend to remove the hydrogen from water to form protonated anions such as hydrogen phosphate ($S_B = -0.22$ vu). In this way strongly bonding anions can reduce their bonding strength until it matches that of water.

Table 7 shows how the bonding strength of an orthosilicate ion, SiO_4^{4-}, changes with successive protonation. The formal ionic charge of the complex is shown in the column labeled V. This is the residual valence that would be expected in the absence of any hydrogen bonding, but since all the hydrogen atoms are expected to form external bonds of valence +0.2 vu, the protonated complexes have a Lewis acid strength of +0.2 vu as well as the expected Lewis base function acting through oxygen. While the total formal charge on the complex is V, the true residual valence, V_R, available to the Lewis base function is found by adding −0.2 vu to V for each hydrogen bond formed by the complex. The expected coordination number, N, is based on the assumption that oxygen will form three external bonds when it is terminal and one when it is part of a hydroxy group. The Lewis base strength is then the ratio of V_R to N. Increasing the degree of protonation lowers the base strength of the silicate anion, with only the fully protonated ion providing a proper match with water. All the protonated complexes have a Lewis acid strength of +0.2 vu through the hydrogen atoms, so the Lewis acid and base strengths of $(HO)_4Si$ are the same, and both are equal to those of water. Like water, $(HO)_4Si$ acts as both an acid and a base of equal strength. Adding further hydrogen atoms would turn the complex into a cation.

The ability of strong Lewis bases such as SiO_4^{4-} to abstract hydrogen atoms from water results in an excess concentration of OH^- ions, and the ability of strong Lewis acids to shed hydrogen atoms from their coordinated water molecules results in an excess concentration of H^+ ions both causing changes in the pH as the respective ions dissolve. It is no surprise that one finds a linear correlation between the pK_a and the logarithm of the anion bonding strength as shown in Fig. 13 [11, p. 47].

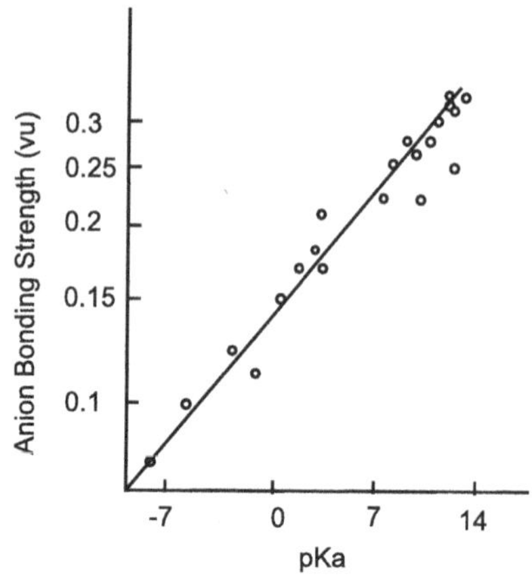

Fig. 13 Correlation between bond valence and pK_a (Fig. 4.3 from [11] by permission of the Oxford University Press)

10 Reactivity

10.1 Lewis Acid and Base Compounds

In the ammonia molecule, NH_3, nitrogen acts as an anion with a valence of −3 vu and the hydrogen as a cation with a valence of +1. In the isolated ammonia molecule, the three N–H bonds have a valence of 1.0 vu with none of the atoms carrying any residual valence. As a result of the strong N–H bond, the lone pair on nitrogen is stereoactive (Sect. 7.1). However, the hydrogen atoms are expected to form hydrogen bonds. The internal donor N–H bonds would have valences of around 0.8 vu, which will tend to reduce the stereoactivity of the nitrogen lone pair. As shown in Table 6, this gives the ammonia complex a Lewis acid strength (at hydrogen) of +0.2 vu and a Lewis base strength (at nitrogen) of −0.6 vu. According to the valence matching rule (9), the nitrogen should bond to a cation with a bonding strength in the range of +0.3 to +1.2 vu, a range typically found in transition metals and organic compounds, which makes ammonia a good complexing agent in coordination chemistry. It also explains why ammonia reacts with BF_3. In a manner complementary to ammonia, BF_3 has a Lewis acid strength of around 0.6 vu at boron and a Lewis base strength of around −0.2 vu at fluorine, the B–F bonds having a valence of around 0.8 vu. However, since both NH_3 and BF_3 have a net residual valence of zero, the N–B bond can only be formed if the H and F atoms also form external bonds. NH_3BF_3 only forms a stable compound because in the solid the molecules are linked by N–H...F hydrogen bonds. This description differs from the usual explanation for the existence of this compound which calls the N–B bond a "dative bond" formed by the lone electron pair on nitrogen. While this seems a plausible explanation, it contains an inherent contradiction since a lone pair is required to be nonbonding. If it is used for bonding, the valence of nitrogen must change from −3 to −5, which is forbidden by the octet rule (11). The bond valence theory provides a more realistic description by pointing

Table 8 Bonding strengths, S, of silicate ions

Complex	V_R (vu)	N_O	$S = V_R/N_O$ (vu)	
SiO_4	4	12	0.33	Isolated ion
Si_2O_7	6	19	0.32	Isolated dimer
SiO_3	2	7	0.29	Chain
Si_2O_5	2	4	0.14	Sheet
SiO_2	0	4	0	Framework

out that the base function on nitrogen can only be activated if the ammonia molecule forms hydrogen bonds which allows it to divert some of the bonding electrons from the N–H to the N–B bond. The lone pair remains nonbonding, but in NH_3BF_3 it is sterically inactive.

Complex ions as well as neutral molecules can sometimes act as both a Lewis acid and a Lewis base. The acetate ion, $H_3CCO_2^-$, being an anion, is primarily a Lewis base with a residual valence of −1.0 vu split between the two oxygen atoms to give a Lewis base strength of −1.00/6 = −0.17 vu, but the methyl hydrogens can also act as weak Lewis acids, typically with a residual valence of around 0.03 vu. If all three hydrogen atoms form hydrogen bonds, the residual valence on the two oxygen atoms increases from −1.00 to −1.09 vu, and the Lewis base strength of the acetate ion increases to −1.09/6 = −0.18 vu. The bonding strength of trifluoroactetate, on the other hand, is smaller (0.16 vu) because the residual negative valence of −1.0 vu must be shared between the oxygen and the fluorine atoms. This difference is reflected in the pK_a values of these two ions.

According to the principle of maximum symmetry, the residual anion valence of the acetate ion should be distributed equally between the two oxygen atoms, allowing it to bond to any cation with a bonding strength in the range of +0.09 to +0.36 vu, but it can bond to more strongly bonding cations, such as silicon (S_{Si} = 1.0 vu) to form $Si(O_2CCH_3)_4$ by distributing its residual valence unequally between the two oxygen atoms [36].

10.2 Stability

In a paper titled "Nonexistent silicates" Dent-Glasser [37] pointed out that no condensed silicates were known with transition metals, and no orthosilicates were known with alkali metals, an observation that is readily understood in terms of valence matching. Table 8 shows the bonding strengths of ortho- and condensed silicates. The bonding strengths of the formula units shown are equal to the residual valence of the complex, V_R, divided by the number of external bonds, N_O, that the formula unit forms. Terminal oxygen atoms are assumed to form three external bonds and bridging oxygen atoms one. The orthosilicate anion, SiO_4^{4-} has a bonding strength of 4/12 = 0.33 vu which matches the bonding strengths of transition metals which range from 0.3 to 0.5 vu, while condensed silicates have bonding strengths in the range of 0.1–0.2 vu which matches the bonding strengths of alkali metals (0.11–0.20 vu). The known silicates are the ones that obey the valence matching rule (9) and Dent-Glasser's nonexistent silicates include all those that do not obey this rule [11, p. 51].

Many biological compounds are required to be stable but at the same time to be sufficiently soluble that they can be assembled or disassembled under aqueous conditions at ambient temperatures. This requirement can be met if the molecule is at its stability limit with the bonding strengths of its components differing by a factor of 2. This condition is satisfied by the hydrolysis of adenosine triphosphate (ATP) into adenosine diphosphate (ADP) and phosphate, the principal method by which energy is transferred in living organisms. ATP can hydrolyze to give either ADP^{3-} and PO_3^- or ADP^- and PO_4^{3-}. In calculating the bonding strength, the coordination number of the bridging oxygen is taken as two, since this reflects its actual coordination number in ATP. With this choice, the Lewis acid or base strength is equal to the residual valence on the atoms that formed the broken bond. The Lewis base strength of ADP^{3-} is −0.6 vu assuming the residual valence of ADP^{3-} is distributed uniformly over all five oxygen atoms. This is to be matched with the Lewis acid strength of P in PO_3^- (+1.25 vu), assuming that all three P–O bonds have the same valence. For this process, the ratio $S_A/S_B = 2.08$. The second possible hydrolysis route yields ADP^- and PO_4^{3-} with the bridging oxygen remaining on the phosphate group. Under the same assumptions ADP^- is the Lewis acid acting through P with $S_{ADP} = +1.4$ vu and PO_4^{3-} is the Lewis base with $S_{phosphate} = -0.75$ vu giving $S_A/S_B = 1.87$. Both routes have ratios close to 2, which is the limit for bond formation, allowing ATP to form or hydrolyze with only a small change in the ambient conditions [11, pp. 201–2].

The mineral apatite, $Ca_3(PO_4)_2$, gives bone its strength. Its components, Ca^{2+} ($S_{Ca} = 0.27$ vu) and PO_4^{3-}, ($S_{phosphate} = -0.25$ vu) are well matched to each other, allowing them to form a strong material, but the bonding strengths are still within the range that matches water, ensuring that both components can be moved through the body to where they are needed, but in this case that the bone, with its better valence match, is more stable against dissolution.

11 Structure Analysis and Prediction

11.1 The Global Instability Index

There are a number of tools that are useful in validating either a measured or a proposed structure. The most widely used is the valence sum rule (2) using bond valences calculated from the observed bond lengths. Experimental uncertainties will mean that the atomic valence and bond valence sum are rarely exactly the same, but for a well-determined structure the difference is usually around 0.05 vu. Larger differences are often found, indicating that some bonds are compressed or stretched by the steric constraints imposed on the structure (Sect. 8). In some cases the differences can be quite large, but for a strained structure to be in equilibrium both stretched and compressed bonds must be present, which is key to verifying the presence of steric strains.

A combined measure of the steric strain over the whole structure is the **global instability index**, G defined in Eq. (23). Experience with this index shows that few stable structures have values of G greater than 0.20 vu. An observed structure with a larger value should be carefully examined to ensure that the structure determination is correct and that the bond lengths have been properly converted to bond valences. Values of G in the range 0.05–0.20 vu indicate the presence of strained bonds; the larger the value of G, the more strained the structure. Garcia-Muñoz and Rordiguez-Carvajál [38] examined an isostructural series of rare earth compounds and showed that those in which G was predicted to exceed 0.20 vu crystallized with a different structure type.

A more interesting index, but one more difficult to apply, is the **bond strain index**, B, which compares the observed and predicted bond valences by summing the squares of the deviation over all m bonds (24):

$$B = \left(\Sigma\{S_{\text{observed}} - S_{\text{ideal}}\}^2/m\right)^{1/2} \tag{24}$$

This calculation requires ideal bond valences to be predicted using the network equations (14a) and (14b). B will not be zero if either electronic anisotropies (Sect. 7) or steric strains (Sect. 8) are present. It measures the deviation from the predictions of the network equations, but does not indicate the origin of these deviations, nor does it measure instability. B will always be large if hydrogen bonds are present, masking the possible presence of other effects.

11.2 Predicting and Mapping the Bond Network into Euclidian Space

In many cases the bond network can be generated from a knowledge of the bonding strengths [11, pp. 134ff]. For a given composition, the first step is to use the electronegativity (Table 1) to identify the anions and the cations while ensuring that the sum of all the atomic valences is zero. The cations and anions with the largest bonding strength are then linked by bonds to form complex ions whose bonding strengths can be matched with the remaining counterions. At each stage, the principle of maximum symmetry is used to decide between alternative choices. This approach works well for binary and ternary compounds, but packing considerations become important if many different elements are present, or if the compound contains weakly bonding cations such as alkali metals, since the number of alternative ways of constructing the network becomes large, and the spatial arrangements of the atoms impose additional constraints. If the bond network has a high symmetry, mapping into three-dimensional (Euclidean) space becomes a straightforward exercise of finding the highest symmetry space group that can accommodate the network [11, pp. 129ff].

11.3 Simulations

A traditional method for predicting complex chemical structure and dynamics is computer simulation, in which atoms are placed at arbitrary, though preferably favorable, positions and are moved in response to the forces acting on them, in such a way as to minimize a cost function, such as the total potential energy [39]. The cost function is usually based on a Coulomb potential, and the total energy of the system is calculated using either quantum or classical mechanics. Whichever method is used, all simulations necessarily involve simplifying assumptions and the adoption of fitted parameters to suit the particular system.

For inorganic compounds, the ionic model (Sect. 5) is the basis of both the classical two-body potential model and the bond valence theory; either can be used, separately or together, in a simulation. The two models are based on identical assumptions, the only difference being in the way in which they describe the repulsion between neighboring atoms. The two-body potential model represents this repulsion (and other factors such as polarization) by adding empirical terms to the cost function; the bond valence theory represents the repulsion through the empirical bond valence parameters, R_0 and b, in Eq. (4). Unlike the repulsive potentials which are optimized for each structure, the bond valence parameters are tabulated, and as they require no special fitting, they are robust. The principal difference between the two approaches lies in the ways in which the calculations are performed: the two-body potential model minimizes a single potential energy, while the bond valence theory ensures that the bond valences obey local rules around each atom. Apart from the choice of empirical parameters, both methods should lead to the same structure. Although bond valence terms are sometimes added to the cost functions, no simulation has yet been performed using only the bond valence rules.

12 Epilogue

Chemical structures are too complex to be described by any single theory without drastically simplifying assumptions, and any quantitative theory requires at least some parameters to be fitted empirically. Consequently we have a variety of models or theories of chemical structure, based on different assumptions, covering different materials and designed to suit different needs, depending on whether one is looking for insights or quantitative predictions, whether one wants simplicity or an understanding of the basic physical principles.

One can identify four desirable characteristics for a theory of chemical structure:

1. It reveals the physical principles involved.
2. It applies to a wide range of compounds.
3. It offers simple insights into the factors that determine chemical structure.
4. It makes quantitative prediction of properties, preferably with simple calculations.

No existing theory meets all of these criteria, which is why structural chemistry has so many different models, but these criteria provide a standard against which each can be evaluated. Since the first pictures of chemical structure appeared in the nineteenth century, many different theories, and variations on theories, have been proposed. The bond valence theory, which is one of these, began a 100 years ago with the ionic model, took form with Pauling's electrostatic valence principle and has evolved into the relatively sophisticated model that we have today. Its ability to give simple quantitative predictions for complex structures appeals to mineralogists and materials scientists, since it provides insights which the extensive computation required by other models tends to obscure. Its greatest weakness in the past has been its apparent lack of any basis in physical theory; its only justification being its surprising success in giving a quantitative account of extended structures.

This chapter has been an attempt to rectify this weakness by taking a new look at the physical basis of the chemical bond. It shows how the rules of the bond valence theory can be derived from established physical principles by making a number of simple assumptions about the properties of atoms and developing the model using the electrostatic field. It provides many insights into chemical structure, but at the cost of losing some of the insights provided by other models. Just as the concept of a chemical bond is not found in the traditional physical models, the concepts of energy and electron density distribution are not found in the bond valence theory. If energy or electron density are important, a different approach is needed.

Even though the bond valence theory does not explain how atoms adhere to each other and its scope is limited to localized bonds, it does provide simple insights into many of the factors that determine chemical structure. It is a model which has been slow to develop, but which has a potential that is far from exhausted. It is a model that will be all the stronger for having the secure base in physical theory presented here.

References

1. Brown ID (2013) Chap. 1 in this volume
2. Brown ID (2013) Chap. 9 in this volume
3. Boeyens JCA, Schutte CJH (2012) In: Putz MV (ed) Chemical information and computational challenges in the 21st century. Nova Science Publishing, New York (Chapter 5)
4. Pritchard H (2012) J Chem Educ 89:301–303
5. Bader RFW (1990) Atoms in molecules, a quantum theory. Oxford University Press, New York
6. Pauling L (1929) J Am Chem Soc 51:1010–1026
7. Bragg WL (1930) Zeit Kristallogr 74:237–305
8. Adams S (2013) Chap. 3 in this volume
9. Pauling L (1947) J Am Chem Soc 69:542–553
10. Brown ID (2012) http://www.iucr.org/resources/data/datasets/bond-valence-parameters
11. Brown ID (2002) The chemical bond in inorganic chemistry, the bond valence model. Oxford University Press, Oxford

12. Allmann R (1975) Monatsh Chem 106:779–793
13. Urusov VS (2003) Z Kristallogr 218:709–719
14. Lufaso MW, Woodward PM (2013) Chap. 4 in this volume
15. Brown ID (1988) Acta Crystallogr B 44:545–553
16. Brown ID (2011) J Phys Chem A 115:12638–12645
17. Preiser C, Lösel J, Brown ID, Kunz M, Skowron A (1999) Acta Crystallogr B 55:698–711
18. Orlov IP, Popov KA, Urusov VS (1998) J Struct Chem 39:575–579 (see also Appendix 2)
19. Kunz M, Brown ID (1995) J Solid State Chem 115:395–406
20. Brown ID (1973) Acta Crystallogr B 29:1979–1983
21. Harvey MA, Baggio S, Baggio R (2006) Acta Crystallogr B 62:1038–1042
22. Zachara J (2007) Inorg Chem 46:9760–9766
23. Bickmore BR, Wander MFC, Edwards J, Maurer J, Shepherd K, Meyer E, Johanson WJ, Frank RA, Andros C, David M (2012) Am Mineral (in press)
24. Waltersson K (1978) Acta Crystallogr A 34:901–905
25. Adams S (2013) Chap. 5 in this volume
26. Brown ID, Gillespie RJ, Morgan KR, Tun Z, Ummat PK (1984) Inorg Chem 23:4506–4508
27. Gillespie RJ, Hargittai I (1991) The VSEPR model of molecular geometry. Allen and Bacon, Needham Heights
28. Alcock NW (1972) Adv Inorg Rad Chem 15:1–57
29. See RF, Kruse RA, Strub WM (1998) Inorg Chem 37:5369–5375
30. Shields GP, Raithby PR, Allen FH, Motherwell WDS (2000) Acta Crystallogr B 56:455–465
31. Petch HE (1957) Can J Phys 35:983–985
32. Brown ID (1976) Acta Crystallogr A 32:786–792
33. Brown ID, Cutforth BD, Davies CG, Gillespie RJ, Ireland PR, Vekris JE (1974) Can J Chem 52:791–793
34. Lufaso MW, Woodward PM (2001) Acta Crystallogr B 57:725–738 (see also Appendix 2)
35. Bickmore BR, Rosso KM, Brown ID, Kerisit S (2009) J Phys Chem A 113:1847–1857
36. Brown ID (1980) J Chem Soc Dalton Trans 1118-1123
37. Dent-Glasser L (1979) Z Kristallogr 149:291–325
38. Garcia-Munoz JL, Rordiguez-Carvajal J (1995) J Solid State Chem 115:324–331
39. Catlow CRA (ed) (1997) Computer modelling in inorganic chemistry. Academic, San Diego

Struct Bond (2014) 158: 59–90
DOI: 10.1007/430_2012_90

Published online: 16 June 2013

Using Bond Valences to Model the Structures of Ternary and Quaternary Oxides

Michael W. Lufaso and Patrick M. Woodward

Abstract The bond valence method is implemented in the modeling of crystal structures with the software program SPuDS. The approach is investigated for the perovskite, pyrochlore, spinel, and garnet structure types. Crystal structures of selected compositions were calculated and compared to experimental structures that were determined using X-ray or neutron diffraction. Bond valence sums (BVSs) of the ions and the global instability indices (G) are investigated to provide insight into the structures of these four structural classes of materials. The predictive ability is examined in the context of understanding the structures of existing compounds. The accuracy and possible uses of the crystal structures obtained from modeling using bond valences are explored.

Keywords Bond valence · Garnet · Perovskite · Pyrochlore · Spinel · SPuDS · Structure modeling · Structure prediction

Contents

M.W. Lufaso (✉)
Department of Chemistry, University of North Florida, 1 UNF Drive, Jacksonville, FL 32224, USA
e-mail: michael.lufaso@unf.edu

P.M. Woodward
Department of Chemistry, The Ohio State University, 100 West 18th Avenue, Columbus, OH 43210-1185, USA
e-mail: woodward@chemistry.ohio-state.edu

Abbreviations

BVS Bond valence sum
SPuDS Structure Prediction Diagnostic Software
v.u. Valence units

1 Introduction

The bond valence model offers an approach to structural modeling that is attractive because of its simplicity, low computational requirements, and intuitive results. Structure types where symmetry constraints limit the number of free positional parameters make particularly attractive targets for structural modeling that is based on bond valence concepts. The software program SPuDS (Structure Prediction Diagnostic Software) is one such example of this approach. It was originally developed to model the structures of perovskites and the various modifications from the aristotype cubic structure including octahedral tilting, cation ordering, and Jahn–Teller distortions. Initial efforts focused on the single octahedral cation structures (ABX_3) [1] and later was extended to ordered double perovskites ($A_2BB'X_6$) [2] and triple perovskites ($A_3BB_2'X_9$) [3]. Applications of SPuDS include generating starting models for structural refinements or more sophisticated calculations, guiding efforts to synthesize new perovskites [4, 5], or to help deconvolute various structural distortions [6].

In this chapter we review the use of SPuDS to model the structures of perovskites and extend this approach to three additional structure types that play a prominent role in solid state materials chemistry: pyrochlore, spinel, and garnet.

2 Cubic ABX_3 Perovskites

The perovskite structure, which is adopted by a large number of mixed-metal oxides [7], has ABX_3 stoichiometry, where A and B are cations and X is an anion. The ideal perovskite structure, shown in Fig. 1, has the A-site cation occupying a 12-coordinate site within a framework of corner-connected $[BX_{6/2}]$ octahedra. The tolerance factor is a geometrical relationship based on a hard sphere model of the atom that gives a measure of the fit of the A-site cation to the octahedral framework. Mathematically the tolerance factor (t) is given by the following expression, $t = (1/\sqrt{2})[(r_A + r_X)/(r_B + r_X)]$, where r_A, r_B, and r_X are the radii of the A, B, and X ions, respectively. When $t = 1$ the A-site cation is a

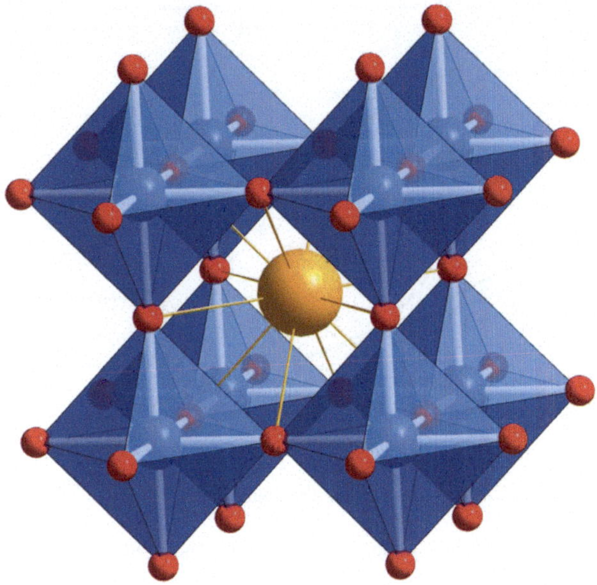

Fig. 1 Crystal structure of a cubic ABX_3 perovskite. *Orange spheres* represent the A cation, *blue spheres* B cations, and *red spheres* X anions

perfect fit, but in most cases t deviates from unity. If the tolerance factor is greater than unity, the A-site ion is too large for the cavity. If the B-sites are d^0 cations in a perovskite with a $t > 1$, the octahedra are prone to undergo an electronically driven distortion as a consequence of a second-order Jahn–Teller distortion [8, section 7.4.1], as observed in $BaTiO_3$ ($t = 1.06$) and $KNbO_3$ ($t = 1.09$). In cases where the A-cation is too small relative to the corner-sharing environment of the $[BX_6]$ octahedra, cooperative octahedral tilting tends to occur. The former distortion effectively increases the volume of the BX_6 octahedra (according to the distortion theorem) (Eq. (5) in [8]) [9] while the latter decreases the volume of the AX_{12} polyhedron [10].

In the aristotype ABX_3 perovskite structure, with $Pm\overline{3}m$ space group symmetry, there is a single free parameter, the cubic lattice parameter, a. A simple method to model the crystal structure is to determine the structure parameter(s), a in this case, that minimize(s) the Global Instability Index, G (Eq. (21) in [8]).

2.1 Equation (1): Global Instability Index

$$G = \sqrt{\frac{\sum_{i=1}^{N} (d_i)^2}{N}} \tag{1}$$

with $d_i = \sum s_{ij} - V_i$, where V_i is the formal valence and $\sum s_{ij}$ is the calculated bond valence sum (BVS) for the ith ion, and N is the number of atoms in the formula unit.

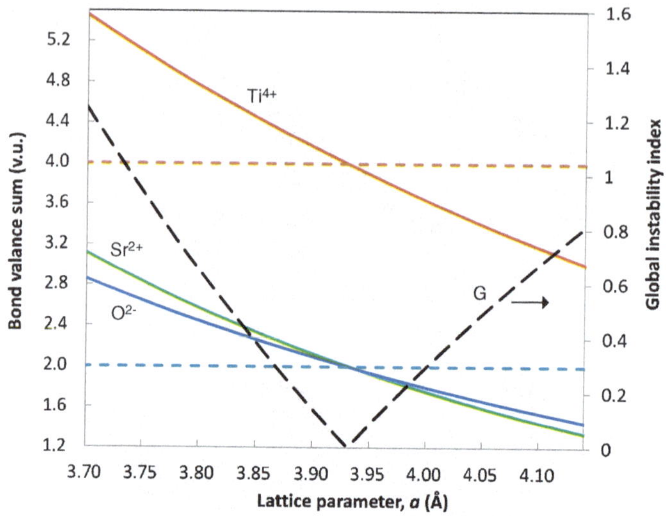

Fig. 2 Bond valence sums of the ions and global instability index versus lattice parameter for the cubic perovskite $SrTiO_3$

The bond valence parameters, calculations of bond valences, and global instability index are more fully described in [8, sections 3, 8 and 11].

With only a single variable it is computationally inexpensive to perform a grid search to find the value of *a* that minimizes *G*. The bond valences are calculated from the A–X and B–X distances and tabulated values of bond valence parameters [11]. Figure 2 shows the *G* and BVSs of the A, B, and X ions in $SrTiO_3$ as a function of lattice parameter. The observed lattice parameter is $a = 3.901(1)$ Å [12] and the predicted lattice parameter is 3.930 Å. For larger lattice parameters the BVS of the ions decreases as the interatomic distances increase, and for smaller lattice parameters the BVS of the ions increase as a result of shorter interatomic distances. The optimized structure and lattice parameter is selected where the minimum of *G* is obtained. Note this is where the BVSs of the ions are near the ideal formal oxidation states, which are shown as horizontal dashed lines.

Figure 3 shows the predicted and observed lattice parameters for eighteen undistorted cubic ABX_3 perovskites: $KTaO_3$ [13], $SrGeO_3$ [14], $BaNbO_3$ [15], $BaMoO_3$ [16], $SrVO_3$ [17], $BaSnO_3$ [18], KUO_3 [19], $BaZrO_3$ [20], $SrTiO_3$ [21], $SrMoO_3$ [22], $KMgF_3$ [23], $KNiF_3$ [24], $KZnF_3$ [25], $KCoF_3$ [26], $KFeF_3$ [27], $BaLiF_3$ [23], $KMnF_3$ [24], $RbCaF_3$ [28]. Two approaches are available in SPuDS: (a) setting the B–X distance at a value that allows the B cation to attain its ideal valence sum and (b) optimizing the *G* of the structure. The former approach is based on the assumption that the larger size and lower valence of the A-site cation makes the A–X bonds more compressible than the B–X bonds [29]. Both approaches are able to reproduce the experimental values with similar accuracy, but for the sake of comparison with modeling employed later in this chapter we only report results based on optimizing *G* here. The predicted lattice parameters are typically within 2% of the observed lattice parameters.

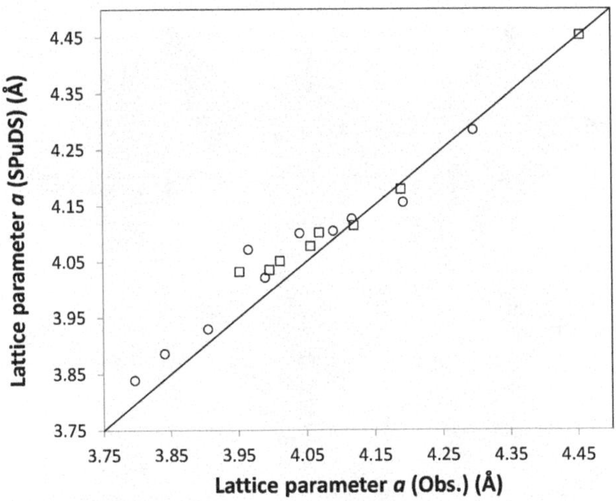

Fig. 3 Predicted and observed lattice parameters for untilted cubic perovskites calculated with optimized B–X distances. *Squares* represent ABF_3 and *circles* represent ABO_3 perovskites

Fig. 4 SPuDS predicted (*filled symbols*) and observed (*open symbols*) global instability index for untilted cubic perovskites. *Squares* represent ABF_3 and *circles* represent ABO_3 perovskites

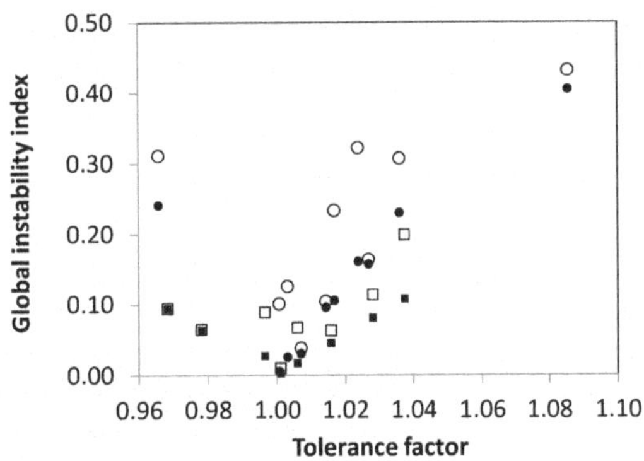

The observed and predicted global instability index versus tolerance factor for untilted cubic ABX_3 perovskites is shown in Fig. 4. As expected the G is smallest for compounds with a tolerance factor near 1. For unstrained structures the normal maximum value of G is 0.2 v.u., but here we see a number of compounds where $G > 0.2$. This can be explained in part by realizing that when t is significantly different than 1 the compounds can be classified as strained. The largest G observed ($G = 0.47$) is for $KTaO_3$ ($t = 1.08$) which has shown on multiple occasions to be a cubic perovskite [13, 30, 31]. These studies used samples prepared with several synthetic techniques and both polycrystalline powders and single crystals have been examined, which support the accuracy of the crystal structure.

Fig. 5 Crystal structure for the cubic $A_2BB'O_6$ ordered double perovskite. *Orange spheres* represent the A cation, *blue* and *green spheres* B and B′ cations, and *red spheres* X anions

3 Ordered Double Perovskites

An even larger family of compounds are the so-called $A_2BB'O_6$ ordered double perovskites, where two B-site cations are present and there is a sufficient size and/or charge between the two to drive a 3D alternation of B and B′ centered octahedra. The crystal structure of an ordered double perovskite is shown in Fig. 5. The ordered perovskites examined here are: Ba_2OsLiO_6 [32], Ba_2NiMoO_6 [33], Ba_2FeNbO_6 [34], Ba_2CoMoO_6 [35], Ba_2CoWO_6 [35], Ba_2MgWO_6 [36], Sr_2AlNbO_6 [37], Ba_2ScTaO_6 [38], Sr_2AlTaO_6 [38], Ba_2PtCeO_6 [39], Ba_2MnWO_6 [40], Ba_2InTaO_6 [38], Ba_2PtPrO_6 [41], Ba_2FeUO_6 [42], Pb_2MgTeO_6 [43], Ba_2RuYbO_6 [44], Ba_2LuRuO_6 [45], Sr_2CrMoO_6 [46], Sr_2CrNbO_6 [47], Ba_2TaYbO_6 [48], Ba_2RuErO_6 [49], Ba_2TaLuO_6 [50], Ba_2OsNaO_6 [32], Ba_2RuTmO_6 [44], Sr_2FeMoO_6 [46], Ba_2ScBiO_6 [38], Ba_2YRuO_6 [45], Ba_2UMnO_6 [51], Ba_2RuHoO_6 [52], Ba_2YNbO_6 [53], Ba_2YIrO_6 [54], Ba_2DySbO_6 [55], Ba_2TlSbO_6 [56], Ba_2CaIrO_6 [57], Ba_2YSbO_6 [58], Ba_2HoSbO_6 [58], Ba_2SmSbO_6 [59], and Ba_2CaMoO_6 [60].

Ordered perovskites with the $A_2BB'X_6$ composition have two free positional parameters, the lattice parameter a and fractional coordinate x, which defines the anion position on the 24e Wyckoff position in space group $Fm\overline{3}m$. Figure 6 shows the predicted (G optimized) and observed lattice parameters for cubic cation ordered $A_2BB'X_6$ perovskites. The predicted lattice parameters are typically within a couple percent of the observed lattice parameters. Interestingly, there is a

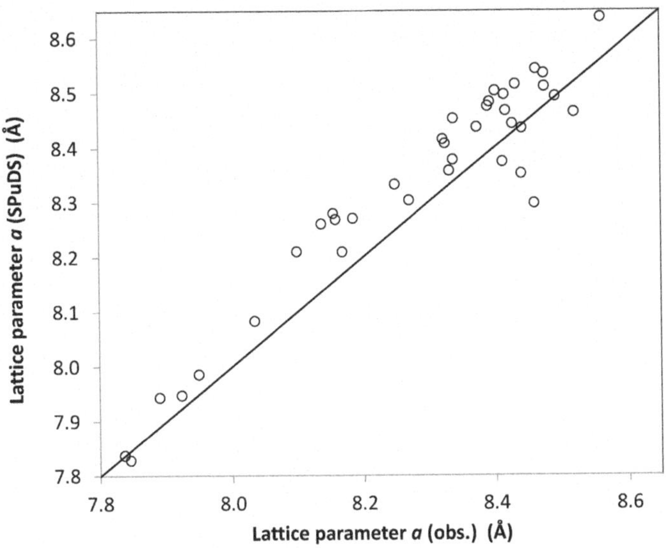

Fig. 6 Predicted and observed lattice parameters for ordered cubic perovskites

systematic tendency for SPuDS to slightly overestimate the lattice parameter. The observed and predicted values of the anion fractional coordinate are shown in Fig. 7. Once again we see agreement within 1–2% for most of the compounds. The observed and predicted global instability index versus tolerance factor for perovskites is shown in Fig. 8. The SPuDS calculated values follow a predictable trend reaching a minimum near $t = 1$, but the G values calculated from experimental structures show considerable scatter, with $G > 0.2$ for a number of entries. There are a couple of factors that could be at work here in addition to the strain effects already discussed for ABX_3 perovskites. Structure determination from X-ray powder diffraction data may not be giving fully accurate values of the oxygen fractional coordinate, x, in some cases. In addition, there is the possibility of some level of B/B′ cation disorder for some compounds. Both effects can lead to errors in the value of G. Table 1 contains the BVSs for Ba_2NiMoO_6, Ba_2LuRuO_6, and Ba_2YNbO_6. For optimized structures of untilted ordered double perovskite with a tolerance factor > 1, the A-site and O-site tend to be overbonded and the B-site is underbonded, whereas for a tolerance factor < 1, the A-site and O-site tend to be underbonded and the B-site is overbonded.

4 Octahedral Tilting in Perovskites

Although the perovskite structure type is found for hundreds of compositions, most of these compounds are distorted from the aristotype cubic structure [1, 61, 62]. Octahedral tilting distortions are the most common type of distortion. They occur as

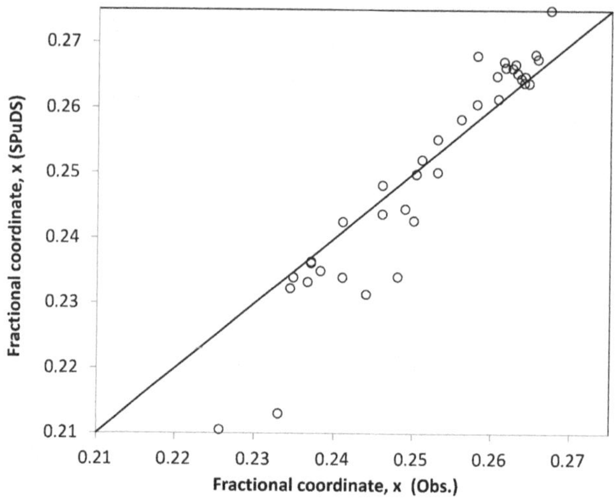

Fig. 7 Predicted and observed lattice fractional coordinates for ordered cubic perovskites

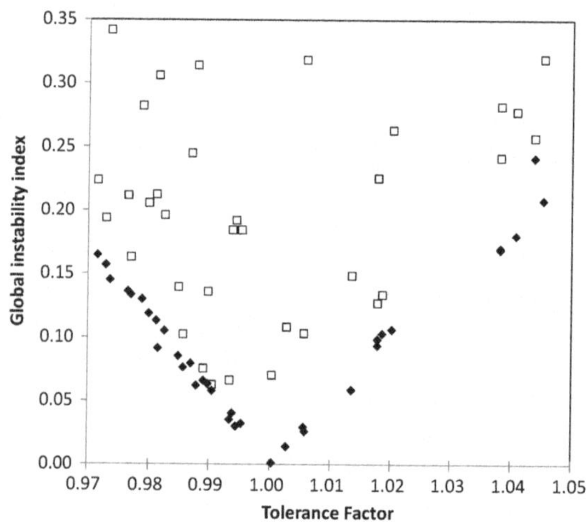

Fig. 8 Global instability index for ordered cubic perovskites with SPuDS predicted with *G* optimized (*diamonds*) and observed (*open squares*) structure

a consequence of the A-site cation being smaller than ideal for the corner-sharing network of [BO_6] octahedra ($t < 1$). Glazer introduced a commonly used notation to describe the octahedral tiling around the three principle axes of the cubic perovskite [63, 64]. Howard and Stokes have shown that by symmetry there are 15 distinct patterns of octahedral tilting, each referred to as a tilt system [65].

The software program POTATO represents an early attempt to model the structures of distorted perovskites [66]. The user is able to generate a structure by inputting the tilt system, the size of the octahedra, and the tilt angles. The program

Table 1 Observed and SPuDS predicted bond valence sums of ions for ordered perovskites Ba_2NiMoO_6 (G_{Obs} = 0.32 v.u., G_{SPuDS} = 0.21 v.u.), Ba_2LuRuO_6 (G_{Obs} = 0.18 v.u., G_{SPuDS} = 0.04 v.u.), and Ba_2YNbO_6 (G_{Obs} = 0.21 v.u., G_{SPuDS} = 0.12 v.u.)

		Observed BVS (v.u.)	*G* Optimized BVS (v.u.)
Ba_2NiMoO_6 (Toler. Fact. = 1.05) [33]	Ba	2.63	2.34
	Ni	1.92	1.60
	Mo	5.67	5.80
	O	2.12	2.02
Ba_2LuRuO_6 (Toler. Fact. = 0.99) [45]	Ba	2.12	1.93
	Lu	3.45	3.07
	Ru	5.10	5.05
	O	2.13	2.00
Ba_2YNbO_6 (Toler. Fact. = 0.98) [53]	Ba	1.80	1.79
	Y	3.49	3.20
	Nb	4.69	5.11
	O	1.97	1.98

generates a structure from these values and the assumption that the octahedra remain rigid (even when space group symmetry does not dictate it). This approach provides flexibility that can be useful for certain applications but at the same time is quite limiting because the program makes no attempt to find the most energetically stable structure. SPuDS builds upon this approach, retaining the assumption of rigid octahedra, but using bond valence concepts to find the most favorable structure.

Because the structures of distorted perovskites can have many degrees of freedom (e.g. there are 10 degrees of freedom in $a^-b^+a^-$, the most common tilt system) a grid search to find the most favorable structure is neither computationally feasible nor guaranteed to find a unique minimum. Therefore, SPuDS first uses bond valence parameters to set the B–O bond distances (B–O and B′–O bond distances for ordered perovskites) so that all six distances are equal and the B-site cation attains its ideal BVS. For perovskites with a tolerance factor of less than unity, this results in an underbonding of the A-site cation(s), a condition that can be remedied by tilting of the octahedra. The positions of the anions are calculated as a function of the tilt angle(s) using equations that have been developed for various tilt systems [1, 67, 68]. In addition to the tilt angle(s) being variables in the optimization process, the positions of the A-site cations are also variable in some tilt systems. For those structures the position of the A-site cation is determined using a valence vector approach, where the ion is located at a position that minimizes the bond valence vector sums [1]. An example valence map of the A-site cation for $CaTiO_3$ (crystal structure in Fig. 9) is shown in Fig. 10, where the *y* fractional coordinate is located on a fixed position while *x* and *z* are free parameters (this atom sits on a mirror plane). Additional details about this approach can be found in the literature [1, 69].

A wide range of tilt angles are examined and for each tilt angle the lattice parameters and fractional coordinates of each of the atoms are determined. The

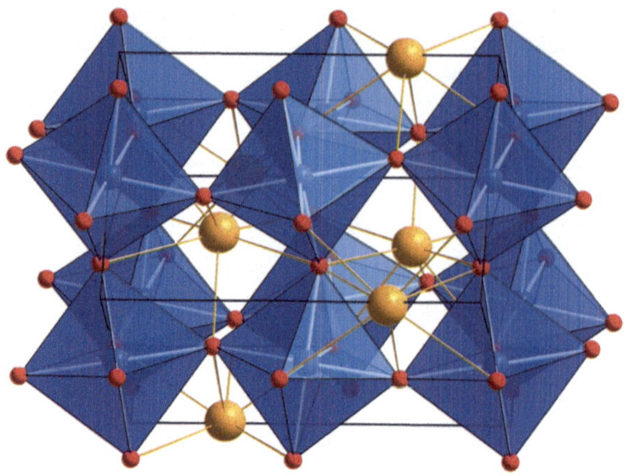

Fig. 9 Crystal structure for the $CaTiO_3$ tilted perovskite. *Orange spheres* represent the Ca^{2+}, *blue spheres* Ti^{4+} cations, and *red spheres* O^{2-} anions

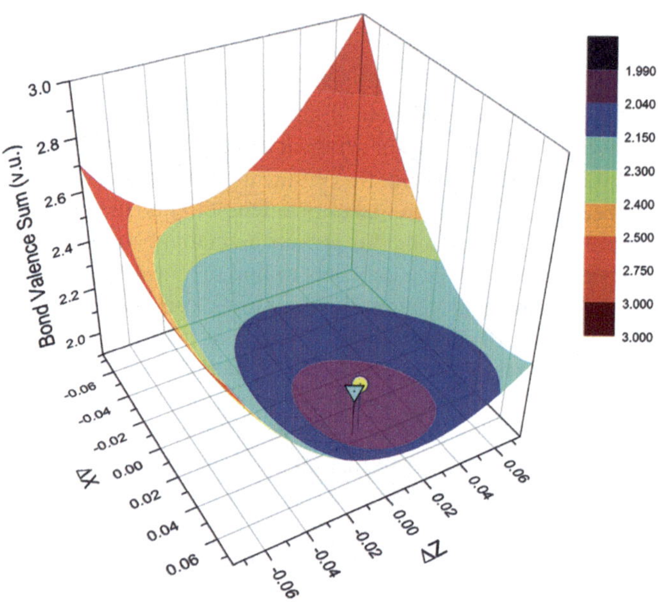

Fig. 10 Valence map plot of A-site cation position for $CaTiO_3$ in tilt system $a^-b^+a^-$ (space group *Pnma*). ΔX and ΔZ are the differences in the fractional position from the high symmetry position located at (1/2, 1/4, 1/2). The valence of A-site cation is shown as the free positional parameters are varied while holding the octahedral tilt angle at 14.60°. The *circle* is the SPuDS predicted position and the *triangle* is the experimental position

bond distances for each symmetry unique atom in the first coordination sphere are calculated. Using these data, the BVSs are calculated and G determined at each tilt angle. The optimized tilt angle is taken as the one that minimizes G. This approach is able to accurately predict the fractional coordinates of all atoms in the structures within a few percent. More details on the accuracy and limitations of SPuDS

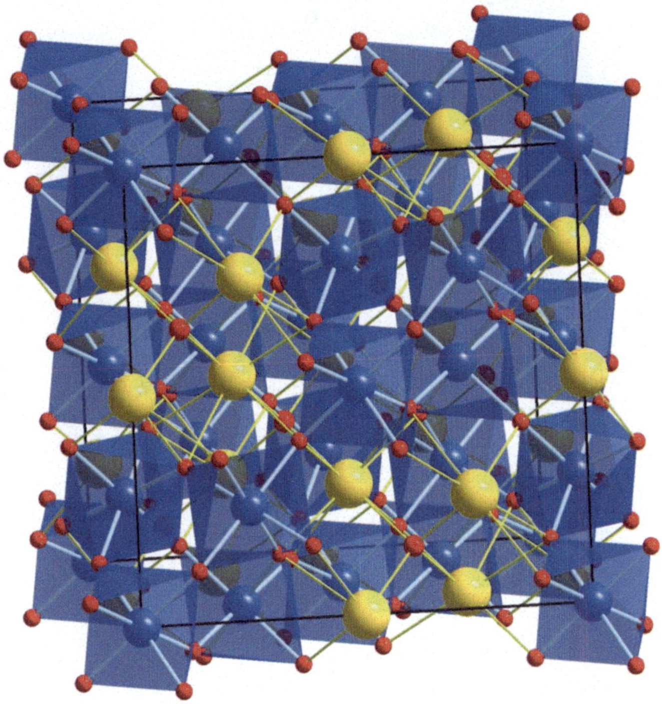

Fig. 11 Crystal structure of an $A_2B_2O_6O'$ pyrochlore. B cations (*blue*) are located in a distorted octahedral coordination and A cations (*yellow*) are in an eight coordinate environments. Anions, in four coordinate environments, are represented by *red spheres*

modeling of octahedral tilting in of single B-site perovskites [1] and ordered B-site perovskites [2, 3] can be found in the literature.

5 Pyrochlores

The pyrochlore structure has the general formula $A_2B_2O_6X$ where A and B are metal cations and X is oxygen or fluorine. The coordination numbers of A[16d], B[16c], O[48f], and X[8b] are 8, 6, 4, and 4, respectively, with Wyckoff sites in brackets. The crystal structure of a pyrochlore is shown in Fig. 11 and can be visualized as an ordered defect fluorite structure. A large number of pyrochlores are known and have been studied for interesting electronic and magnetic properties as well as for radioactive waste disposal [70]. The crystal structure accepts X-site vacancies and A-site vacancies to a certain extent; however, the focus of this section is on stoichiometric pyrochlores where X is oxygen. To distinguish the two crystallographically distinct oxygen sites, the oxide ions that occupy the X-site and are coordinated only by A-site cations will be referred to as O′ while the oxide ions

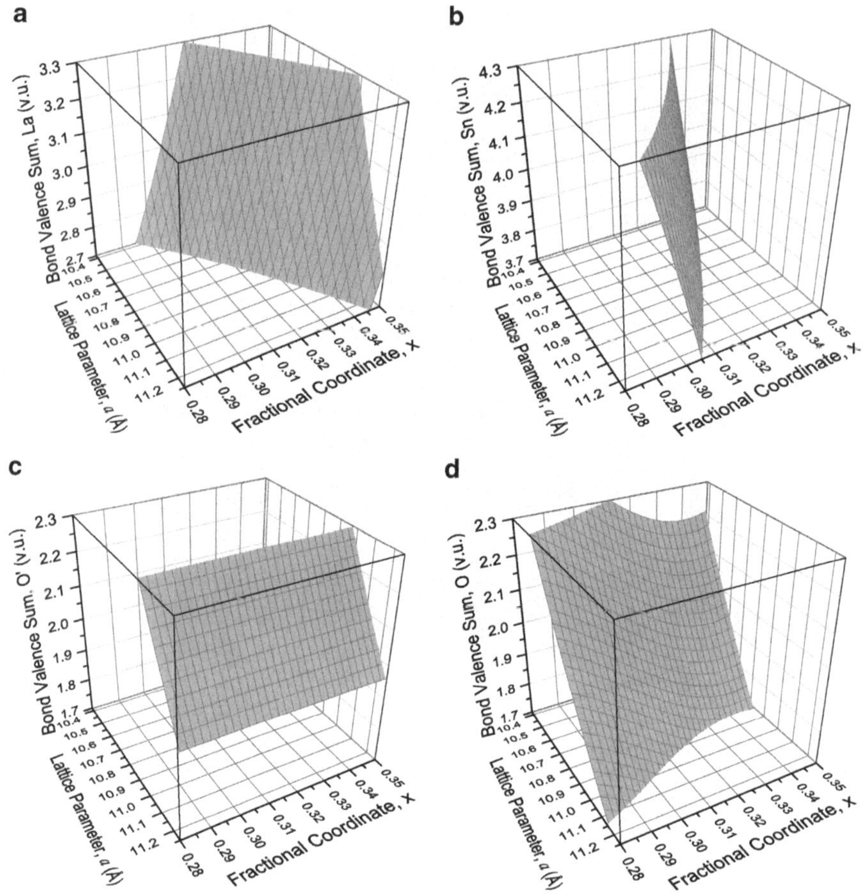

Fig. 12 3D graphs showing the bond valence sums for the four symmetry unique ions: (**a**) La^{3+}, (**b**) Sn^{4+}, (**c**) O'^{2-}, and (**d**) O^{2-}, versus the lattice parameter and fractional coordinate x of the $La_2Sn_2O_7$ pyrochlore. The optimized fractional coordinate $x = 0.3250$ and lattice parameter $a = 10.7040$ Å

that make up the B-site centered octahedra will be referred to as O. In the simple pyrochlore there are two free parameters within the structure, the lattice parameter a in the $Fd\overline{3}m$ space group and x for the O anion on the 48f Wyckoff site (x, 0, 0).

A detailed analysis of the $La_2Sn_2O_7$ pyrochlore follows. A graphical representation of the BVSs for La^{3+}, Sn^{4+}, O^{2-}, and O'^{2-} as a function of the lattice parameter and the x coordinate of the O on the 48f site is shown in Fig. 12. There is a range of fractional coordinates that results in a BVS of 3 v.u. for the La^{3+} ion. The change of the BVS of the Sn^{4+} ion is more strongly affected by the change in lattice parameter and fractional coordinates than any of the other ions. The BVS of the oxygen on the 8b site only depends on the lattice parameter. In $La_2Sn_2O_7$ ideal BVS of the O′ on the 8b site may only be achieved when $a = 11.216$ Å, but the O on the 48f site cannot achieve its ideal BVS at this value. A 2d projection of these 3d plots,

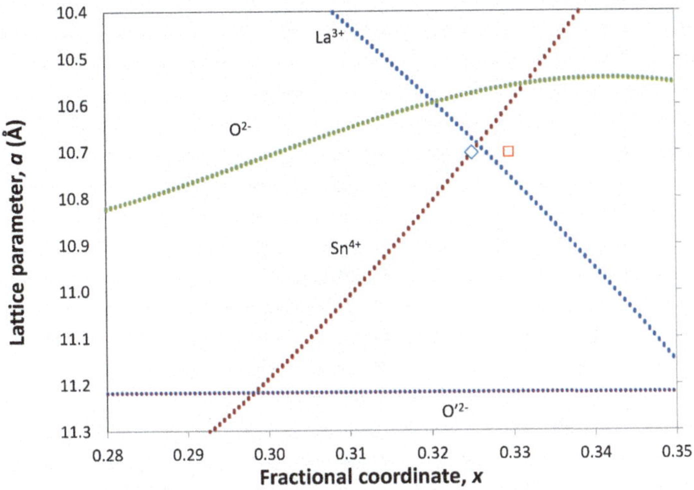

Fig. 13 Ideal bond valence sums for the four symmetry unique ions for the lattice parameter and fractional coordinate x of the $La_2Sn_2O_7$ pyrochlore. The optimized (*diamond*) fractional coordinate $x = 0.3250$ and lattice parameter $a = 10.7040$ Å and the observed values (*square*) are $x = 0.3294$ and lattice parameter $a = 10.7026$ Å

Fig. 13, shows the values in which the ideal valence is obtained. It is noteworthy the structure does not have a unique combination of lattice parameter a and fractional coordinate x that simultaneously optimizes the BVSs of each ion. In both the calculated and observed structures the values of a and x are those that come close to optimizing the valences of the La^{3+} and Sn^{4+}, while the O ion is slightly underbonded (1.863 v.u. SPuDS; 1.848 v.u. obs.) and the O′ ion is significantly overbonded (2.700 v.u. SPuDS; 2.703 v.u. obs.).

Figure 14 contains a 3d surface and 2d contour plot of the global instability index as a function of the lattice parameter and fractional coordinate for the $La_2Sn_2O_7$ pyrochlore. The predicted structure has a lattice parameter and fractional coordinate that are close to the values for the experimentally determined structure. The G for the experimental structure is 0.258 v.u., whereas the predicted structure has a slightly lower G equal to 0.235 v.u. The relatively large value of G seen for $La_2Sn_2O_7$ is something that will be repeated for many other pyrochlores. It is a result of the topological constraints associated with this structure type that prevent distortions that could alleviate the lattice induced strain (e.g. octahedral tilting in perovskites) [69].

The lattice parameters and fractional coordinates, for SPuDS predicted and experimentally observed from neutron powder diffraction, of several representative pyrochlores are shown in Table 2. The optimized structure is taken as the lattice parameter and fractional coordinate where the G is minimized. The predicted structures are similar to the experimentally observed structures for a variety of oxide pyrochlores.

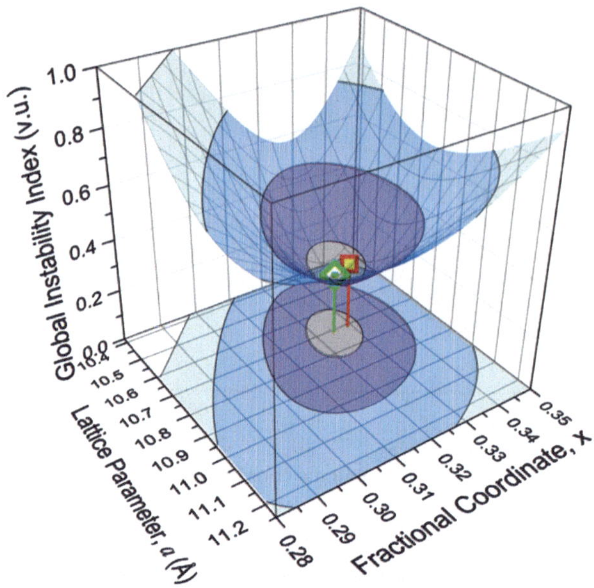

Fig. 14 Contour plot of the SPuDS calculated global instability index with the lattice parameter *a* and fractional coordinate *x* for the $La_2Sn_2O_7$ pyrochlore. The *white diamond* with *green outline* represents the SPuDS predicted lattice parameter and fractional coordinate, respectively. The *yellow square* with *red outline* represents the observed lattice parameter and fractional coordinate, respectively [71]. *Contour lines* are at 0.29, 0.5, 0.75, and 1 valence units

Table 2 Observed and SPuDS predicted structural information for selected pyrochlores

		SPuDS	Observed
$La_2Sn_2O_7$ [71]	*a* (Å)	10.7040	10.7026(1)
	x	0.3250	0.32943(7)
$Y_2Ru_2O_7$ [72]	*a* (Å)	10.1660	10.1429(2)
	x	0.3328	0.33536(3)
$Y_2Ti_2O_7$ [73]	*a* (Å)	10.1260	10.0947(1)
	x	0.3298	0.32918(9)

In space group $Fd\bar{3}m$, the fixed atomic fractional coordinates, with Wyckoff sites in brackets, are A[16d] (1/2, 1/2, 1/2), B[16c] (0, 0, 0), O′[8b] (3/8, 3/8, 3/8), O[48f] (*x*, 1/8, 1/8)

In order to investigate the robustness of the calculation approach a representative sample of stoichiometric $A_2^{3+}B_2^{4+}O_7$ pyrochlore compounds were investigated. The chemical compositions and references are as follows: $Pr_2Ru_2O_7$ [74], $Sm_2Ti_2O_7$ [75], $Sm_2Ru_2O_7$ [74], $Eu_2Ti_2O_7$ [76], $Gd_2Ti_2O_7$ [75], $Gd_2Ru_2O_7$ [77], $Y_2Mo_2O_7$ [78], $Lu_2Mn_2O_7$ [79], $Nd_2Ru_2O_7$ [77], $Er_2Mn_2O_7$ [80], $Tm_2V_2O_7$ [81], $Y_2Ti_2O_7$ [73], $Y_2Ru_2O_7$ [72], $Yb_2Ge_2O_7$ [82], $Pr_2Sn_2O_7$ [71], $Lu_2V_2O_7$ [81], $La_2Sn_2O_7$ [71], $Tb_2Ru_2O_7$ [83], $In_2Mn_2O_7$ [79], $Yb_2V_2O_7$ [81], $Er_2Ti_2O_7$ [84], $Nd_2Sn_2O_7$ [71], $Dy_2Ru_2O_7$ [74], $Yb_2Ru_2O_7$ [83], $Y_2Sn_2O_7$ [71], $Dy_2Sn_2O_7$ [71], $Tm_2Sn_2O_7$ [85], $Lu_2Sn_2O_7$ [71], $Er_2Sn_2O_7$ [85], and $Yb_2Sn_2O_7$ [85]. There are

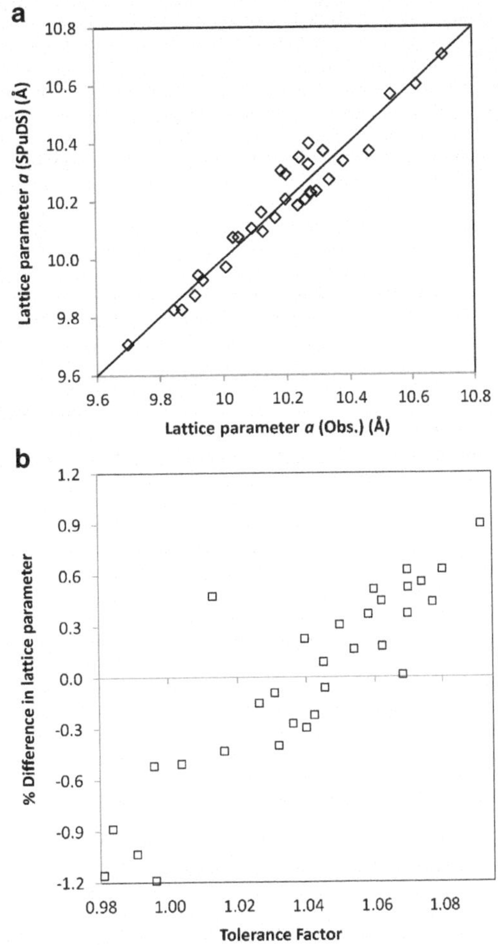

Fig. 15 The (**a**) predicted and observed lattice parameters and (**b**) percent difference in lattice parameters versus tolerance factor for $A_2^{3+}B_2^{4+}O_7$ pyrochlores

known $A_2^{2+}B_2^{5+}O_7$ pyrochlores, but those are fewer in number and not considered in the present analysis. To simplify comparison, the predicted structural variables are plotted as a function of the pyrochlore tolerance factor. The pyrochlore tolerance factor is defined differently than for perovskites, although both attempt to give a measure of the fit of the A-site cation to the [BO_6] octahedral framework. Here we have used the definition that assumes perfect octahedra [86], even though that condition is seldom realized [87]. The observed and SPuDS predicted lattice parameters and percent difference in lattice parameters versus tolerance factor are shown in Fig. 15. There is a systematic tendency to overestimate the lattice parameter for pyrochlores with large tolerance factors and underestimate the lattice parameter for pyrochlores with small tolerance factors. The predicted lattice parameter is typically within ±1% of the observed experimental value.

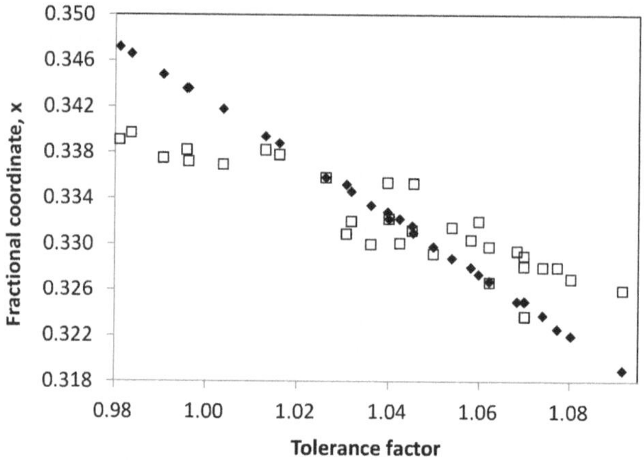

Fig. 16 SPuDS predicted (*diamonds*) and observed (*open squares*) fractional coordinate of the O-site for $A_2^{3+}B_2^{4+}O_7$ pyrochlores

The observed and SPuDS predicted fractional coordinate of the O ion versus tolerance factor is shown in Fig. 16. The fractional coordinate decreases as the tolerance factor increases, but the magnitude of this change is more pronounced for the calculated values than seen in the observed structures. The difference between observed and calculated is most pronounced for $t \leq 1$. The divergence of the calculated and observed values for small tolerance factors may be a result of the fact that the bond valence modeling does not consider the repulsive forces arising from cation–cation and anion–anion interactions. For small tolerance factor pyrochlores ($t < 1.00$), this results in O–O distances that are about 0.1 Å shorter in the predicted structures than they are in the observed structures.

The observed and predicted global instability index versus tolerance factor for pyrochlores is shown in Fig. 17. Unlike the perovskites a relatively large G is observed over the entire range of tolerance factors investigated. Despite the rigidity imposed by the topology of the pyrochlore structure, this result is a little surprising, since the pyrochlore structure is known as a "chemical garbage can" and has an amazing variety of chemical compositions [88]. For tolerance factors above 1.02, the G tends to increase for both predicted and observed pyrochlore structures. For $\tau < 1.02$, the G of the observed and predicted structures begins to deviate. In order to understand the source of this deviation we examine the individual BVSs of each of the ions.

The site-specific BVSs are plotted versus tolerance factor for $A_2^{3+}B_2^{4+}O_6O'$ pyrochlores in Fig. 18. In general the BVSs for the B-site cation and O are reasonably close to their formal valences. This is particularly true for the calculated valences of the B-site cations. While the BVS of O decreases systematically as the tolerance factor increases. In sharp contrast the BVS of the O′ site is close to its expected value of 2 only for the smallest tolerance factors. Both calculated

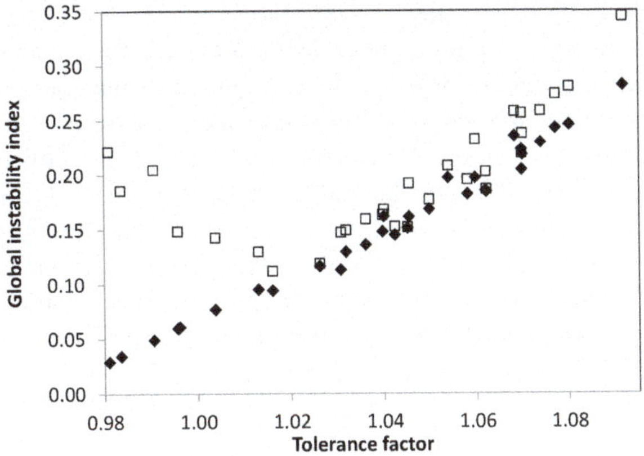

Fig. 17 SPuDS predicted (*blue diamonds*) and observed (*open red squares*) global instability index for $A_2^{3+}B_2^{4+}O_7$ pyrochlores

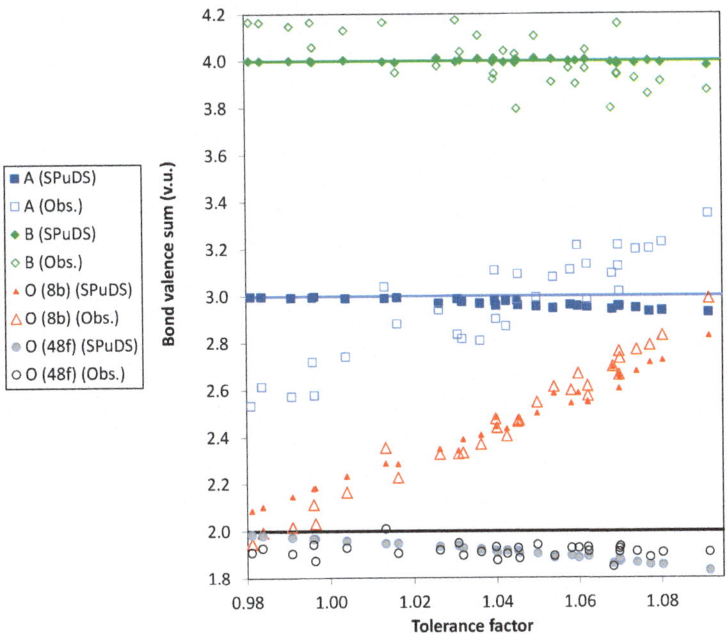

Fig. 18 Observed and SPuDS predicted bond valence sums individual ions for $A_2^{3+}B_2^{4+}O_7$ pyrochlores. Ideal values are indicated by *horizontal lines*

and observed values increase sharply with increasing tolerance factor, approaching 3 v.u. for the largest tolerance factors.

While the SPuDS predicted BVSs do a good job of mirroring the tolerance factor dependence of the observed values for the B-site cation and both oxygen sites, the same cannot be said for modeling the valence of the A-site cation. The predicted structures show a BVS near the formal charge of 3+; however, the observed structure BVSs display a strong increase as the tolerance factor increases, similar to O′. The discrepancy factor for the A-site cation varies from about 0 to −0.07 v.u. for predicted structures, whereas the observed structures show a substantially larger deviation, +0.35 to −0.46 v.u. This observation indicates the lattice parameter is largely dictated by the $[BO_6]$ octahedral network and the $[A_2O']$ network must accommodate lattice strains imposed upon it. The fact that SPuDS is able to find structures with considerably lower G values than those observed experimentally suggests that it is more favorable for the $[A_2O']$ network to accommodate the tensile strain than it is to compress the $[BO_6]$ octahedra beyond a certain point.

6 Spinels

The spinel structure has the general formula $A^{2+}B_2{}^{3+}O_4$ and may be described as a cubic close packed arrangement of oxide anions with the A and B cations occupying 1/8 of the tetrahedral and 1/2 of the octahedral holes, respectively. The normal spinel is typified by $MgAl_2O_4$ which has Mg^{2+} ions on the tetrahedral site and Al^{3+} ions on the octahedral site. In an inverse spinel, typified by Fe_3O_4, the B cations (Fe^{3+} in this case) occupy the tetrahedral sites, which means that the octahedral sites are occupied by a 50:50 mixture of A (Fe^{2+}) and B cations. There are also intermediate structures with an arrangement of the A and B-cations between the end limits of the normal and inverse spinel, including a random arrangement $(A_{1/3}B_{2/3})(A_{2/3}B_{4/3})O_4$. In our modeling we have excluded inverse spinels and those prone to intermediate degrees of site mixing. The spinel structure has a face-centered cubic unit cell and adopts the $Fd\overline{3}m$ space group. The cations occupy the special positions 8a and 16d, whereas the anion is located on the 32e Wyckoff site (x, x, x). So like pyrochlore and ordered double perovskites, the spinel structure has two degrees of freedom, the lattice parameter a and the x coordinate of the anion. The crystal structure of an AB_2O_4 spinel is shown in Fig. 19.

A series of spinels were examined and the chemical compositions and references are as follows: $CoAl_2O_4$ [89], $MgAl_2O_4$ [90], $ZnAl_2O_4$ [91], $FeAl_2O_4$ [92], $NiCr_2O_4$ [93], $ZnCr_2O_4$ [94], $ZnGa_2O_4$ [95], $FeCr_2O_4$ [96], $CoCr_2O_4$ [97], $CoCo_2O_4$ [98], $MgCr_2O_4$ [91], FeV_2O_4 [99], MgV_2O_4 [100], ZnV_2O_4 [100], $ZnFe_2O_4$ [101], $CdGa_2O_4$ [102], CdV_2O_4 [99], $HgCr_2O_4$ [103], $CdFe_2O_4$ [104], $CdRh_2O_4$ [105], $SiFe_2O_4$ [106], $SiCo_2O_4$ [107], $GeCo_2O_4$ [97], $GeFe_2O_4$ [108], $GeMg_2O_4$ [109], $GeNi_2O_4$ [96], and $SiNi_2O_4$ [110]. The SPuDS predicted and observed lattice parameters and fractional coordinates for three representative spinels are shown in Table 3.

The predicted values of the fractional coordinates agree with observed values typically within 0.004, as shown in Fig. 20. The agreement between observed and calculated lattice parameters is shown in Fig. 21. Unlike the previous cases here we see that SPuDS systematically underestimates the size of the cubic unit cell. The origin of this systematic error is not clear. One possibility is that cation–cation

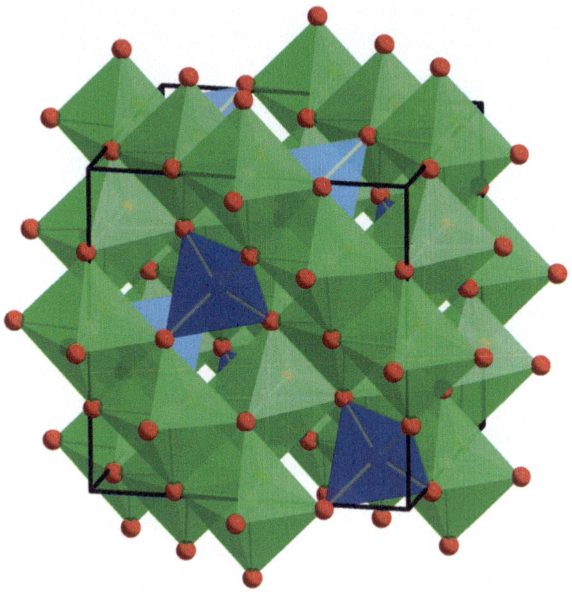

Fig. 19 Crystal structure of an AB_2O_4 spinel. *Blue spheres* represent A cations, *green spheres* B cations, and *red spheres* represent O anions

Table 3 Observed and SPuDS predicted structural information for selected representative spinels

		SPuDS	Observed
$MgAl_2O_4$ [90]	a (Å)	7.9807	8.08360(6)
	x	0.2660	0.26201(3)
$HgCr_2O_4$ [103]	a (Å)	8.6555	8.658(1)
	x	0.2736	0.2714(2)
$GeFe_2O_4$ [108]	a (Å)	8.3972	8.4127(7)
	x	0.2452	0.2466(1)

In space group $Fd\bar{3}m$, the fixed atomic fractional coordinates, with Wyckoff sites in brackets, are A[8a] (1/8, 1/8, 1/8), B[16d] (1/2, 1/2, 1/2), O1[32e] (x, x, x)

repulsions across the shared edges of the octahedral network expand the octahedra and result in larger values of a than predicted by bond valence modeling.

The predicted structures obtained by optimizing the bond valences of the ions have BVSs close to their formal valences. The degrees of freedom in the spinel structure enable the simultaneous optimization of the BVS for each of the A-, B-, and X-site ions. Within calculation limitations, a zero global instability index (G) is found for each of the predicted structures. The experimentally determined structures have a larger G, ranging from about 0.04 to 0.27 v.u. with an average of 0.15 v.u. for the examined structures. Unlike the cubic perovskite and pyrochlore structure, the spinel is not a strained structure. Regardless of the sizes of the A and B

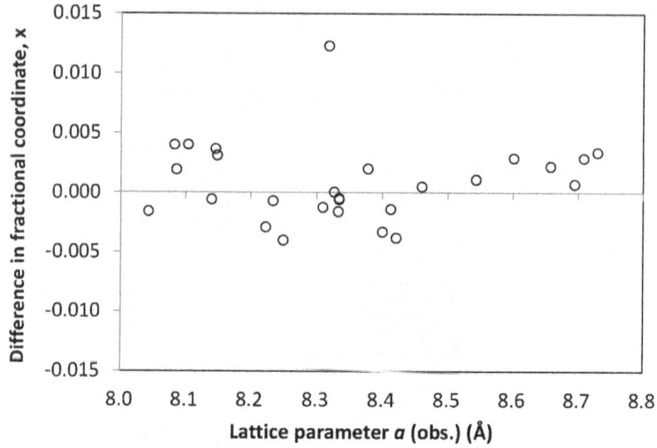

Fig. 20 Difference in SPuDS predicted and observed fractional coordinate versus observed lattice parameter for $A^{2+}B_2{}^{3+}O_4$ spinels

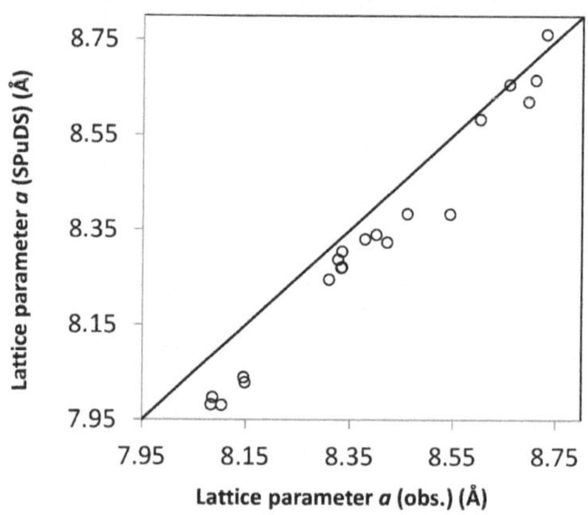

Fig. 21 Predicted and observed lattice parameters for $A^{2+}B_2{}^{3+}O_4$ spinels

cations all of the ions can achieve their ideal valences. A majority of the observed structures have $G > 0.1$ v.u., so the view that a good structure has a $G < 0.1$ v.u. appears to be questionable for the simple spinel structure type. It should be noted that structures of oxides determined by X-ray powder diffraction are not generally as accurate as those structures determined from neutron powder or single crystal diffraction methods, which may be a source of some scatter. Structures with large deviations between predicted and observed values could indicate the structure is not accurate, there is undetected interchange between A-site and B-site ions, or the bond valence parameters are not correct.

Examination of the individual BVSs was undertaken to determine if any trends were evident. The site-specific BVSs of the ions in $A^{2+}B_2{}^{3+}O_4$ spinels are shown in Fig. 22. As expected the BVSs for all ions in the calculated structures attain their formal valences. The B-site cations and the oxide anions are systematically

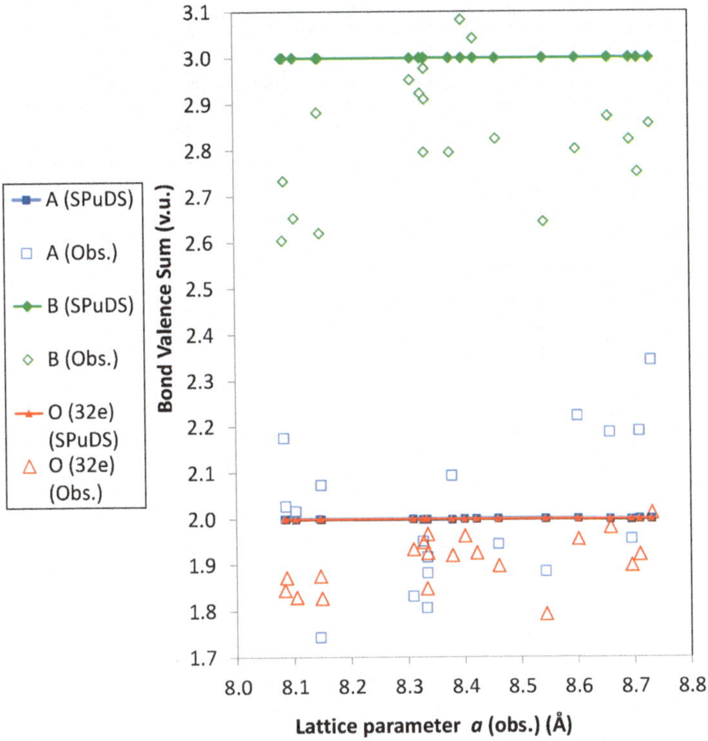

Fig. 22 Observed and SPuDS predicted bond valence sums for $A^{2+}B_2^{3+}O_4$ spinels

underbonded. The discrepancy factors for the B-site range from +0.083 to −0.395 v.u., with an average of −0.173 v.u., while for the oxide ions the average discrepancy factor is −0.092 v.u. This underbonding could have been expected from the earlier observation that the lattice parameter is always larger than predicted by SPuDS. The fact that B-site cations are systematically underbonded while the A-site cations are not lends further credence to the idea that B–B repulsions across the shared edges of the octahedra play a role in the SPuDS underestimation of the lattice parameter. The divalent A-site cations located on a tetrahedral site tend to exhibit the largest variance in BVSs. The observed structures display an A-site discrepancy factor of +0.34 to −0.41 v.u., with an average discrepancy factor of −0.021 v.u., indicating the presence of a slight underbonding on average.

The observed and SPuDS predicted BVSs $A^{4+}B_2^{2+}O_4$ spinels are shown in Fig. 23. Although at first glance similar to the $A^{2+}B_2^{3+}O_4$ spinels there is an interesting twist. Now the BVS of the octahedrally coordinated B-site ions are close to the formal valences, with an average discrepancy factor of only +0.05 v.u., but the tetrahedrally coordinated A-site ions are substantially underbonded, with an average discrepancy factor of −0.35 v.u. The X anion is once again slightly

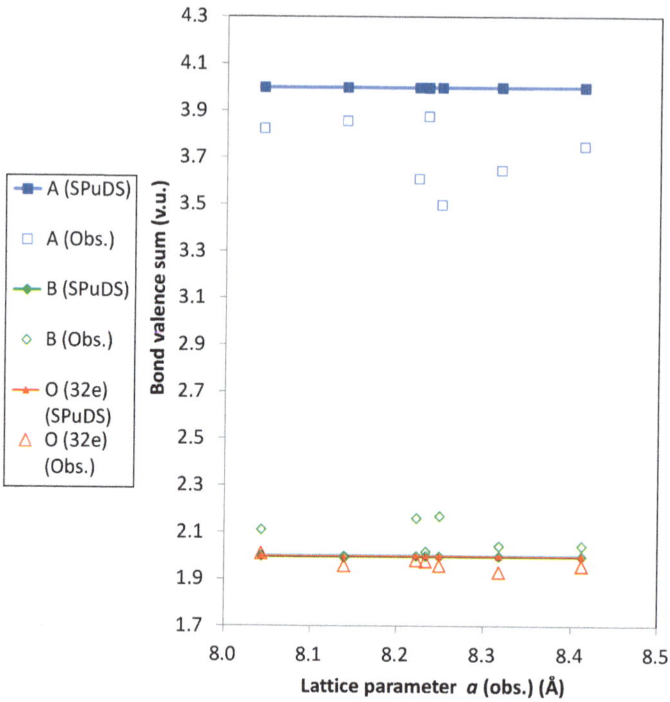

Fig. 23 Observed and SPuDS predicted bond valence sums for $A^{4+}B_2^{2+}O_4$ spinels

underbonded, with an average discrepancy factor of −0.06 v.u. The reversal of the location of the underbonded cations to the site occupied by the highest valent cation is consistent with the hypothesis that cation–cation repulsions are responsible for the underestimation of the lattice parameter.

7 Garnets

The garnet structure has the general structure $A_3B_2C_3O_{12}$ where A, B, and C are cations [111]. Garnets crystallize in the $Ia\overline{3}d$ space group. There are four free parameters within the structure, the lattice parameter a, and three positional parameters x, y, and z for the anion on the 96h Wyckoff site. The crystal structure is shown in Fig. 24 and is drawn in such a way to highlight the octahedra and tetrahedra present in the structure.

A representative subset of garnet structures was examined and the predicted structures compared to the experimental structures. The garnets chemical formulas and references are as follows: $Y_3Al_5O_{12}$ [112], $Y_3Fe_5O_{12}$ [113], $Yb_3Fe_5O_{12}$ [114], $Lu_3Fe_5O_{12}$ [114], $Gd_3Fe_5O_{12}$ [115], $Dy_3Fe_5O_{12}$ [114], $Eu_3Fe_5O_{12}$ [116], $Tb_3Fe_5O_{12}$ [116], $Pr_3Fe_5O_{12}$ [117], $Nd_3Fe_5O_{12}$ [117], $Y_3Ga_5O_{12}$ [118], $Sm_3Ga_5O_{12}$ [119],

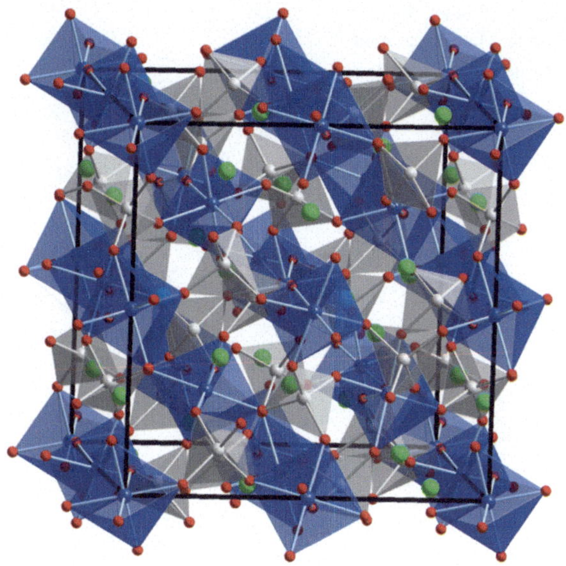

Fig. 24 Crystal structure of $A_3B_2C_3O_{12}$ garnet. *Green spheres* represent A cations and *red spheres* represent O anions. B and C cations are located in the octahedra (*blue*) and tetrahedra (*gray*), respectively

Table 4 Observed and SPuDS predicted structural information for selected $A_3B_2C_3O_{12}$ garnets

		SPuDS	Observed
$Y_3Fe_2Fe_3O_{12}$ [113]	*a* (Å)	12.2515	12.3723(6)
	x	−0.0355	−0.0270(2)
	y	0.0664	0.0566(2)
	z	0.1525	0.1505(2)
$Y_3Ga_2Ga_3O_{12}$ [118]	*a* (Å)	12.3313	12.273(1)
	x	−0.0239	−0.0274(5)
	y	0.0482	0.0546(5)
	z	0.1519	0.1493(5)
$Y_3Al_2Al_3O_{12}$ [112]	*a* (Å)	11.8567	12.0031(4)
	x	−0.0353	−0.03042(9)
	y	0.0489	0.05095(9)
	z	0.1463	0.1495(1)

In space group $Ia\bar{3}d$, the fixed atomic fractional coordinates, with Wyckoff sites in brackets, are A [24c] (1/8, 0, 1/4), B[16a] (0, 0, 0), C[24d] (3/8, 0, 1/4), O[96h] (*x*, *y*, *z*)

$Gd_3Ga_5O_{12}$ [119], $Tb_3Ga_5O_{12}$ [119], $Lu_3Ga_5O_{12}$ [114], $Ho_3Ga_5O_{12}$ [120], $Yb_3Ga_5O_{12}$ [114], $Er_3Ga_5O_{12}$ [116], $Dy_3Ga_5O_{12}$ [120], $Mn_3Cr_2Ge_3O_{12}$ [121], $Ca_3Al_2Si_3O_{12}$ [122], and $Na_3Cr_2Li_3F_{12}$ [123]. The $A_3M_5O_{12}$ compounds have the same ion on the tetrahedral and octahedral (B) and tetrahedral (C) sites.

Table 4 contains a comparison of predicted and experimental parameters for selected garnet structures. The predicted structures are in reasonable agreement with the experimental structures and could be useful as a starting model in the process of refining the crystal structures of garnets.

Fig. 25 Predicted and observed lattice parameters for $A_3B_2C_3O_{12}$ garnets

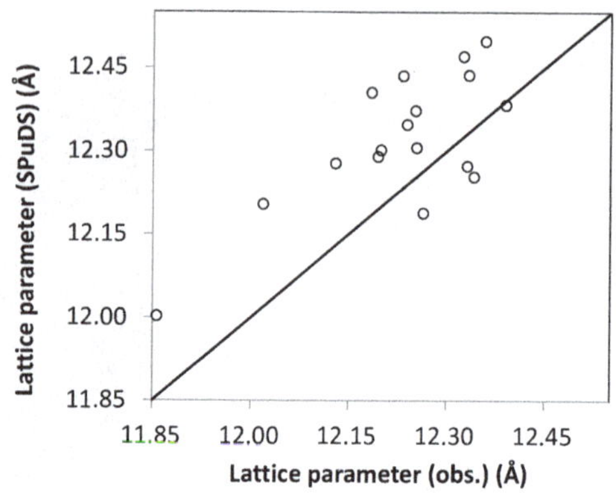

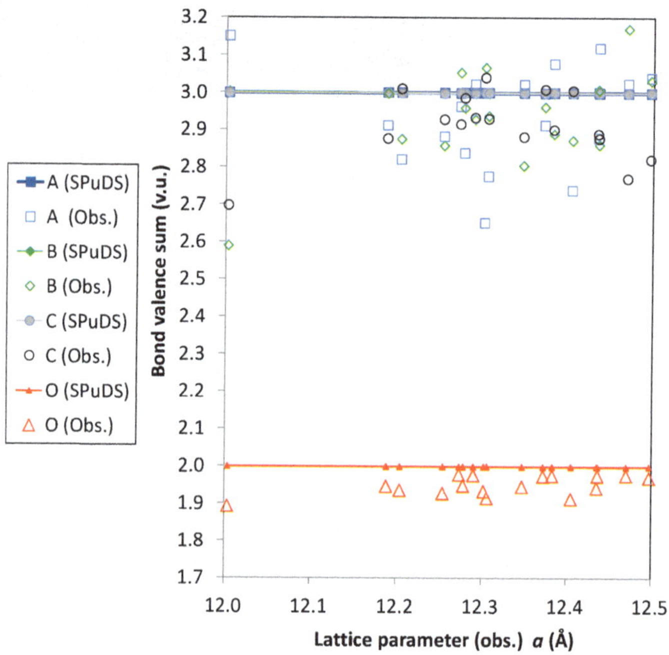

Fig. 26 Observed and SPuDS predicted bond valence sums for $A_3^{3+}B_5^{3+}O_{12}$ garnets

The predicted and observed lattice parameters are shown in Fig. 25. The predicted structures tend to have lattice parameters that are smaller than experimentally observed lattice parameters by an average of 1.5%, but with significant scatter. Similar to the spinels, the predicted structures have BVSs that approach their formal valences. The degrees of freedom in the garnet structure enable the simultaneous optimization of the BVS for each of the A-, B-, C-site cations and oxygen. Within calculation limitations, a zero global instability index (G) is found for each of the predicted structures. The garnets with a small lattice parameter

Table 5 Observed and SPuDS predicted structural information for selected $A_3^{2+}B_2^{3+}C_3^{4+}O_{12}$ and $A_3^{1+}B_2^{3+}C_3^{5+}O_{12}$ oxide garnets and $A_3^{1+}B_2^{3+}C_3^{1+}F_{12}$ fluoride garnets

		SPuDS	Observed
$Na_3Cr_2As_3O_{12}$ [124]	a (Å)	12.3334	12.188(2)
	x	−0.0471	−0.03819(14)
	y	0.0444	0.04743(12)
	z	0.1470	0.15186(13)
$Ca_3Al_2Si_3O_{12}$ [122]	a (Å)	11.8630	11.845(1)
	x	−0.0465	−0.03808(11)
	y	0.0434	0.04493(11)
	z	0.1448	0.15140(9)
$Na_3Cr_2Li_3F_{12}$ [123]	a (Å)	12.2578	12.328(2)
	x	−0.0370	−0.035(5)
	y	0.0588	0.050(5)
	z	0.1398	0.140(5)

In space group $Ia\bar{3}d$, the fixed atomic fractional coordinates, with Wyckoff sites in brackets, are A [24c] (1/8, 0, 1/4), B[16a] (0, 0, 0), C[24d] (3/8, 0, 1/4), O[96h] (x, y, z)

exhibit the largest value of G, indicating the structures have the highest amount of strain.

The BVSs of the ions in $A_2^{3+}B_2^{3+}C_3^{3+}O_{12}$ garnets are shown in Fig. 26. The BVSs of the ions for the predicted structures are equal to the formal atomic valences. The A-, B-, and C-site cations in the experimentally observed structures typically have an average BVS that is smaller than the formal atomic valence. There are similarities to the spinel structure. The structure is not strained and the oxide anions tend to be slightly underbonded for each experimental structure examined in this study, exhibiting an average discrepancy factor of −0.05 v.u.

The most commonly observed non-silicate garnets are the $A_3^{3+}B_2^{3+}C_3^{3+}O_{12}$ ($A_3^{3+}B_5^{3+}O_{12}$) compositions, but there are also $A_3^{2+}B_2^{3+}C_3^{4+}O_{12}$ and $A_3^{1+}B_2^{3+}C_3^{5+}O_{12}$ oxide garnets and $A_3^{1+}B_2^{3+}C_3^{1+}F_{12}$ fluoride garnets. The evaluations of several representative compounds are given in Table 5 to illustrate the accuracy in the prediction of the structures of those garnets.

8 Summary

The use of the bond valence concept to model the structures of perovskites is already well established and has been put to a number of practical uses over the last decade. The crystal structures obtained from modeling using bond valence concepts can be used for a variety of purposes including, but not limited to: (1) starting points for Rietveld refinements, (2) an initial structure model for density functional theory (DFT) calculations, (3) estimate the structure stability of target compositions for the synthesis of new materials, and (4) investigating of the nature of the chemical bonding present in extended solids. The extension of this approach to pyrochlores, spinels, and garnets provides some interesting insights into the crystal chemistry of these structure types.

If one describes the $A_2B_2O_6O'$ pyrochlore structure as interpenetrating $[B_2O_6]^{n-}$ and $[A_2O']^{n+}$ networks, the bond valence modeling reveals a clear size mismatch between the two networks that leads to a significant amount of lattice induced strain [9]. In general the $[A_2O']^{n+}$ network would be stabilized by a smaller unit cell than is optimal for the $[B_2O_6]^{n-}$ network. In this tug of war the size of the unit cell is largely dictated by the $[B_2O_6]^{n-}$ network. As a consequence the O′ ion is significantly overbonded in almost all of the pyrochlores modeled here. Nonetheless, there is a slight expansion of the $[B_2O_6]^{n-}$ network in response to the lattice induced strain, which leads to a slight underbonding of the ions in this network.

Unlike perovskites and pyrochlores, the spinel and garnet structures do not exhibit lattice induced strain. In other words, it is always possible to find a structure where the BVSs of all ions are equal to their formal valences and the global instability index goes to zero. Despite this fact bond valence modeling systematically underestimates the lattice parameter of cubic spinels by approximately 1% on average. It is hypothesized that this systematic error comes from the neglect of cation–cation repulsions in the bond valence modeling. Nonetheless, the ability to predict the lattice parameter of spinels with an accuracy of 1%, and the fractional coordinate of the anion with an accuracy of ± 0.005, is impressive given the simplicity of the approach. Garnets represent a more complex problem with 4 structural degrees of freedom and 160 atoms in the unit cell. It is shown here that they can also be modeled effectively, but the deviations between predicted and observed values of the lattice parameter are somewhat larger ($\pm 3\%$) than for the other structure types modeled here.

In the examination of past and current uses of the bond valence approach to structure modeling, we offer a few comments looking towards possible future developments in the use of bond valences to model the crystal structures of ternary and quaternary oxides. Despite its relative simplicity, the bond valence approach is able to model the structures of complex oxides with a reasonably high level of accuracy. The discrepancies between calculated and observed values of the structural parameters, ± 1–3% are comparable to much more sophisticated and computationally expensive approaches, like density functional theory. The advantage of bond valence modeling is the fact that it requires minimal computational resources, which means that calculations can often be run in a few seconds on a typical desktop computer. Perhaps the biggest limitation is the constraint that this approach cannot be applied in a general sense, algorithms must be developed for each new structure type that is to be modeled. Furthermore, it is not clear how many structural degrees of freedom can be present before this approach breaks down. In garnets and perovskites that undergo octahedral tilting a reasonably high degree of accuracy can be achieved when there are 4 degrees of freedom. There are many more oxide structures that can be described with 4–5 degrees of freedom (dof) [e.g. $KAlF_4$ (3 dof), lithium ferrite (3 dof), K_2MgF_4 (4 dof), zircon(4 dof), tetragonal (4 dof) and hexagonal bronzes (5 dof), scheelite (5 dof), bixbyite (5 dof)], which makes it quite likely that this approach can be extended much further.

References

1. Lufaso MW, Woodward PM (2001) Prediction of the crystal structures of perovskites using the software program SPuDS. Acta Crystallogr B 57:725–738
2. Lufaso MW, Barnes PW, Woodward PM (2006) Structure prediction of ordered and disordered multiple octahedral cation perovskites using SPuDS. Acta Crystallogr B 62:397–410
3. Lufaso MW (2004) Crystal structures, modeling, and dielectric property relationships of 2:1 ordered $Ba_3MM_2'O_9$ (M = Mg, Ni, Zn; M′ = Nb, Ta) perovskites. Chem Mater 16: 2148–2156
4. Byeon SH, Lufaso MW, Parise JB, Woodward PM, Hansen T (2003) High-pressure synthesis and characterization of perovskites with simultaneous ordering of both the A- and B-site cations, $CaCu_3Ga_2M_2O_{12}$ (M = Sb, Ta). Chem Mater 15:3798–3804
5. Byeon SH, Lee SS, Parise JB, Woodward PM, Hur NH (2005) A new ferrimagnetic oxide $CaCu_3Cr_2Sb_2O_{12}$: high-pressure synthesis, structure, and magnetic properties. Chem Mater 17:3552–3557
6. Zhou JS, Goodenough JB (2005) Universal octahedral-site distortion in orthorhombic perovskite oxides. Phys Rev Lett 94:065501
7. Mitchell RH (2002) Perovskites: modern and ancient. Almaz Press, Ontario
8. Brown D (2013) Bond valence theory. Struct Bond. doi:10.1007/430_2012_89
9. Brown ID (2006) The chemical bond in inorganic chemistry: the bond valence model, International Union of Crystallography monographs on crystallography; 12. Oxford University Press, Oxford, p 278
10. Thomas NW (1998) A new global parameterization of perovskite structures. Acta Crystallogr B 54:585–599
11. Brown ID (2012) Bond valence parameters. http://www.iucr.org/resources/data/datasets/bond-valence-parameters. Accessed 30 Nov 2012
12. Abramov YA, Tsirelson VG, Zavodnik VE, Ivanov SA, Brown ID (1995) The chemical bond and atomic displacements in $SrTiO_3$ from X-ray diffraction analysis. Acta Crystallogr B B51:942–951
13. Mann M, Jackson S, Kolis J (2010) Hydrothermal crystal growth of the potassium niobate and potassium tantalate family of crystals. J Solid State Chem 183:2675–2680
14. Shimizu Y, Syono Y, Akimoto S (1970) High-pressure transformations in $SrGeO_3$, $SrSiO_3$, $BaGeO_3$, and $BaSiO_3$. High Temperatures – High Pressures 2:113–120
15. Casais MT, Alonso JA, Rasines I, Hidalgo MA (1995) Preparation, neutron structural study and characterization of $BaNbO_3$: a Pauli-like metallic perovskite. Mater Res Bull 30:201–208
16. Brixner LH (1960) X-ray study and electrical properties of the system $Ba_xSr_{(1-x)}MoO_3$. J Inorg Nucl Chem 14:225–230
17. Rey MJ, Dehaudt P, Joubert JC, Lambert-Andron B, Cyrot M, Cyrot-Lackmann F (1990) Preparation and structure of the compounds $SrVO_3$ and Sr_2VO_4. J Solid State Chem 86:101–108
18. Smith AJ, Welch AJE (1960) Some mixed metal oxides of perovskite structure. Acta Crystallogr 13:653–656
19. Dickens PG, Powell AV (1991) Powder neutron diffraction study of potassium uranate(V), KUO_3. J Mater Chem 1:137–138
20. Levin I, Amos TG, Bell SM, Farber L, Vanderah TA, Roth RS, Toby BH (2003) Phase equilibria, crystal structures, and dielectric anomaly in the $BaZrO_3$–$CaZrO_3$ system. J Solid State Chem 175:170–181
21. Hutton J, Nelmes RJ, Scheel HJ (1981) Extinction corrections for a highly perfect crystal ($SrTiO_3$). Acta Crystallogr A 37:916–920
22. Liu G, Zhao X, Eick HA (1992) The synthesis, structure and characterization of $SrMoO_{2.6}{}^{15}N_{0.4}$. J Alloys Compd 187:145–156

23. Zhao C, Feng S, Chao Z, Shi C, Xu R, Ni J (1996) Hydrothermal synthesis of the complex fluorides $LiBaF_3$ and $KMgF_3$ with perovskite structures under mild conditions. Chem Commun 1641–1642
24. Kijima N, Tanaka K, Marumo F (1983) Electron-density distributions in crystals of $KMnF_3$ and $KNiF_3$. Acta Crystallogr B 39:557–561
25. Buttner RH, Maslen EN (1988) Electron difference density in potassium zinc fluoride perovskite. Acta Crystallogr C 44:1707–1709
26. Kijima N, Tanaka K, Marumo F (1981) Electron density distribution in crystals of potassium trifluorocobaltate(II). Acta Crystallogr B 37:545–548
27. Miyata N, Tanaka K, Marumo F (1983) Electron density distribution in crystals of iron(II) potassium trifluoride. Acta Crystallogr B 39:561–564
28. Hutton J, Nelmes RJ (1981) High-resolution studies of cubic perovskites by elastic neutron diffraction II: $SrTiO_3$, $KMnF_3$, $RbCaF_3$ and $CsPbCl_3$. J Phys C: Solid State Phys 14: 1713–1736
29. Brown ID, Klages P, Skowron A (2003) Influence of pressure on the lengths of chemical bonds. Acta Crystallogr B 59:439–448
30. Tavora Weber I, Audebrand N, Bouquet V, Guilloux-Viry M, Perrin A (2006) $KTaO_3$ powders and thin films prepared by polymeric precursor method. Solid State Sci 8:606–612
31. Zhurova EA, Ivanov Y, Zavodnik V, Tsirelson V (2000) Electron density and atomic displacements in $KTaO_3$. Acta Crystallogr B 56:594–600
32. Stitzer KE, Smith MD, zur Loye HC (2002) Crystal growth of Ba_2MOsO_6 (M = Li, Na) from reactive hydroxide fluxes. Solid State Sci 4:311–316
33. Martinez-Lope MJ, Alonso JA, Casais MT (2003) Synthesis, crystal and magnetic structure of the double perovskites A_2NiMoO_6 (A = Sr, Ba): a neutron diffraction study. Eur J Inorg Chem 2839–2844
34. Tezuka K, Henmi K, Hinatsu Y, Masaki NM (2000) Magnetic susceptibilities and Mossbauer spectra of perovskites A_2FeNbO_6 (A = Sr, Ba). J Solid State Chem 154:591–597
35. Martinez-Lope MJ, Alonso JA, Casais MT, Fernandez-Diaz MT (2002) Preparation, crystal and magnetic structure of the double perovskites Ba_2CoBO_6 (B = Mo, W). Eur J Inorg Chem 2463–2469
36. Patwe SJ, Achary SN, Tyagi AK (2005) Synthesis, phase transition and thermal expansion studies on M_2MgWO_6 ($M = Ba^{2+}$ and Sr^{2+}) double perovskites. J Alloys Compd 390: 100–105
37. Iwakura H, Einaga H, Teraoka Y (2010) Relationship between cation arrangement and photocatalytic activity for Sr–Al–Nb–O double perovskite. Inorg Chem 49:11362–11369
38. Woodward PM (1996) Structural distortions, phase transitions, and cation ordering in the perovskite and tungsten trioxide structures. Oregon State University, Corvallis
39. Ouchetto K, Archaimbault F, Pineau A, Choisnet J (1991) Chemical and structural characterization of a new barium ceroplatinate – Ba_2CePtO_6 a double perovskite mixed-oxide. J Mater Sci Lett 10:1277–1279
40. Azad AK, Ivanov SA, Eriksson SG, Eriksen J, Rundlof H, Mathieu R, Svedlindh P (2001) Synthesis, crystal structure, and magnetic characterization of the double perovskite Ba_2MnWO_6. Mater Res Bull 36:2215–2228
41. Amador U, Hetherington CJD, Moran E, Alario-Franco MA (1992) Ba_2PrPtO_6 – a novel double perovskite. J Solid State Chem 96:132–140
42. Dianoux AJ, Poix P (1968) ETude cristallographique et paramagnetique de l'oxyde mixte Ba_2UFeO_6 comportant de l'uranium U^{5+}. C R Hebd Seances Acad Sci 266:283–285
43. Baldinozzi G, Grebille D, Sciau P, Kiat JM, Moret J, Berar JF (1998) Rietveld refinement of the incommensurate structure of the elpasolite (ordered perovskite) Pb_2MgTeO_6. J Phys Condens Matter 10:6461–6472
44. Doi Y, Hinatsu Y, Nakamura A, Ishii Y, Morii Y (2003) Magnetic and neutron diffraction studies on double perovskites A_2LnRuO_6 (A = Sr, Ba; Ln = Tm, Yb). J Mater Chem 13:1758–1763

45. Battle PD, Jones CW (1989) The crystal and magnetic-structures of Sr_2LuRuO_6, Ba_2YRuO_6, and Ba_2LuRuO_6. J Solid State Chem 78:108–116
46. Arulraj A, Ramesha K, Gopalakrishnan J, Rao CNR (2000) Magnetoresistance in the double perovskite Sr_2CrMoO_6. J Solid State Chem 155:233–237
47. Choy JH, Hong ST, Choi KS (1996) Crystal structure, magnetism and phase transformation in perovskites A_2CrNbO_6 (A = Ca, Sr, Ba). J Chem Soc-Faraday Trans 92:1051–1059
48. Taira N, Hinatsu Y (2000) Magnetic susceptibility of Ba_2YbTaO_6 with the ordered perovskite structure and electron paramagnetic resonance of Yb^{3+} doped in Ba_2LuTaO_6. J Solid State Chem 150:31–35
49. Izumiyama Y, Doi Y, Wakeshima M, Hinatsu Y, Nakamura A, Ishii I (2002) Magnetic and calorimetric studies on ordered perovskite Ba_2ErRuO_6. J Solid State Chem 169:125–130
50. Doi Y, Hinatsu Y (2001) Magnetic properties of ordered perovskites Ba_2LnTaO_6 (Ln = Y, lanthanides). J Phys Condens Matter 13:4191–4202
51. Grenet JC, Poix P, Michel A (1972) Determinations crystallographiques et magnetiques sur l'oxyde mixte de formule Ba_2MnUO_6. Ann Chim (Paris) 1972:231–234
52. Hinatsu Y, Izumiyama Y, Doi Y, Alemi A, Wakeshima M, Nakamura A, Morii Y (2004) Studies on magnetic and calorimetric properties of double perovskites Ba_2HoRuO_6 and Ba_2HoIrO_6. J Solid State Chem 177:38–44
53. Barnes PW, Lufaso MW, Woodward PM (2006) Structure determination of $A_2M^{3+}TaO_6$ and $A_2M^{3+}NbO_6$ ordered perovskites: octahedral tilting and pseudosymmetry. Acta Crystallogr B B62:384–396
54. Fu WT, Ijdo DJW (2005) Re-examination of the structure of Ba_2MIrO_6 (M = La, Y): space group revised. J Alloys Compd 394:L5–L8
55. Karunadasa H, Huang Q, Ueland BG, Schiffer P, Cava RJ (2003) Ba_2LnSbO_6 and Sr_2LnSbO_6 (Ln = Dy, Ho, Gd) double perovskites: lanthanides in the geometrically frustrating fcc lattice. Proc Natl Acad Sci USA 100:8097–8102
56. Fu WT, Ijdo DJW (1997) On the structure of $BaTl_{0.5}Sb_{0.5}O_3$: an ordered perovskite. J Solid State Chem 128:323–325
57. Jung D, Gravereau P, Demazeau G (1993) Stabilization of six-coordinated iridium(VI) in a perovskite oxygen lattice Ba_2MIrO_6 (M = Ca, Sr. Eur J Solid State Inorg Chem 30: 1025–1037
58. Alonso JA, Cascales C, Casado PG, Rasines I (1997) On characterization of barium rare-earth antimonates: ordered perovskites suitable as substrates for superconducting films. J Solid State Chem 128:247–250
59. Fu WT, Ijdo DJW (2005) X-ray and neutron powder diffraction study of the double perovskites Ba_2LnSbO_6 (Ln = La, Pr, Nd and Sm). J Solid State Chem 178:2363–2367
60. Day BE, Bley ND, Jones HR, McCullough RM, Eng HW, Porter SH, Woodward PM, Barnes PW (2012) Structures of ordered tungsten- or molybdenum-containing quaternary perovskite oxides. J Solid State Chem 185:107–116
61. Woodward PM (1997) Octahedral tilting in perovskites. 1. Geometrical considerations. Acta Crystallogr B 53:32–43
62. Woodward PM (1997) Octahedral tilting in perovskites. 2. Structure stabilizing forces. Acta Crystallogr B 53:44–66
63. Glazer AM (1972) The classification of tilted octahedra in perovskites. Acta Crystallogr B 28:3384–3392
64. Glazer AM (2011) A brief history of tilts. Phase Transitions 84:405–420
65. Howard CJ, Stokes HT (1998) Group-theoretical analysis of octahedral tilting in perovskites. Acta Crystallogr B 54:782–789
66. Woodward PM (1997) POTATO – a program for generating perovskite structures distorted by tilting of rigid octahedra. J Appl Crystallogr 30:206–207
67. O'Keeffe M, Hyde BG (1977) Some structures topologically related to cubic perovskite (E21), ReO_3 (D09) and Cu_3Au (L12). Acta Crystallogr B B33:3802–3813

68. Lufaso MW (2002) Perovskite synthesis and analysis using structure prediction diagnostic software. The Ohio State University, Columbus
69. Brown ID (2009) Recent developments in the methods and applications of the bond valence model. Chem Rev 109:6858–6919
70. Subramanian MA, Aravamudan G, Rao GVS (1983) Oxide pyrochlores – a review. Prog Solid State Chem 15:55–143
71. Kennedy BJ, Hunter BA, Howard CJ (1997) Structural and bonding trends in tin pyrochlore oxides. J Solid State Chem 130:58–65
72. Kennedy BJ (1995) Structure refinement of Y2Ru2O7 by neutron powder diffraction. Acta Crystallogr C 51:790–792
73. Haile SM, Wuensch BJ, Prince E (1990) Neutron Rietveld analysis of anion and cation disorder in the fast-ion conducting pyrochlore system yttrium zirconium titanium oxide ($Y_2(Zr_xTi_{1-x})_2O_7$). In: Materials Research Society symposium proceedings: neutron scattering for materials science, vol. 166, Department of Materials Science & Engineering, Massachusetts Institute of Technology, Cambridge, pp 81–86
74. Yamamoto T, Kanno R, Takeda Y, Yamamoto O, Kawamoto Y, Takano M (1994) Crystal structure and metal–semiconductor transition of the $Bi_{2-x}Ln_xRu_2O_7$ pyrochlores (Ln = Pr-Lu). J Solid State Chem 109:372–383
75. Knop O, Brisse F, Castelliz L (1969) Pyrochlores. V. Thermoanalytic, X-ray, neutron, infrared, and dielectric studies of $A_2Ti_2O_7$ titanates. Can J Chem 47:971–990
76. Chtoun E, Hanebali H, Pierre G (2001) X-ray Rietveld analysis of (1-x)$A_2Ti_2O_7$-(x)Fe_2TiO_5 (A = Eu, Y) solid solutions. Ann Chim (Paris) 26:27–32
77. Kobayashi H, Kanno R, Kawamoto Y, Kamiyama T, Izumi F, Sleight AW (1995) Synthesis, crystal structure, and electrical properties of the pyrochlores $Pb_{2-x}Ln_xRu_2O_{7-y}$(Ln = Nd, Gd). J Solid State Chem 114:15–23
78. Reimers JN, Greedan JE, Sato M (1988) The crystal structure of the spin-glass pyrochlore, $Y_2Mo_2O_7$. J Solid State Chem 72:390–394
79. Shimakawa Y, Kubo Y, Hamada N, Jorgensen JD, Hu Z, Short S, Nohara M, Takagi H (1999) Crystal structure, magnetic and transport properties, and electronic band structure of $A_2Mn_2O_7$ pyrochlores (A = Y, In, Lu, and Tl). Phys Rev B 59:1249–1254
80. Subramanian MA, Torardi CC, Johnson DC, Pannetier J, Sleight AW (1988) Ferromagnetic $R_2Mn_2O_7$ pyrochlores (R = Dy-Lu, Y). J Solid State Chem 72:24–30
81. Soderholm L, Greedan JE (1982) Relationship between crystal structure and magnetic properties of $(RE)_2V_2O_7$; RE = Lu, Yb, Tm. Mater Res Bull 17:707–713
82. Dem'yanets LN, Radaev SF, Mamin BF, Maksimov BA (1988) Synthesis and atomic structure of pyrochlore-type $Yb_2Ge_2O_7$ crystals. J Struct Chem 29:485–487
83. Kennedy BJ, Vogt T (1996) Structural and bonding trends in ruthenium pyrochlores. J Solid State Chem 126:261–270
84. Knop O, Brisse F, Castelliz L (1965) Determination of the crystal structure of erbium titanate, $Er_2Ti_2O_7$, by x-ray and neutron diffraction. Can J Chem 43:2812–2826
85. Kennedy BJ (1996) Structural trends in pyrochlore oxides. Mater Sci Forum 228–231: 753–758
86. Isupov VA (1958) Geometric criteria of structures of the pyrochlore type. Kristallografiya 3:99–100
87. Cai L, Arias AL, Nino JC (2011) The tolerance factors of the pyrochlore crystal structure. J Mater Chem 21:3611–3618
88. Vanderah TA, Levin I, Lufaso MW (2005) An unexpected crystal-chemical principle for the pyrochlore structure. Eur J Inorg Chem 14:2895–2901
89. Nakatsuka A, Ikeda Y, Yamasaki Y, Nakayama N, Mizota T (2003) Cation distribution and bond lengths in $CoAl_2O_4$ spinel. Solid State Commun 128:85–90
90. Redfern SA, Harrison RJ, O'Neill HSC, Wood DRR (1999) Thermodynamics and kinetics of cation ordering in $MgAl_2O_4$ up to 1600 C from in situ neutron diffraction. Am Mineral 84:299–310

91. O'Neill HSC, Dollase WA (1994) Crystal structures and cation distributions in simple spinels from powder XRD structural refinements: $MgCr_2O_4$, $ZnCr_2O_4$, Fe_3O_4 and the temperature dependence of the cation distribution in $ZnAl_2O_4$. Phys Chem Minerals 20:541–555
92. Harrison RJ, Redfern SAT, O'Neill HSC (1998) The temperature dependence of the cation distribution in synthetic hercynite ($FeAl_2O_4$) from in-situ neutron structure refinements. Am Mineral 83:1092–1099
93. Ueno G, Sato S, Kino Y (1999) The low-temperature phase of $NiCr_2O_4$. Acta Crystallogr C 39:1963–1966
94. Sawada H (1997) Electron density study of spinels: zinc chromium oxide. Mater Res Bull 32:873–879
95. Wendschuh-Josties M, O'Neill HSC, Bente K, Brey G (1995) Lattice and oxygen parameters of $ZnGa_2O_4$ as a function of equilibriation temperature. Neus Jahrbuch Mineralogie 1995 (6):273–280
96. Shirane G, Cox DE (1964) Magnetic structure in $FeCr_2S_4$ and $FeCr_2O_4$. J Appl Phys 35: 954–955
97. Hirota K, Inoue T, Mochida N, Ohtsuka A (1990) Study of germanium spinels (Part 3). J Ceram Soc Japan (Nippon Seramikkusu Kyokai Gakujutsu Ronbunshi) 98:976–986
98. Liu X, Prewitt CT (1990) High-temperature X-ray diffraction study of Co_3O_4: transition from normal to disordered spinel. Phys Chem Minerals 17:168–172
99. Reuter B, Riedel E, Hug P, Arndt D, Geisler U, Behnke J (1969) Zur kristallochemie der vanadin(III)-spinelle. Z Anorg Allgemeine Chem 369:306–312
100. Rueedorff W, Reuter B (1947) Die struktur der magnesium- und zink-vanadinspinelle. Z Anorg Allgemeine Chem 253:194–208
101. Schaefer W, Kockelmann W, Potzel W, Martin A, Campbell SJ (1997) Neutron diffraction data on preparation dependent variations of structural and magnetic order in $ZnFe_2O_4$ spinels. Z Kristallogr 15:167
102. Stone SF, Arean CO, Diez Vinuela JS, Platero EE (1985) Structural characterization of cadmium-copper gallium oxide ($Cd_xCu_{1-x}Ga_2O_4$) spinels. J Chem Soc-Faraday Trans 81: 1255–1261
103. Wessels AL, Czekalla R, Jeitschko W (1998) Structure of the mercury(II) chromate(III) $HgCr_2O_4$ and lattice constants of isotypic mercury(I) compounds Hg_2MoO_4 and Hg_2WO_4. Mater Res Bull 33:95–101
104. Arean CO, Diaz EG, Gonzales JMR, Garcia MAV (1988) Crystal structure of cadmium-zinc ferrites. J Solid State Chem 77:275–280
105. Agarwala RP (1961) Structure of cadmium rhodite, $CdRh_2O_4$. Z Anorg Allgemeine Chem 307:205–207
106. Yagi T, Marumo F, Akimoto SI (1974) Crystal structures of spinel polymorphs of Fe_2SiO_4 and Ni_2SiO_4. Am Mineral 59:486–490
107. Marumo F, Isobe M, Akimoto S (1977) Electron-density distributions in crystals of gamma-Fe_2SiO_4 and gamma-Co_2SiO_4. Acta Crystallogr B 33:713–716
108. Welch MD, Cooper MA, Hawthorne FC (2001) The crystal structure of brunogeierite, Fe_2GeO_4. Mineral Mag 65:441–444
109. von Dreele RB, Navrotsky A (1977) Refinement of the crystal structure of Mg_2GeO_4. Acta Crystallogr B 33:2287–2288
110. Marumo F, Isobe M (1974) Electron-density distribution in crystals of gamma Ni_2SiO_4. Acta Crystallogr B 30:1904–1906
111. Geller S (1967) Crystal chemistry of the garnets. Z Kristallogr 125:1–47
112. Rodic D, Mitric M, Tellgren R, Rundlof H (2001) The cation distribution and magnetic structure of $Y_3Fe_{(5-x)}Al_xO_{12}$. J Magn Magn Mater 232:1–8
113. Rodic D, Mitric M, Tellgren R, Rundlof H, Kremenovic A (1999) True magnetic structure of the ferrimagnetic garnet $Y_3Fe_5O_{12}$ and magnetic moments of iron ions. J Magn Magn Mater 191:137–145

114. Euler F, Bruce JA (1965) Oxygen coordinates of compounds with garnet structure. Acta Crystallogr 19:971–978
115. Weidenborner JE (1961) Least squares refinement of the structure of gadolinium-iron garnet, $Gd_3Fe_2Fe_3O_{12}$. Acta Crystallogr 14:1051–1056
116. Dukhovskaya EL, Saksonov YG, Titova AG (1973) Oxygen parameters of certain compounds with the garnet structure. Izvestiya Akad Nauk SSSR Neorganicheskie Mater 9:809–813
117. Guo L, Huang K, Chen Y, Li G, Yuan L, Peng W, Yuan H, Feng S (2011) Mild hydrothermal synthesis and ferrimagnetism of $Pr_3Fe_5O_{12}$ and $Nd_3Fe_5O_{12}$ garnets. J Solid State Chem 184:1048–1053
118. Nakatsuka A, Yoshiasa A, Takeno S (1995) Site preference of cations and structural variation in $Y_3Fe_{5-x}Ga_xO_{12}$ ($0 < x < 5$) solid solutions with garnet structure. Acta Crystallogr B 51:737–745
119. Sawada H (1997) Electron density study of garnets: $Z_3Ga_5O_{12}$; Z = Nd, Sm, Gd, Tb. J Solid State Chem 132:300–307
120. Patzke G, Wartchow R, Binnewies M (1999) Crystal structure of triholmium pentagallium dodecaoxide, $Ho_3Ga_2(GaO_4)_3$ and of tridysprosium pentagallium dodecaoxide, $Dy_3Ga_2(GaO_4)_3$. New Cryst Struct (Z Kristallogr) 214:143–144
121. Lipp C, Strobel S, Lissner F, Niewa R (2012) Garnet-type $Mn_3Cr_2(GeO_4)_3$. Acta Crystallogr E 68:35
122. Novak GA, Gibbs GV (1971) The crystal chemistry of the silicate garnets. Am Mineral 56:791–825
123. De Pape R, Portier J, Gauthier G, Hagenmuller P (1967) Fluorinated garnets of the transition elements, $Na_3Li_3M_2F_{12}$ (M = titanium, vanadium, chromium, iron, or cobalt). C R Hebd Seances Acad Sci 265:1244–1246
124. Bouzemi B, Boughzala H, Jouini T (2002) $Na_3Cr_2(AsO_4)_3$ trisodium dichromium(III) triarsenate. Acta Crystallogr E 58:117–118

Struct Bond (2014) 158: 91–128
DOI: 10.1007/430_2013_96

Published online: 4 June 2013

Practical Considerations in Determining Bond Valence Parameters

Stefan Adams

Abstract Based on an investigation of empirical links of the bond valence method to observable quantities, especially the electron density at the bond critical point as well as absolute electronic potential and hardness values in the frame of the hard and soft acids and bases concept, it is ascertained that bond valence can be understood as a functional of valence electron density. Therefrom a systematic approach for deriving bond valence parameters and related quantities such as coordination numbers and bond breaking energies is discussed that together allow for a conversion of the bond valence method to a simple effective atomistic forcefield.

Keywords Atomistic forcefield · Bond critical point · Bond valence parameters · Coordination numbers · Electron density

Contents

S. Adams (✉)
Department of Materials Science and Engineering, National University of Singapore, Singapore 117576, Singapore
e-mail: mseasn@nus.edu.sg

Abbreviations

b	Bond valence parameter defining the compliance of a bond length R to external forces
$b_{average}$	Average b value for the interactions of all anions in a unit cell to a given cation M
$b_{effective}$	b value to be used in bond valence calculations for compounds containing several anion types (derived from partial b-averaging)
BVS	Bond valence sum
BVSE	Bond valence site energy
D_0	Bond dissociation energy
EA	Electron affinity
$E_{repulsion}$	Energy penalty due to Coulomb repulsions among cations or among anions
G	Global Instability Index
HSAB	Hard and soft acids and bases
IE	Ionization energy
k	Force constant of a bond at the equilibrium distance
n	Principal quantum number = period number in periodic table of the elements
N_C	Coordination number
N_{RCN}	Running coordination number
R_0	Bond valence parameter (distance corresponding to a bond valence value of 1 v.u.)
R_1	Radius of first coordination shell
R_{min}	Equilibrium distance M–X for a given coordination number
$<R(M–X)>$	Expected M–X bond length
s_{min}	Bond valence corresponding to $R = R_{min}$
V	Bond valence sum
V_{id}	Oxidation state
α	Bond stiffness parameter in the Morse interaction potential, here identified with $1/b$
ΔE_{EVR}	Energy penalty for a deviation from the equal valence rule among bonds to the same central cation M
$\rho(r)$	Electron density as a function of distance r
ρ_{BCP}	Electron density at the bond critical point

1 Introduction

Empirical relationships between bond length R and bond valence $s_{M-X} = \exp[(R_0 - R(M - X))/b]$ are widely used in crystal chemistry to identify plausible equilibrium sites for an atom in a structure as sites where the bond valence sum of the atom matches its oxidation state. In our earlier work, we suggested a systematic adjustment of bond valence parameters to the bond softness [1–3], which together with the inclusion of interactions beyond the first coordination sphere was intended to allow for more adequate estimates of nonequilibrium site energies and used the approach to model ion transport in solids [4, 5]. In this context the decision on whether to include weak interactions to more distant counterions beyond the first coordination shell in the determination of bond valence parameters mostly depends on the purpose of the application and the treatment of the bond softness parameter b. For modeling ion transport pathways (cf. [6]) a description of bond valence mismatches or the related bond valence site energies are required for low-symmetry arrangements and thus a self-consistent cutoff will be advantageous, as it avoids artifacts in the bond valence variation, when an ion moves across the border of its coordination shell. Where the application is the description of equilibrium sites, the computationally simpler first coordination shell cutoff criterion will be sufficient, but one should still be aware that this choice not only affects the values of the bond valence parameter R_0 but also requires different (lower) values of the bond valence parameter b (the larger the value of b the higher will be the contribution of the more distant counterions compared to the nearest neighboring counterions, see Sect. 3.4). In this work we therefore prefer to deviate from the first-coordination shell convention while looking into factors that should be considered when determining bond valence parameters.

In the course of the discussion of bond valence parameter determination we will also link the energy of an atom M in a given structural environment to deviations of its bond valence sum $V(A)$ from the absolute value of its oxidation state and to a (also bond valence-based) penalty function ΔE_{EVR} that penalizes deviations in the bond arrangement from the equal valence rule:

$$\mathrm{BVSE(M)} = D_0 \cdot |\Delta V(\mathrm{M})|^g + \Delta E_{\mathrm{EVR}} + E_{\mathrm{repulsion}}. \quad (1)$$

Here, the scaling factor D_0 of the bond valence sum term can be understood as a measure of the bond breaking energy. While the first two terms in (1) describe the attractive and repulsive interaction with counterions, an additional term $E_{\mathrm{repulsion}}$ has to account for cation–cation or anion–anion interactions on the total energy for M at a given site. A model yielding explicit values for D_0, the exponent g and the penalty terms will be derived in Sect. 2.2. To emphasize the approximative nature of such approaches, BVSE(M) is called the *bond valence site energy*. The boundary conditions chosen in the systematic determination of bond valence parameters as discussed in this chapter are chosen to facilitate a translation into such BVSEs. The possibility to reproduce dynamic structural features when using the BV-based atomistic forcefields resulting from such an approach in molecular dynamics simulations provide an avenue for further parameter validation.

2 A Physical Basis of the Bond Valence Approach

2.1 Bond Valence: A Functional of Electron Density

Before turning to the practical determination of bond valence parameters, it appears appropriate to briefly discuss connections between experimentally observable quantities and the bond valence parameters. Qualitatively, it is obvious that the closer two atoms of opposite charge approach each other, the more electron density will be found in the bonding region and the stronger their interaction will be. So it appears natural to link bond valence to the electron density, $\rho(r)$, and several such approaches can be found in the literature.

$\rho(r)$ measures the probability of an electron occupying an infinitesimally small element of space. For values of r on the scale of interatomic distances, the electron density $\rho(r)$ for individual atoms follows an exponential decay function

$$\rho(r) \sim \exp\left[-2\sqrt{2I}r\right], \tag{2}$$

where I is the ionization energy of the system [7].

The topological analysis of $\rho(r)$ as formulated in the "quantum theory of atoms in molecules" (QTAIM) theory of Bader and co-workers [8] led to the concepts of *bond path* (BP) and *bond critical point* (BCP). The electron density along BP(r) descends steeply from each nucleus toward the unique intermediate stationary point BCP [9]. Because $\rho(r = r_{\mathrm{BCP}}) = \rho_{\mathrm{BCP}}$ at the BCP as well as its Laplacian $\nabla^2\rho_{\mathrm{BCP}}$ are directly observable quantities from both experimental diffraction studies and ab initio calculations, ρ_{BCP} may be the natural starting point when aiming to establish links of the bond valence approach to the physical world.

If in a simple *gedankenexperiment* we assume that electron densities of overlapping atoms do not affect each other but simply add up, then the electron density as a function of the position x along the bond path[1] between the two atoms at a distance R $(0 < r < R)$ varies as the linear combination of electron densities

$$\rho(r) = a_1 \exp[-c_1 r] + a_2 \exp[-c_2(R - r)] \ . \tag{3}$$

Here the electron density ρ_{BCP} at the bond critical point, i.e. the minimum electron density along this bond path for a fixed value of R can be found from the minimum condition $\mathrm{d}\rho/\mathrm{d}r = 0$ (as $\nabla^2\rho_{\mathrm{BCP}}$ is automatically >0 in this simplified model and will be typically >0 for bonds with significant ionicity):

[1] For this simple two-atom case the straight connecting line between the atoms is the *bond path*, i.e. the connecting line between the atoms for which each point on the line is a maximum of the electron density within the perpendicular plane. This does not necessarily remain true for multi-atom configurations or solids.

$$- a_1c_1 \exp[-c_1r_{BCP}] + a_2c_2 \exp[c_2r_{BCP} - c_2R] = 0, \tag{4}$$

$$\frac{a_1c_1}{c_2} \exp[-c_1r_{BCP}] = a_2 \exp[c_2r_{BCP} - c_2R], \tag{5}$$

so that at the position r_{BCP} of the BCP for a given interatomic distance R

$$r_{BCP} = \frac{c_2R - \ln\left(\frac{a_2c_2}{a_1c_1}\right)}{(c_1 + c_2)}, \tag{6}$$

we can express ρ_{BCP} as

$$\rho_{BCP} = \left(\frac{c_2 + c_1}{c_2}\right) a_1 \exp[-c_1r_{BCP}], \tag{7}$$

and hence

$$\rho_{BCP} = \exp\left[\left(\ln\left(a_1\frac{c_1 + c_2}{c_2}\right) + \frac{c_1 \ln\left(\frac{a_2c_2}{a_1c_1}\right)}{(c_1 + c_2)}\right) - \frac{c_1c_2}{(c_1 + c_2)}R\right]. \tag{8}$$

Equation 8 can be rewritten into the form

$$\rho_{BCP} = \exp\left[\frac{\left(\frac{c_1+c_2}{c_1c_2}\ln\left(a_1\frac{c_1+c_2}{c_2}\right) + \frac{\ln\left(\frac{a_2c_2}{a_1c_1}\right)}{c_2}\right) - R}{\frac{c_1+c_2}{c_1c_2}}\right] = \exp\left[\frac{A - R}{B}\right], \tag{9}$$

that simplifies a comparison to the corresponding bond valence. While the assumption that the two electron densities will add up without otherwise affecting each other is of course oversimplifying, it may be plausible (at least for the Lewis acid–Lewis base–type interactions in the focus of the bond valence model) to assume that the perturbation of the electron density at the BCP will be a simple function of ρ_{BCP} itself, so that the parameters A, B in (9) will change, but the functional form of the correlation is preserved.[2] This includes electron density variations of the generalized type $a_1 \exp[-c_1r] + a_2 \exp[-c_2(R - r)] + a_3 \exp[-c_1r]$

[2] The simple expression $\rho_{BCP} = \exp\left[\frac{\frac{2}{c}\ln(2a) - R}{\frac{2}{c}}\right] = \exp\left[\frac{\frac{\ln(2a)}{\sqrt{2I}} - R}{\frac{1}{\sqrt{2I}}}\right]$ would follow in the same way for the superposition of two identical atoms, but it may be questionable whether the electron density variation in such a perfectly covalent bond can still be thought of as a minor perturbation of the additive linear combination of the electron densities of the contributing atoms.

$\exp[-c_2(R-r)]$ that to a good approximation can be described using the form of (4) $[a_1' \exp[-yc_1 r] + a_2' \exp[-yc_2(R-r)]$ with a factor y slightly below 1 for the relevant case of positive a_3 values. A more detailed discussion of such a similar approach can be found in Mohri's [10] work, though his consequence of suggesting a redefinition of bond valence will be of limited helpfulness for our task.

The correlation $\rho_{\mathrm{BCP}}(R) = \exp\left[\frac{A-R}{B}\right]$ can be directly compared to the bond valence $s(R) = \exp\left[\frac{R_0-R}{b}\right]$ for the same interatomic distance R suggesting a simple power law relationship between (valence) electron density and bond valence:

$$R = R_0 - b \ln s = A - B \ln \rho_{\mathrm{BCP}} \Rightarrow s = \rho_{\mathrm{BCP}}^{\frac{B}{b}} \cdot \exp\left[\frac{R_0 - A}{b}\right]. \tag{10}$$

So the bond valence can be thought of as a *functional* of (valence) electron density, but it should be kept in mind that so far the coefficients of this functional relationship appear to be specific to the type of atom pair.

A more generally applicable correlation would require that ρ_{BCP}, s and R(M–X) can be (at least approximately) predicted for a wider range of atom types if one of the three quantities is known. There have been various attempts to establish such correlations, most notably by Gibbs and co-workers [11,12], who inspired by the pioneering description of Brown and Shannon [13] of bond valences of oxides in terms of a power law dependence on the bond length reported a general power law dependence on the expected bond length $\langle R(\mathrm{M{-}{-}O})\rangle$ in a wide range of oxides

$$\langle R(\mathrm{M-O})\rangle = 1.39\left(\frac{\mathrm{e.b.s}}{n-1}\right)^{-0.22}, \tag{11}$$

(based on Shannon and Prewitt's dataset of coordination number-dependent radii sums) [14] using for the sake of simplicity Pauling's electrostatic bond strength e.b.s.[3] as an approximation to the bond valence, and practically the same correlation was found for a much smaller dataset available at that time when e.b.s. was replaced by bond valences as calculated from the Brown and Shannon [13] parameter set. Here, n refers to the principal quantum number of the metal ion M (i.e. its row number in the periodic table).[4] Complementary studies on fluorides, sulfides, and nitrides found analogous correlations. Later the same group [15] reported an analogous correlation of the electron density

$$\langle R(\mathrm{M-O})\rangle = 1.42\left(\frac{\rho_{\mathrm{BCP}}}{n-1}\right)^{-0.19}, \tag{12}$$

---

[3] The electrostatic bond strength e.b.s. of an ionic bond is defined as the nominal charge (oxidation state) of the cation $V_{\mathrm{id}}(\mathrm{M})$ divided by the coordination number $N_{\mathrm{C}}(\mathrm{M})$ of the cation. For a cation M^{m+} symmetrically coordinated by $N_{\mathrm{C}}(\mathrm{M})$ anions X^{z-}, the numerical value of e.b.s.(M–X) is thus identical to the expectation value of the (conventional) bond valence s(M–X).

[4] Note that Gibbs et al. refer to row numbers in the sense row number = n–1.

although in the meanwhile an exponential expression was found to be more appropriate for expressing bond length-bond valence correlations.

To test whether a power-law expression of the type shown in (11) and (12) is superior to an exponential correlation as expected from (10), here the same data used in deriving these relationships are analyzed in a variety of ways. In Figs. 1, 2, and 3 results for double-logarithmic fits of the scaled s, e.b.s. or ρ_{BCP} vs. the bond distance that yield the power law relationships are compared to single-logarithmic plots for the same data that yield the exponential relationships in (10). For hydrogen a scaling by $n - 1 = 0$ is obviously inapplicable, so $n - 1$ is arbitrarily set to 0.5 for this case. Moreover the fits are generalized in Figs. 2. and 3 by substituting the scaling factor $(n - 1)$ by $(n - 1)^x$ or n^x with an adjustable exponent x.

The double-logarithmic plot of Pauling's electrostatic bond strength divided by $(n - 1)$ vs. the radii sum based on coordination number-dependent bonding radii of cations M tabulated by Shannon and Prewitt and a fixed radius of $R(\mathrm{O}) = 1.4$ Å for oxygen (see top graph of Fig. 1) essentially reproduces (11) by Gibbs et al. [11] though with slightly different parameter values (prefactor 1.58, exponent −0.20) with a correlation coefficient of $R^2 = 0.968$, but a polynomial fit to the same data (broken line) suggests that data show a slightly curved correlation instead of randomly scattering around a linear trend. If the same data are plotted with just replacing the ln[R(M–O)] axis by a linear R(M–O) axis, then the correlation coefficient improves marginally and the linear and polynomial fits become nearly congruent (see insert of lower graph in Fig. 1). So while a clear distinction of the two fitting approaches based on the achieved correlation coefficient is not feasible (due to the small range of bond distances), an exponential fit appears to be equally well suited. In the following paragraph only results for the main group (s and p block) cations are briefly discussed, and it should be noted that for all graphs d-block (transition metal) cations follow a trend with a slightly different slope. The trend for f-block elements seems to harmonize more to that of the main group elements, but the limited range of cation bond distances makes it discriminate the slopes.

As seen in Fig. 1 the correlation of the scaled bond valence for main group cations can as well be expressed by the exponential form ($R^2 = 0.970$)

$$\frac{s_{\mathrm{conv}}}{n-1} \approx \frac{\mathrm{e.b.s}}{n-1} = \exp\left(\frac{1.4598\ \text{Å} - \langle R(\mathrm{MO})\rangle}{0.479\ \text{Å}}\right), \tag{13a}$$

if conventional bond valence parameters or Pauling's e.b.s. approach are used, or

$$\frac{s_{\mathrm{softBV}}}{n-1} = \exp\left(\frac{1.2478\ \text{Å} - \langle R(\mathrm{MO})\rangle}{0.476\ \text{Å}}\right), \tag{13b}$$

if *softBV* parameters are used ($R^2 = 0.981$; *softBV* parameters assume that a part of the total bond valence arises from interactions beyond the first coordination shell and thus necessarily assign a slightly lower bond valence to a bond of a given bond length).

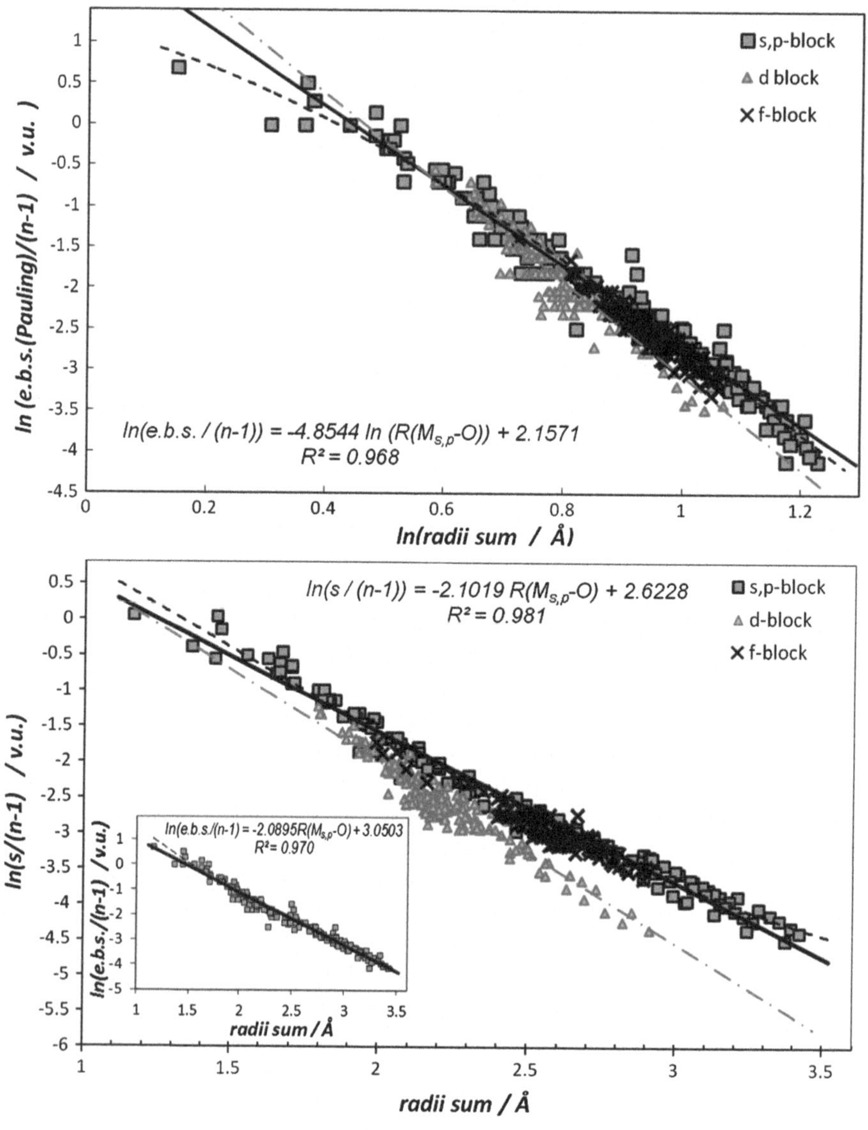

Fig. 1 *Top*: Double-logarithmic plot of Pauling's electrostatic bond strength for M–O bonds divided by $n - 1$ (where n is the cation period number) vs. the radii sum based on coordination number dependent bonding radii of M tabulated by Shannon and Prewitt and a fixed radius of $R(\mathrm{O}) = 1.4$ Å for oxygen. *Bottom*: Logarithmic plot of bond valence divided by $(n - 1)$ vs. the radii sum (using a linear radii sum axis) based on the same R(M–O) distances. In the main graph the bond valence is calculated using *softBV* bond valence parameters, while the inset shows the fit when Pauling's e.b.s. values are used. The two main graphs show data separately for s or p-block cations (*squares*), d-block cations (*triangles*), and f-block cations (*crosses*), while the *inset* shows main group cations only. In each graph the linear regression line for main group cations is shown as a *bold black line* and the resulting parameters analysis are listed. The *broken black line* indicates a third order polynomial fit to the same data. The *gray dash-dot line* represents a linear fit to the data for d-block cations

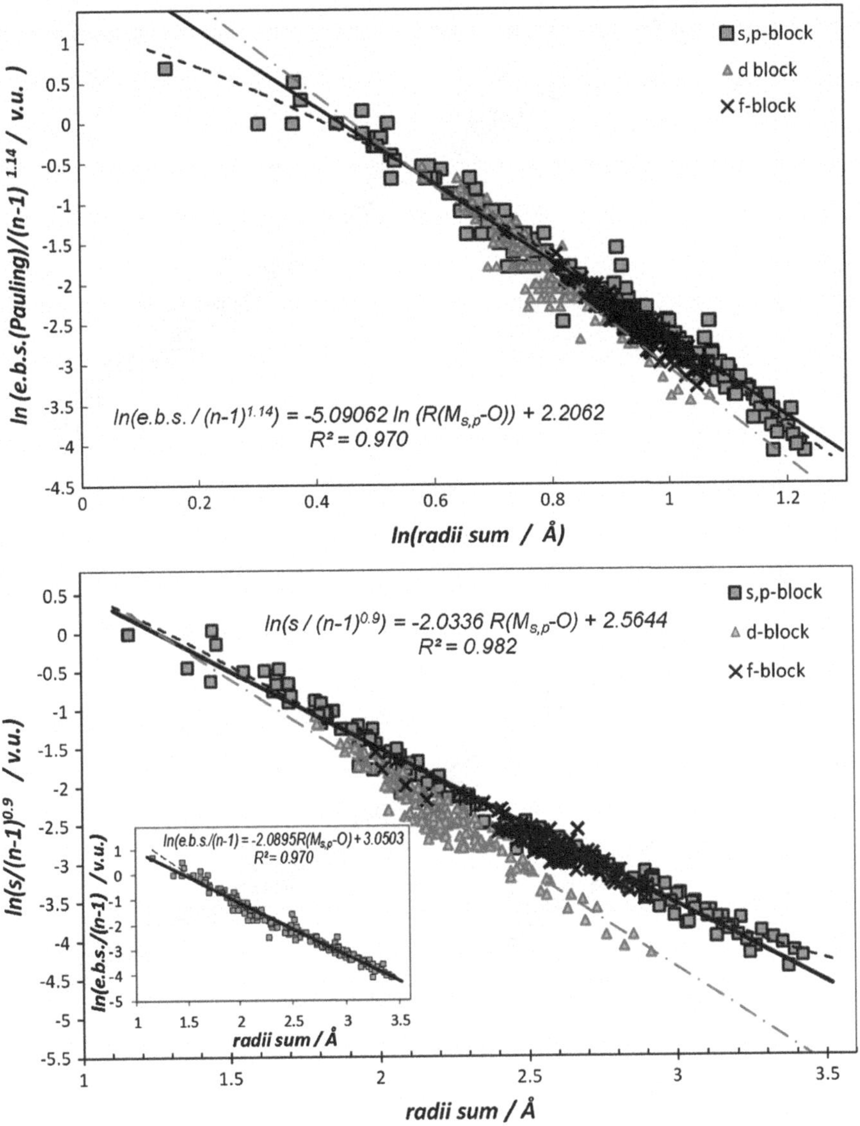

Fig. 2 *Top*: Double-logarithmic plot of Pauling's electrostatic bond strength divided by an optimized exponent of $(n - 1)$ vs. the radii sum. *Bottom*: Logarithmic plot of bond valence (for main graph, e.b.s. for inset) divided by an optimized exponent of $(n - 1)$ vs. the radii sum. Data (and the chosen symbols) are identical to Fig. 1. The only change is that the scaling factor $(n - 1)$ is replaced by $(n - 1)^x$. The respective value of the exponent x is chosen in each graph separately to maximize the correlation coefficient R^2

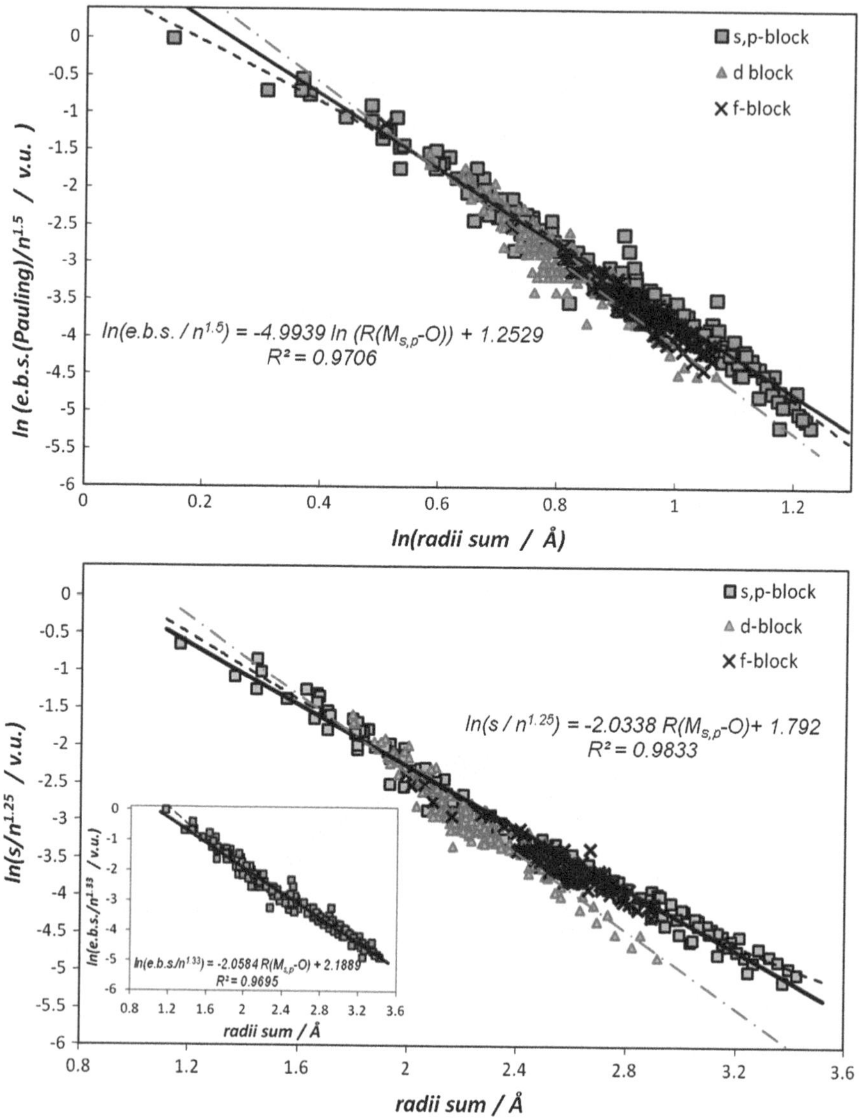

Fig. 3 *Top*: Double-logarithmic plot of Pauling's electrostatic bond strength divided by an optimized exponent of the row number n of the periodic table vs. the radii sum. *Bottom*: Logarithmic plot of bond valence (for main graph, e.b.s. for inset) divided by an optimized exponent of the row number of the periodic table vs. the radii sum. Data (and the chosen symbols) are identical to Figs. 1 and 2. The only change is that the scaling factor $(n - 1)$ is replaced by n^x. The respective value of the exponent x is chosen in each graph separately to maximize the correlation coefficient R^2

Though the exponential correlations have marginally higher correlation coefficients than the power law model (upper graph of Fig. 1), the difference appears to be too small to be conclusive. A generalization of the scaling factor $(n - 1)$ to $(n - 1)^x$ with an optimized exponent yields only a minor improvement (see Fig. 2) and all optimized exponents are close to 1.[5] Thus the added parameter does not significantly improve the refinement. This is a bit different when using n as the scaling factor which yields clearly inferior correlations (especially for the power law correlation), while as shown in Fig. 3 n^x with an optimized x slightly larger than one yields (again only by a slim margin) the highest correlation coefficients for both the power law and the exponential correlation, but the exact value varied from $x = 1.5$ for the power law fit of e.b.s. data ($R^2 = 0.971$ for linear regression, significantly reduced curvature in polynomial fit)

$$\langle R(\mathrm{M-O})\rangle = 1.285\left(\frac{\mathrm{e.b.s.}}{n^{3/2}}\right)^{-0.20}, \tag{14a}$$

via $x = 1.33$ for the exponential fit of e.b.s. data ($R^2 = 0.970$)

$$s_{\mathrm{conv}} \approx \mathrm{e.b.s} = n^{1.33}\exp\left(\frac{1.0634\ \text{Å} - \langle R(\mathrm{MO})\rangle}{0.485\ \text{Å}}\right), \tag{14b}$$

to $x = 1.25$ for the exponential fit to s_{softBV} data ($R^2 = 0.983$) leading to

$$s_{\mathrm{softBV}} = n^{1.25}\exp\left(\frac{0.8811\ \text{Å} - \langle R(\mathrm{MO})\rangle}{0.4917\ \text{Å}}\right). \tag{14c}$$

In the same way the electron density at the bond critical point ρ_{BCP} can be related to the bond length either via a power law or an exponential law. To establish such relationships obviously ρ_{BCP} must be known from experimental or ab initio simulation studies. From systematic determinations of ρ_{BCP} values for a range of oxide minerals [16] and sulfide systems [17] Gibbs et al. [15–17] suggested power law relationships for individual anions. As demonstrated in Fig. 4 it is also possible to fit correlations for the 303 M–O bonds [15, 16] [that were used to derive (12)] and data [17] for 108 M–S bonds of main group cations M by a single masterplot. When using the period number scaling in the form $(n_\mathrm{M} - 1)^x(n_\mathrm{X} - 1)^y$ as shown in Fig. 4a, b the refinement results correspond to

$$\langle R(\mathrm{M-X})\rangle = 1.445\left(\frac{\rho_{\mathrm{BCP}}}{(n_\mathrm{M}-1)(n_\mathrm{X}-1)^{1.5}}\right)^{-0.179}, R^2 = 0.977, \tag{15a}$$

or

[5] $x = 1.14$ ($R^2 = 0.970$) for the power law correlation, $x = 1.0$ ($R^2 = 0.970$) or $x = 0.9$ ($R^2 = 0.982$) for the exponential correlation when using e.b.s, or *softBV* parameters.

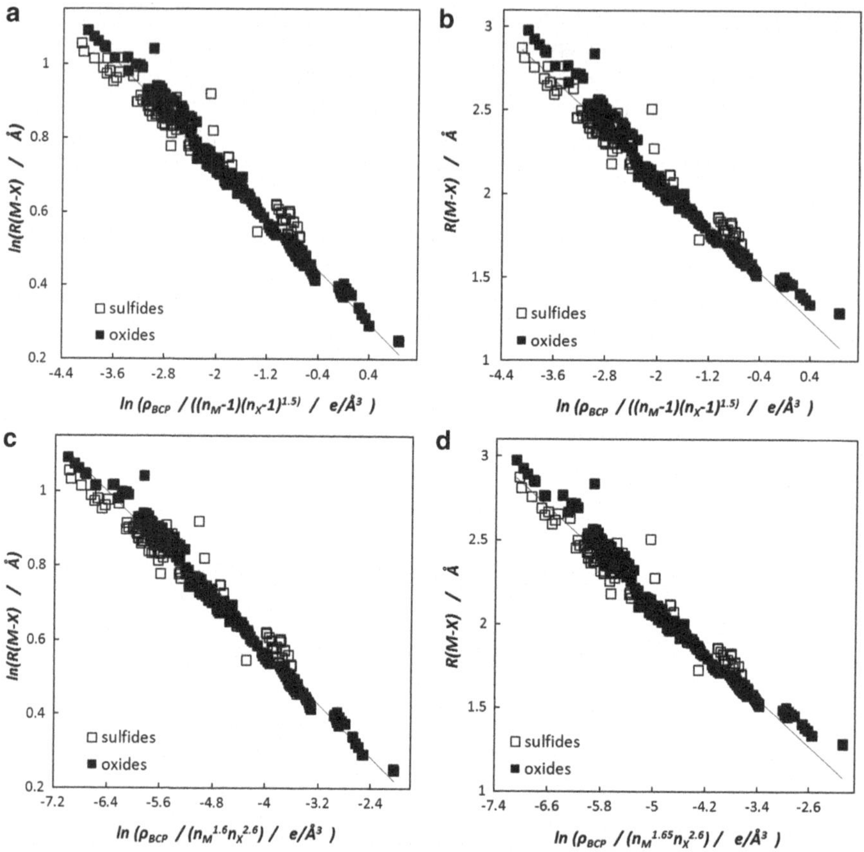

Fig. 4 Correlation between the scaled electron density ρ_{BCP} at the bond critical point and the cation–anion distance R(M–X) for bonds between main group cations M and oxides (*filled symbols*, 303 data points) and sulfides (*filled symbols*, 108 data points). The double-logarithmic plots (**a**, **c**) yields the power law relationship in (15a) and (15c), while the linear regression of the r.h.s. single-logarithmic plots (**b**, **d**) result in the exponential correlations of (15b) and (15d). The *solid line* indicates the fitted linear regression. The respective values x and y of the exponents in the scaling factors $n_{\mathrm{M}}{}^{x}n_{\mathrm{X}}{}^{y}$ are chosen to maximize the correlation coefficient R^2

$$\rho_{\mathrm{BCP}} = (n_{\mathrm{M}} - 1)(n_{\mathrm{X}} - 1)^{1.5} \exp\left(\frac{1.386\ \text{Å} - \langle R(\mathrm{M} - \mathrm{X})\rangle}{0.359\ \text{Å}}\right),\ R^2 = 0.969. \quad (15\mathrm{b})$$

Alternatively scaling by $n_{\mathrm{M}}{}^{x}\, n_{\mathrm{X}}{}^{y}$ in Fig. 4c, d yields

$$\langle R(\mathrm{M} - \mathrm{X})\rangle = 0.864 \left(\frac{\rho_{\mathrm{BCP}}}{n_{\mathrm{M}}^{1.6} n_{\mathrm{X}}^{2.6}}\right)^{-0.179},\ R^2 = 0.979, \quad (15\mathrm{c})$$

or

$$\rho_{BCP} = n_M^{1.65} n_X^{2.6} \exp\left(\frac{0.343\ \text{Å} - \langle R(\text{M}-\text{X})\rangle}{0.356\ \text{Å}}\right),\ R^2 = 0.973. \quad (15d)$$

Again the differences among the correlation coefficients which yield nearly the same correlation coefficient appear inconclusive. Despite the slightly lower R^2 values the optimized exponents for (15a) and (15b), 1 and 3/2, let these correlations appear to be more natural choices than (15c) and (15d) that lead to fractional exponents. It may also be noted that in this case the deviations from the exponential fit appear to be not only marginally larger but also more systematic.

From a pragmatic point of view it does not matter, whether the two observed correlations s vs. R(M–X) and ρ_{BCP} vs. R(M–X) are both described by the exponential or both by the power law function: their combination will in both approaches yield the same type of power law relationship between bond valence s and the electron density at bond critical point ρ_{BCP} (that can be thought of as a generalization of (10). Indeed, as shown in Fig. 5, such a power law can be observed when fitting the same set of literature data on R(M–X) and ρ_{BCP} used in Fig. 4, yet converting R(M–X) into the corresponding bond valence $s(R(\text{M–X}))$. To minimize the number of refineable parameters the exponents x and y of the scaling factor $(n_M - 1)^x(n_X - 1)^y$ were fixed to $y = 2x$ in the optimization in accordance with results of free refinements. For the above-mentioned reference data set of main group oxides and sulfides

$$s(R(\text{M}-\text{X})) = A \cdot (n_M - 1)^x (n_X - 1)^y \rho_{BCP}^z, \quad (16a)$$

with the coefficients $A = 0.832$, $x = y/2 = 0.2$, $z = 0.825$ and a correlation coefficient $R^2 = 0.984$ results when using *softBV* parameters. For conventional bond valence parameters taken from Brown's compilation [18] the optimized parameters $A = 0.846, x = y/2 = 0.25, z = 0.829$ yield a slightly lower correlation coefficient $R^2 = 0.946$. Unsurprisingly the *softBV* bond valence parameters (with two refined bond valence parameters R_0 and b) are somewhat more closely correlated to ρ_{BCP} than the conventional bond valence parameters (that assume a fixed value of b and refine only R_0) and hence allow for a more precise prediction of ρ_{BCP}. An almost equally close correlation can be derived using a scaling by $n_M^x n_X^y$

$$s(R(\text{M}-\text{X})) = A \cdot n_M^x n_X^y \rho_{BCP}^z, \quad (16b)$$

and the coefficients $A = 0.418, x = y/2 = 1/3, z = 0.826$ ($R^2 = 0.983$) for *softBV* parameters or $A = 0.346$, $x = y/2 = 0.43$, $z = 0.832$ ($R^2 = 0.945$) for conventional bond valence parameters.

As noted before in the literature the Laplacian $\nabla^2\rho_{BCP}$ of the electron density ρ_{BCP} at the BCP is correlated to ρ_{BCP} (and hence to the BV), but obviously also to the degree of bond ionicity (which can e.g. be expressed by the electronegativity or hardness differences of cation and anion). Using absolute electronegativity differences χ_{diff} (in eV) based on Mulliken's electronegativity definition [19]

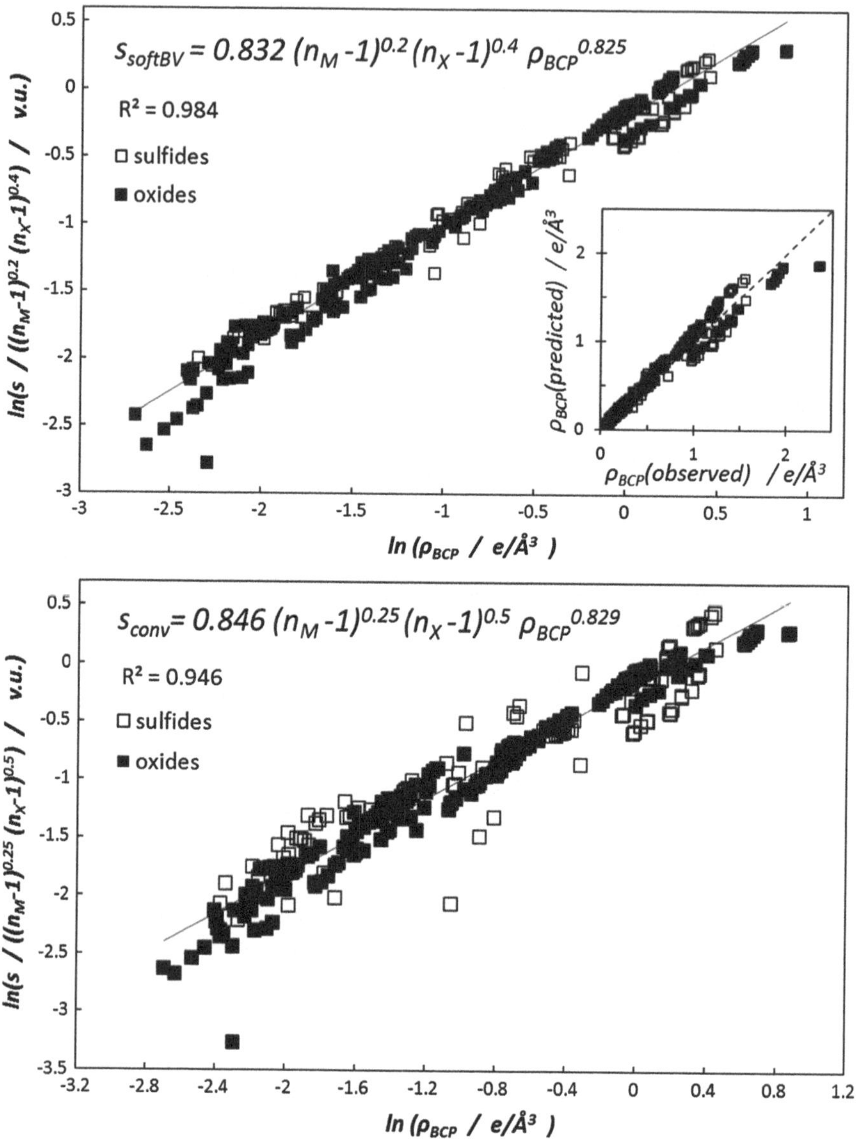

Fig. 5 *Top*: Double-logarithmic plots of the bond valence divided by the scaling factor $(n_M - 1)^x$ $(n_x - 1)^y$ vs. ρ_{BCP} , where n_M (n_X) is the period number of the cation (anion). (*Top*: using *softBV* bond valence parameters, bottom using conventional bond valence parameters). The respective values of the period number exponents are chosen to maximize the correlation coefficient R^2 for the linear regression over all data points (*solid line*), while maintaining a 1:2 ratio between x and y. The inset of the top graph compares the observed ρ_{BCP} to values predicted from the correlation derived in the main graph

cf. ref. and Sect. 3.2) calculated from the electronegativities of the corresponding ions (not elements) according to

$$\chi_{\text{diff}} = \sqrt{|\ \chi^2_{\text{M}^{m+}} - \chi^2_{\text{O}^{2-}}|}. \tag{17}$$

As Fig. 6 demonstrates, a significant correlation of the form

$$s(R(\text{M}-\text{O})) = C \cdot n^x_{\text{M}} n^{2x}_{\text{X}} \left(\chi_{\text{diff}}\ \nabla^2 \rho_{\text{BCP}}\right)^z, \tag{18}$$

with $C = 5.46 \cdot 10^{-3}$, $x = 0.5$, $z = 0.509$ (for χ_{diff} in eV and $\nabla^2\rho_{\text{BCP}}$ in e/Å^5) and $R^2 = 0.980$ is found for the 303 M–O bonds of the reference data set. It should however be mentioned that the sulfide reference data [17] only roughly follow the same trend. A similar quality of correlation for the same oxide data set is reached when using the average of cation and anion absolute hardness

$$\overline{\eta} = \sqrt{\frac{\eta^2_{\text{M}^{m+}} + \eta^2_{\text{O}^{2-}}}{2}}, \tag{19}$$

as the scaling parameter for the Laplacian, so that the bond valence can be expressed as a function of Laplacian (in e/Å^5), the conceptually closely related average atomic hardness (in eV) and atomic row number:

$$s(R(\text{M}-\text{O})) = C \cdot (n_{\text{M}} n_{\text{X}})^x \left(\overline{\eta}\ \nabla^2 \rho_{\text{BCP}}\right)^z, \tag{20}$$

with $C = 4.13 \times 10^{-3}$, $x = 0.85$, $z = 0.577$ and $R^2 = 0.984$. In this case a rough comparison with the more scattered data for the sulfide data set suggested that equal exponents of the row numbers are more appropriate than the 1:2 ratio applied in other correlations. Similar correlations with insignificantly lower correlation coefficients result for a scaling based on $(n_{\text{M}} - 1)^x (n_{\text{X}} - 1)^y$.

Overall the detailed parameter values listed in this section should be understood as approximate values only. This is not only due to imperfections of the reference data sets, but largely due to strong fundamental correlations among the parameters. The absolute electronegativity χ of a cation is, e.g., approximately determined by its formal charge q and total number N of electrons: As discussed more in detail in the literature [2, 20, 21] there is a nearly perfect and fairly general linear relation

$$\chi \sim \sqrt{\frac{q^3}{N}}, \tag{21}$$

(except for highly charged transition metal cations). As N obviously depends on the atomic number Z and thereby on the cation mass M, and – with a lower correlation coefficient – on the atomic row number n, differences between scalings based on M

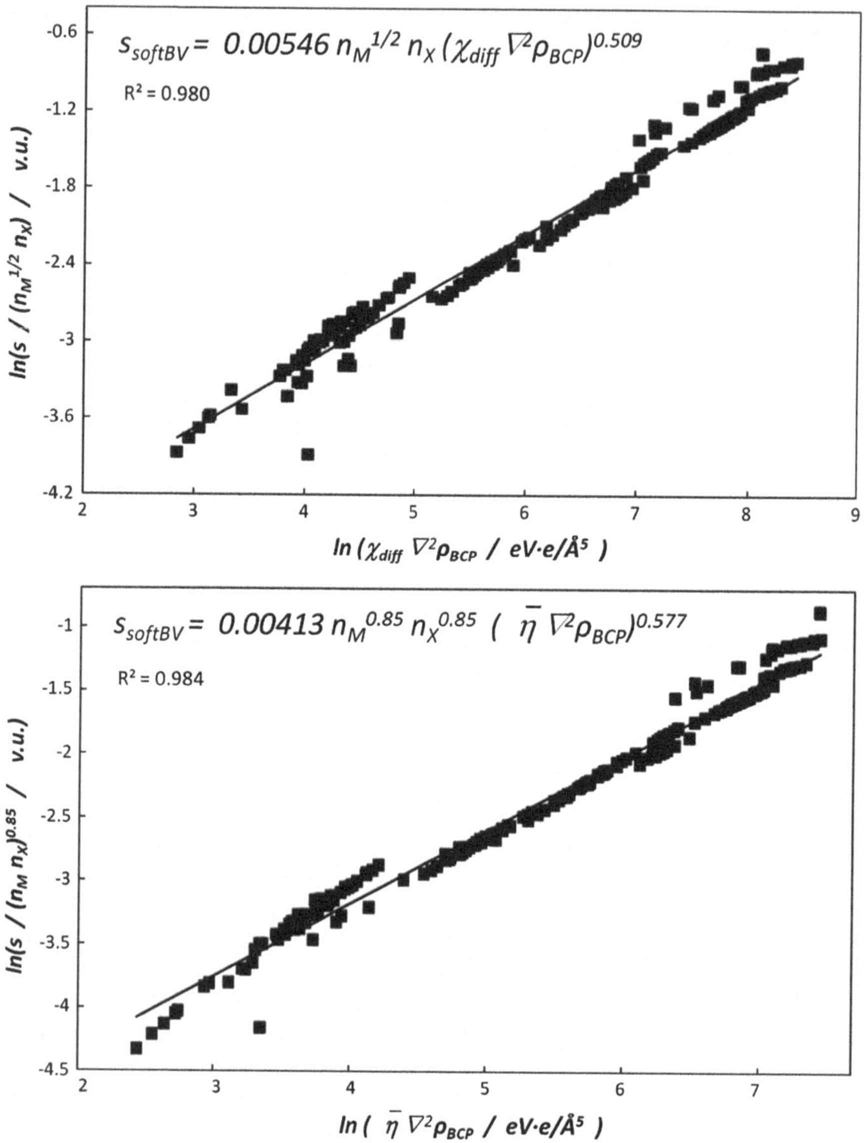

Fig. 6 Double-logarithmic plots of the bond valence (using *softBV* bond valence parameters) divided by an optimized exponent of the row numbers n of cation M and anion X in the periodic table vs. the product of Laplacian of electron density at the bond critical point $\nabla^2\rho_{BCP}$ and either electronegativity difference χ_{diff} (*top*) or average atomic hardness $\bar{\eta}$(*bottom*). In the top graph the exponent in the scaling factor $n_M^x n_X^y$ $x = 1/2$ is chosen to maximize the correlation coefficient R^2, while maintaining a 1:2 ratio between the exponents of n_M and n_X. For the bottom graph the exponent in the scaling factor $x = 0.85$ is chosen to maximize the correlation coefficient R^2. Based on a comparison with data for sulfides (not shown) a 1:1 ratio between the exponents of n_M and n_X has been chosen in this case

or χ become significant for highly charged light cations only and even an empirical distinction between χ and the atomic row number n as a scaling factor is often not straightforward for scattered data.

Still it should have become obvious from the above discussion that there is a close functional relationship between bond valence and electron density at the bond critical point (and in the same way between bond valence and the Laplacian of $\nabla^2\rho_{BCP}$) and that this correlation involves a scaling based on the principal quantum number (row number) of the atoms involved or a closely correlated quantity,[6] and at least for the case of the Laplacian to a measure of atomic polarizability (such as the atomic hardness or its inverse the atomic softness). This fundamental correlation should thus be taken into account when fine-tuning approaches to determine bond valence parameters and BV-related forcefields.

2.2 Bond Valence-Based Atomistic Forcefields

Another entirely empirical approach to reveal correlations between the bond valence scale and the energy scale that has been explored by several groups including ours is to try and derive energy-scaled atomistic forcefields starting from the bond valence parameters. Validating that these forcefields can reproduce experimental observables (lattice parameters, thermal expansion, compressibilities, static and dynamic structural features, etc.) then also indirectly validates links between the underlying bond valence parameters and the terms controlling the energetics of the simulated structure models.

Early approaches include the works of Eck and Dronskowski [22, 23], who in their *aixCCAD* software integrated bond valence terms to derive fictional extra charges, while the interaction is essentially treated as a Coulomb interaction. Shin et al. [24] suggested a bond valence mismatch term in combination with standard Coulombic, short-range repulsion and angle bending terms to describe, e.g., the ferroelectric transition in $PbTiO_3$

$$E_{\text{Shin}} = \sum_{A=1}^{N} S_A \sum_{i=1}^{n(A)} |V(a_i) - V_{id}(a)| + E_{\text{Coulomb}} + E_{\text{Born repulsion}} + E_{\text{octahedral tilt angle}} \quad (22)$$

Here the double sum of the bond valence term runs over the $n(A)$ atoms of type A and over the N atom types in the system. Besides the unclear distribution of potential energy between the Coulomb and bond valence contributions, the freely refined additional scaling parameter S_A for the bond valence mismatches of each atom type A and the adjustment of the fractional charge in the independent

[6] This correlation with the atomic row number n or $(n - 1)$ is as mentioned hardly distinguishable from a correlation with atomic number or atomic mass and has been preferred here more in line with the existing proposal in the literature.

Coulomb terms makes it difficult to take this as the basis for a transferable potential. From a mathematical point of view moreover the choice of the absolute value of the bond valence mismatch [i.e., of an exponent $g = 1$ in (1)] is disadvantageous as this potential contribution is not differentiable at $\Delta V = 0$.

A different empirical approach to assess the coefficients D_0 and g in (1) as well as a suitable functional form for the influence of the equal valence rule (E_{EVR}) have been derived in our earlier work [25, 26] from comparing the distance dependence of the bond valence sum mismatch with the distance dependence of interaction energies in various empirical interatomic potentials: The variation of an individual bond valence can be straightforwardly translated into the variation of a Morse-type interaction potential

$$E = D_0\left\{(\exp[\alpha(R_{\min} - R)] - 1)^2 - 1\right\}$$
$$= D_0\left\{\left(\frac{\exp\left[\frac{R_0-R}{b}\right] - s_{\min}}{s_{\min}}\right)^2 - 1\right\} \quad (23a)$$

This description implies that the interaction energy E can be approximated as a quadratic function of the deviation of the bond valence from its value $s_{\min} = \exp[(R_0 - R_{\min})/b]$ for the energy minimum distance ($R = R_{\min}$) and hence g in (1) assumes the value 2, so that in contrast to Shin's approach (23a) is continuously differentiable.[7] The bond valence parameter b is simply identified with the reciprocal of the stiffness parameter α of the Morse potential. Note that the bond valence parameter R_0 (i.e., the distance corresponding to a bond valence value of $s = 1$) in general differs from $R_{\min}$ (the bond distance for which the interaction potential yields an energy minimum). By introducing the dimensionless relative bond valence $s_{rel} = s/s_{\min}$ the Morse potential can be expressed concisely as

$$E = D_0\left\{\frac{(s - s_{\min})^2}{s_{\min}^2} - 1\right\} = D_0\left(s_{rel}^2 - 2s_{rel}\right) \quad (23b)$$

In contrast to the conventional bond valence sum mismatch description, such bond valence interaction potentials of the type described in (23a) or (23b) fulfill formal requirements for an anharmonic diatomic interaction potential, allowing for molecular dynamics simulation based on bond valence parameters. The Morse-type interaction potential is characterized by three parameters D_0, $R_{\min}$ (or $s_{\min}$) and $b = 1/\alpha$ representing the bond breaking energy, the equilibrium bond distance and the elastic compliance of the bond. Constraining b to a universal value is thus analogous to approximating the three-parameter Morse-type interaction by a

[7] The quadratic dependence may also be derived from equating bond valence to bond fluxes in a point charge model. For details, see Brown [3].

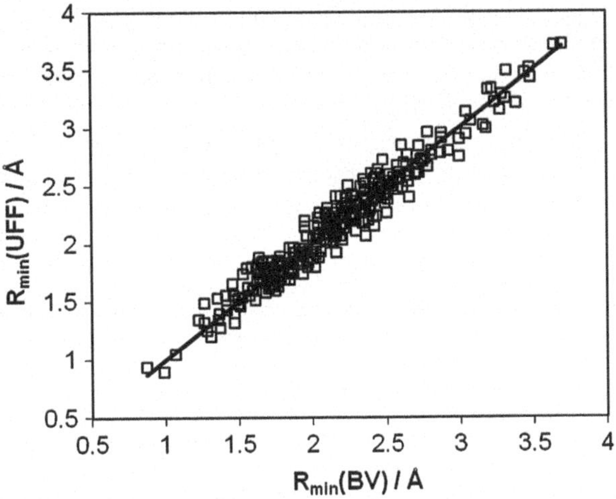

Fig. 7 Comparison of the equilibrium bond distances R_{min}(UFF) in the widely employed empirical forcefield UFF [27] and R_{min}(BV) in our BV-based Morse-type forcefield for 283 cation–anion pairs according to (23a) and (23b). The *solid line* indicates the fit target $R_{min}(\text{UFF}) = R_{min}(\text{BV})$

two-parameter description in which the stiffness of the bond is predetermined by the remaining refineable parameters of bond length and bond energy (as in the two-parameter Lennard–Jones potential).

A consistent set of R_{min} (and hence s_{min}) values [25] was approximated as

$$R_{min} \approx R_0 \times [f_1 + f_2 \cdot |\sigma_A - \sigma_X|] - b \cdot \ln\left(\frac{V_{id}}{N_C}\right). \tag{24}$$

Here, N_C refers to the preferred coordination number of the central ion and the empirically determined term in square brackets accounts for the effect of polarization (σ_A, σ_X refer to the absolute softness values of the cation and anion, respectively) as well as the influence of higher coordination shells. Practically, the coefficients $f_1 = 0.9185$ and $f_2 = 0.2285$ eV were derived by comparison with the parameters used in other empirical force fields such as the universal force field (UFF) [27]. The level of agreement for the resulting R_{min} values with the established UFF forcefield can be judged from Fig. 7 ($R^2 \approx 0.95$).

For a Morse-type interaction potential the bond dissociation energy is $D_0 = b^2\,k/2$, k being the force constant at the distance $R = R_{min}$. We have thus approximated D_0 for a wide range of main group cations as

$$D_0 = \frac{kb^2}{2} = c \cdot 14.4 \frac{\text{eV}}{\text{Å}} \frac{[V_{id}(\text{M}) \cdot V_{id}(\text{X})]^{1/c} b^2}{2R_{min}\sqrt{n_M n_X}}, \tag{25}$$

with $c = 1$ if A is an s or p block element, or $c = 2$ if A is a d block element. Here n_A, n_X represent the principal quantum numbers of cation A and anion X, respectively. Thus the principle quantum numbers are used as scaling factors for the bond valence term in an analogous way to their use in Sect. 2.1. The terms V_{id}(M), V_{id}(X) in (25) refer to the absolute values of the nominal charges of M or X. Table 1 gives examples of parameters derived by the above formalism from our respective *softBV* bond valence parameters [26]. A wider range of parameters is available from the electronic appendix.

Starting from this definition of the energy of a single bond, the total *bond valence site energy* BVSE(A) of a cation M can then be determined as the sum over bond valence terms for the interactions with each of the N_X adjacent anions X_j:

$$\mathrm{BVSE(M)} = D_0\left[\sum_{j=1}^{N_X}\frac{\left(s_{\mathrm{M-X}_j}-s_{\min}\right)^2}{s_{\min}^2}-N\right]+\sum_{i=1}^{N_M}E_{\mathrm{Coulomb}}(\mathrm{M}-\mathrm{M_i}) \qquad (26)$$

and vice versa. The Coulomb repulsion term runs over all N_M cations M_i Rewriting (26) then reveals how this bond valence site energy BVSE(M) (i.e., the total potential energy of cation M in the bond valence approximation) varies with both the mismatch of the bond valence sum and the asymmetry of the coordination. This provides an avenue to quantify the correlation between (1) the *bond valence sum rule* [28] stating that that the sum of the bond valences around an atom is equal to its atomic valence and (2) the *equal valence rule* [28], which states that the sum of the bond valences around any loop is zero, is that the most isometric distribution of atomic valence among the bonds is energetically preferable.

In this brief summary of the derivation we assume for the sake of simplicity only contributions from the $N_X = N_C$ counterions of type X in the 1st coordination shell around M. Then the correlation takes the simple form of (27):

$$\mathrm{BVSE(M)} = D_0\left[N_C\left\{\frac{(V(\mathrm{M})-V_{\min}(\mathrm{M}))^2}{V_{\min}^2}-1\right\}+\sum_{i=1}^{N_C}\frac{\left(s_{\mathrm{M-X}_i}-\bar{s}_{\mathrm{M-X}_i}\right)^2}{s_{\min}^2}\right]$$
$$+\sum_{i=1}^{N_M}E_{\mathrm{Coulomb}}(\mathrm{M}-\mathrm{M}_i), \qquad (27)$$

where $V_{\min}(\mathrm{M}) = N_C \cdot s_{\min}$ in the first term (the BVS rule term), while the second term (the equal valence rule term) quantifies the effect of the deviation of individual bond valences from their average value $s_{\mathrm{M-X}} = V(\mathrm{M})/N_C$ [25].

A major advantage of such an energy-scaled bond valence mismatch is that it allows for straightforward combinations of the bond valence sum term as an effective (attractive or repulsive) short-range interaction term with suitably weighted penalty functions for coordinations with differing bond lengths and particularly a Coulombic cation–cation repulsion. To model Coulomb repulsion we commonly use simple fractional charges q_M, q_X that are calculated based on

Table 1 Bond valence parameters R_0, b of selected cations in oxides, the corresponding cutoff distance R_{cutoff} and the resulting Morse potential parameters D_0, R_{min} ($\alpha = 1/b$)

Cation M	R_0/Å	b/Å	N_C	R_{cutoff}/Å	D_0/eV	R_{min}/Å
Ag(I)	1.78239	0.457	4.438	5	0.63519	2.22578
Al(III)	1.59901	0.516	5.327	5	1.80346	1.75806
As(III)	1.76706	0.541	3	5	1.51493	1.64554
As(V)	1.76689	0.385	4.029	5	2.71934	1.58127
Au(III)	1.81761	0.415	4	5.5	1.96967	1.81312
B(III)	1.35761	0.45	3.417	4.5	2.38924	1.34003
Ba(II)	2.15998	0.448	10.32	6	0.57994	2.73769
Be(II)	1.20903	0.442	4	5.5	2.76882	1.52217
Bi(III)	2.03677	0.482	6.058	5.5	0.97904	2.18321
Bi(V)	2.04498	0.512	6	5	1.4405	1.98599
Br(VII)	1.83658	0.424	4	5.5	4.24339	1.50274
C(II)	1.41368	0.432	1	5	2.40553	1.03098
C(IV)	1.39826	0.437	3	5	4.79187	1.20089
Ca(II)	1.79519	0.402	7.544	5.5	0.99429	2.32032
Cd(II)	1.83926	0.441	6.176	5.5	0.98346	2.1694
Ce(III)	2.03118	0.427	9.147	5.5	1.22048	2.37861
Ce(IV)	2.02821	0.443	7.867	5.5	1.48412	2.19872
Cl(V)	1.69552	0.491	3	5.5	4.29089	1.35653
Cl(VII)	1.67946	0.436	4	5	5.991	1.34801
Co(II)	1.59773	0.476	5.506	5.5	1.51476	1.93362
Co(III)	1.59234	0.494	6	5.5	1.87024	1.7762
Cr(III)	1.66198	0.503	6	5.5	1.77335	1.83887
Cr(VI)	1.82471	0.426	4	5.5	3.68751	1.53251
Cs(I)	2.25899	0.439	11.79	6.5	0.23307	3.13121
Cu(I)	1.58730	0.476	2.56	5	0.66417	1.78269
Cu(II)	1.57422	0.402	5.087	5	1.85341	1.56633
Dy(III)	1.96029	0.43	7.828	5.5	1.1735	2.22689
Er(III)	1.95608	0.52	7.135	5.5	1.12394	2.17477
Eu(III)	2.00469	0.413	7.743	5.5	1.19545	2.26888
Fe(II)	1.57911	0.402	5.743	5.5	1.69269	1.96005
Fe(III)	1.70840	0.437	5.733	5	1.66681	1.86647
Ga(III)	1.71606	0.41	4.905	5	1.18456	1.79391
Ge(IV)	1.73939	0.48	4.305	5	1.91375	1.66872
H(I)	0.87045	0.414	1.923	4	1.8858	1.12768
Hg(II)	1.81276	0.451	6.966	5.5	1.12852	2.25275
In(III)	1.90305	0.403	6.024	5	0.84076	2.02471
K(I)	1.94117	0.411	8.846	6	0.34985	2.76636
La(III)	2.06392	0.404	9.83	5.5	1.18587	2.46989
Li(I)	1.17096	0.416	5.021	5.5	0.98816	1.94001
Lu(III)	1.91728	0.427	6.83	5.5	1.19488	2.136
Mg(II)	1.48398	0.423	5.897	5.5	1.57554	1.95627
Mn(II)	1.62758	0.413	5.91	5.5	1.64143	2.02969
Mn(IV)	1.73272	0.478	5.923	5	1.85886	1.77045
Mn(VII)	1.87362	0.526	4	6.5	4.9163	1.48171
Mo(III)	1.78933	0.501	5.7	5.5	1.42826	1.92974

(continued)

Table 1 (continued)

Cation M	R_0/Å	b/Å	N_C	R_{cutoff}/Å	D_0/eV	R_{min}/Å
Mo(IV)	1.72390	0.391	6	6.5	3.10807	1.85099
Mo(VI)	1.90934	0.562	4.764	5	1.9915	1.71254
N(III)	1.40795	0.418	2	5	3.81089	1.13758
N(V)	1.46267	0.425	3	5	6.27677	1.16142
Na(I)	1.56225	0.436	6.52	6	0.57523	2.37433
Nb(III)	1.74581	0.449	6	6	2.02848	1.9519
Nb(IV)	1.78543	0.403	6	6	2.7096	1.85989
Nb(V)	1.86588	0.478	6.044	5.5	2.72326	1.85459
Nd(III)	2.02425	0.449	8.647	5.5	1.13205	2.33016
NH_4(I)	2.03380	0.498	3.467	6	0.40537	2.45364
Ni(II)	1.55920	0.394	5.933	5.5	1.46841	1.92452
Ni(III)	1.64888	0.407	6	5.5	1.66191	1.81887
P(III)	1.51555	0.4	3	4.5	2.02062	1.41051
P(V)	1.62038	0.423	4	5	3.89635	1.44066
Pb(II)	2.01825	0.419	7.541	5.5	0.63833	2.44191
Pb(IV)	2.03293	0.424	5.74	5	1.02719	2.02857
Pd(II)	1.62359	0.412	4	5.5	1.7391	1.83671
Pd(IV)	1.80500	0.401	5.333	5.5	2.04218	1.79813
Pr(III)	2.03652	0.419	9.067	5.5	1.17041	2.37113
Pt(II)	1.51205	0.437	4	5.5	2.14999	1.80179
Pt(IV)	1.82198	0.451	6	5.5	2.03825	1.87174
Rb(I)	2.08597	0.443	10.02	6.5	0.26813	2.89683
Re(III)	2.20710	0.449	6	6	0.81067	2.33218
Re(V)	1.82664	0.439	6	6	2.41099	1.76914
Re(VII)	1.97792	0.433	4.098	6	3.55593	1.59634
Rh(III)	1.67013	0.434	6	5.5	1.92826	1.86915
Ru(IV)	1.79363	0.415	6	5.5	1.99513	1.84053
Ru(V)	1.87442	0.494	6	5.5	2.13208	1.81571
S(IV)	1.64282	0.426	3	5.5	3.03672	1.41188
S(VI)	1.64220	0.415	4	5	4.96726	1.38102
Sb(III)	1.92036	0.412	6	5	1.17786	2.07526
Sb(V)	1.89768	0.421	6	5.5	1.95523	1.86318
Sc(III)	1.73220	0.426	6.255	5.5	2.1561	1.99615
Se(IV)	1.80095	0.421	3	5.5	2.38082	1.55957
Se(VI)	1.79866	0.478	4	5.5	3.44865	1.53287
Si(IV)	1.60817	0.486	4.1	5	2.8572	1.53594
Sm(III)	2.01168	0.401	8.119	5.5	1.17622	2.29536
Sn(II)	1.87499	0.498	3.325	5.5	0.97261	1.9642
Sn(IV)	1.89019	0.52	6.069	5	1.35268	1.93422
Sr(II)	1.95311	0.508	9.4	5.5	0.74351	2.53589
Ta(IV)	1.75632	0.498	5.5	6	2.75655	1.79826
Ta(V)	1.86816	0.479	6.09	5.5	2.36669	1.85532
Tb(III)	1.95675	0.401	7.958	5.5	1.20764	2.23563
Tb(IV)	1.96244	0.512	6	6	1.70132	2.38506
Te(IV)	1.95290	0.479	3.396	5.5	1.67169	1.75208
Te(VI)	1.91343	0.463	6	5.5	2.56406	1.80876

(continued)

Table 1 (continued)

Cation M	R_0/Å	b/Å	N_C	R_{cutoff}/Å	D_0/eV	R_{min}/Å
Ti(III)	1.69766	0.498	6	5.5	1.97851	1.88619
Ti(IV)	1.72394	0.479	6	5.5	2.81333	1.83144
Tl(I)	1.91752	0.436	8.03	6	0.34999	2.77086
Tl(III)	2.06297	0.479	5.22	5	0.67637	2.10642
Tm(III)	1.94901	0.574	6.912	5.5	1.18138	2.16042
V(III)	1.67799	0.498	6	5.5	1.82936	1.85797
V(IV)	1.74932	0.441	5.738	5	2.08047	1.77638
V(V)	1.79445	0.465	4.166	5.5	3.69533	1.60258
W(IV)	1.74558	0.465	6	6	2.47114	1.81945
W(V)	1.81975	0.338	6	6	2.6157	1.76261
W(VI)	1.90641	0.483	5.688	5	1.84267	1.77713
Y(III)	1.90384	0.354	7.285	5.5	1.62701	2.21523
Yb(III)	1.92872	0.433	6.875	5.5	1.21989	2.1422
Zn(II)	1.65344	0.371	4.718	5	1.24031	1.88557
Zr(IV)	1.84505	0.414	6.765	5.5	2.19103	1.99602

$$q_{M_i} = \frac{V_{id}(M_i)}{\sqrt{n_{M_i}}}\sqrt{\frac{\sum_j \frac{V_{id}(X_j)N_{X_j}}{\sqrt{n_{X_j}}}}{\sum_i \frac{V_{id}(M_i)N_{A_i}}{\sqrt{n_{M_i}}}}}, \quad q_{X_j} = \frac{V_{id}(X_j)}{\sqrt{n_{X_j}}}\sqrt{\frac{\sum_i \frac{V_{id}(M_i)N_{M_i}}{\sqrt{n_{M_i}}}}{\sum_j \frac{V_{id}(X_j)N_{X_j}}{\sqrt{n_{X_j}}}}}, \tag{28}$$

in which N_{Mi}, (N_{Xj}) refer to the occupancies of the ith cation M_i (jth anion X_j in the structure model). This scaling of fractional charges ensures that the model is overall charge-neutral. Obviously, fractional charges from quantum mechanical calculations could be used instead and may improve the quality of the fit, but at the expense of suitability of the approach for the fast and automatic generation of forcefields for screening of a wide range of compounds.

The Coulomb repulsions e.g. between two different cations M_1 and M_2 cations are then taken into account in a screened $E_{Coulomb}(M_1{-}M_2)$:

$$E_{Coulomb}(M_1 - M_2) = \frac{q_{M_1} q_{M_2}}{R_{M_1-M_2}} \operatorname{erfc}\left(\frac{R_{M_1-M_2}}{\rho_{M_1-M_2}}\right). \tag{29}$$

The screening factor $\rho_{M1-M2} = f\,(r_{M1} + r_{M2})$ therein is assumed to equal the sum of the covalent radii r_{M_i} of the two ions involved times a factor f that depends on the average absolute cation electronegativity and the average cation charge in the compound. Typical values of f for oxides fall into the range 0.75 ± 0.1 and thus typical values of ρ are of the order of 2–3 Å. While this simplification restricts long-range interactions to the real part of the Ewald sum, molecular dynamics simulations employing such a localized interaction model yield realistic activation energies of diffusion, e.g., for a range of Li conducting oxyacid salts cf. [6]. It should be emphasized that the attractive Coulomb interactions between cations and anions are already covered by the bond valence terms in (27) and thus should not be included a second time in $E_{Coulomb}$.

3 Practical Steps of Bond Valence Parameter Determination

3.1 Data Mining

The determination of new bond valence parameters typically involves as the first and crucial step the data mining: a database of reliable reference crystal structure data has to be selected. In our work the main source is the Inorganic Crystal Structure Database (ICSD) complemented by structures extracted from the recent literature. The selection of suitable structures has to take into account the quality of datasets in terms of small residual factor values (typical values were $R \leq 0.055$) and global instability index G values < 0.2 valence units (v.u.). Though in general a smaller R value should indicate a more reliable structure model, inconsistencies in the type of R value reported in databases as well as intrinsic weaknesses of the R values as a quality criterion for structures (e.g., the small influence of light atoms on R values from X-ray diffraction data) limit its significance and so it should not be used as the only criterion. Using the global instability index, G, as a selection criterion obviously leads to an iterative process, as at least starting values of bond valence parameters are required to calculate the G.

Typically it is also advisable to limit the complexity of reference structures, e.g. by restricting the number of elements in the compound [e.g., to 2–4 of which only one should be an anion (see Sect. 3.3)] and by excluding structure models for modulated structures, polytypes, or structures with disordered cation arrangement (for which a bond valence analysis requires more complex models of the actual local structure), structures determined under high pressures or at extreme temperatures, or structure models based on theoretical predictions (that tend to overestimate the unit cell volume). Focusing on high symmetry compounds to further reduce structural complexity is, however, not viable, as the size of smaller ions in simple high symmetry structures often deviates somewhat from the ideal radius ratio. In contrast, it should be ascertained that the reference data set contains a sufficient number of low symmetry cation environments and a representative mixture of different coordination numbers. In a significant number of cases binary high symmetry compounds turn out to be outliers and were eliminated from the reference databases in the course of the parameter refinement.

Obviously the reference data set should consist only of structures for which the assignment of oxidation states to the atoms is straightforward. Particular care should be taken when including compounds that contain the same element in different oxidation states and for transition metal compounds with short cation–cation distances, where significant metal–metal interactions have to be expected.

For some cation–anion pairs automatic data extraction routines lead to reference data sets containing multiple structure refinements of the same compound (e.g., a compound of high technological or scientific relevance) or to data sets dominated by a group of isostructural compounds. In both cases it is advisable to either reduce the number of representatives of this structure type or to introduce a weighting factor that scales down their influence on the refined parameters.

As mentioned above the refinement process may involve the need to eliminate outliers that would strongly bias the refined parameters. Still for each such eliminated structure it should be made sure that the reason for a deviating bond distance is understood (or that there are further indications for a problem of the structure refinement) and that the elimination does not unduly bias the balance between different coordination polyhedra in the surviving reference data set.

3.2 The Role of the Bond Softness Parameter *b*

As briefly mentioned in Sect. 2.2, the bond valence parameter b represents the compliance of a bond to external forces. Approximating b by a universal value therefore eliminates the crystal-chemical information on elastic behavior from bond valence parameters (or more precisely reduces the information from an approximation that takes into account structure type and atomic properties to a crude estimate solely based on the coordination type). Whether such information is relevant for a given application purpose and available for specific cation–anion pair may depend on individual circumstances. Here it will be assumed that retaining this information available is desired, and thus it is necessary to elaborate suitable procedures to systematically determine the respective b values.

This aim of deriving b values that preserve the information on the softness (compliance, polarizability) of a bond obviously implies the need for an independent measure of this bond softness from experimentally observable or ab initio accessible quantities. Parr and Pearson [29] proposed to characterize individual particles in equilibrium by their constant site-independent *electronic chemical potential* μ

$$-\mu = -\left(\frac{\partial E}{\partial \rho}\right)_{\nu} \approx \frac{\mathrm{IE} + \mathrm{EA}}{2} = \chi, \tag{30}$$

and the global average of the (site dependent) *absolute hardness* η

$$\eta = \frac{1}{2}\left(\frac{\partial \mu}{\partial \rho}\right)_{\nu} \approx \frac{\mathrm{IE} - \mathrm{EA}}{2}, \tag{31}$$

or its inverse the *absolute softness* $\sigma = \eta^{-1}$. Here, ρ represents the electron density, while the subscript ν indicates the potential of the nucleus and external influences. In this approximation $-\mu$ becomes identical to Mulliken's definition of the absolute electronegativity χ [19]. This approximate identification with the independently accessible quantities ionization energy IE and electron affinity EA was originally only derived for neutral particles, but according to Pearson [30] the corresponding properties of a cation M^{m+} may be calculated in the same manner using the ionization energy of M^{m+} (i.e. the $(m + 1)$th ionization energy of M) as IE and replacing EA by the ionization energy of $M^{(m-1)+}$.

For anions we need a somewhat different approach as electron affinities of anions are not always accessible and their relevance for the determination of bond softnesses appears questionable. Pearson suggested to use the values of IE and EA for the neutral elements as a rough approximation for the anions. As shown in our earlier work [1], an empirical correlation between the anion radius and anion softness may be utilized to obtain a more precise estimate of the softness values for anions with different charges: To eliminate a shift of the softness-versus-anion radius relationships for halides and chalcogenides, we use – in line with Pearson's suggestion – the softness values of neutral atoms for the monovalent anions, but reduce the softnesses of the divalent chalcogenide anions (as calculated from IE and EA of the neutral atoms) by 0.017 eV^{-1}. It may be noted that the true softness values will still be slightly overestimated by this crude approximation, but there is good indication that the use of this modified softness definition is sufficient to achieve comparability at least among chalcogenides and halides [1].

To derive a measure for the softness of an M–X bond, the softness values of the interacting species M^{m+} and X^{x-} need to be combined. The empirical HSAB (hard and soft acids and bases) concept [31, 32] suggests that reactions occur most readily between species that match each other in softness. From the empirically observed formation of strong bonds between anions and cations of equal softness it appears straightforward to conclude that the interatomic potentials for these bonds will be steeper (and thus correspond to a smaller value of b) than the potentials for the weaker bonds between particles of mismatched softness values. The low b values proposed by Pauling [33, 34] for bonds between the same type of atoms from his early investigations of bond length bond order relationships ($b = 0.30$ Å for metals, $b = 0.26$ for C–C bonds) may be tentatively interpreted as a first hint that bonds between particles of equal softness are characterized by a low value of b. This hypothesis was further supported by a vague trend in a survey of literature bond valence parameters that were refined without postulating a universal b value, but following the convention of limiting bond valence contributions to the first coordination shell [1]. The large scatter in this correlation can mainly be attributed to the short range of bond distances R usually found in the first coordination shell, so that the shape of the $s(R)$ correlation can hardly be fitted. When the limitation to the first coordination shell is lifted so that the weak interactions with more distant counterions are included, a much wider range of R values becomes available allowing for more accurate fits of $s(R)$. Thus a comprehensive determination of freely refined bond valence parameters that we conducted in the frame of deriving the *softBV* parameter set shows a much clearer correlation of the bond valence parameter to the softness difference. As seen in Fig. 8 the lowest b values are found for softness differences of ca. 0.05 eV^{-1}, whereas for cation–anion pairs with higher softness difference (as well as for the pairs with smaller or even negative softness difference)

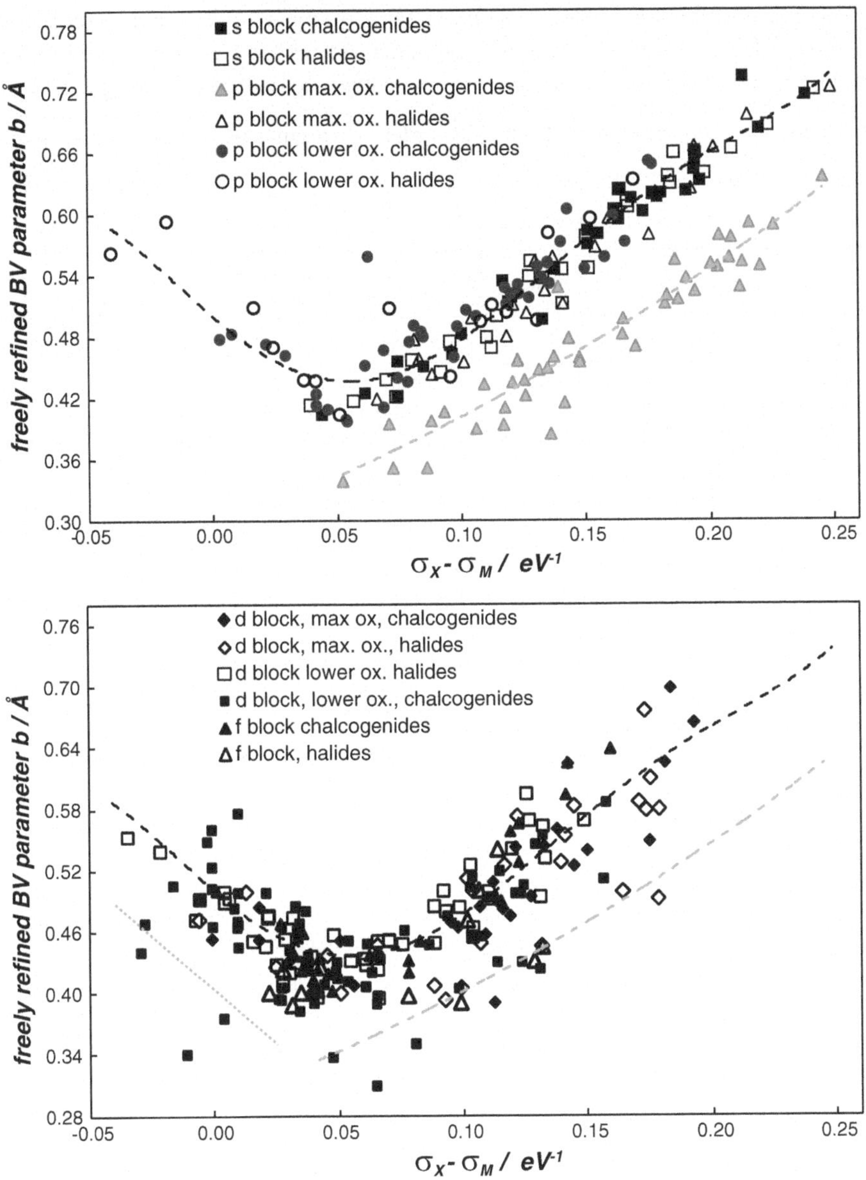

Fig. 8 Correlation of b values found in freely refined bond valence parameter sets for the interaction of main group cations (*upper graph*) or transition element and rare earth cations (*lower graph*) with halide (*open symbols*) or chalcogenide anions (*filled symbols*). The *broken black line* represents a polynomial fit to all data in the upper graph except for the bonds of p block cations in their maximum oxidation state to chalcogenide anions (*gray triangles*). For the latter a polynomial fit (*broken gray line*) yields a parallel trend with somewhat lower b values. In the lower graph the same polynomial fits to the main group cation data are shown again, now along with data for d and f block cations. Therefrom it is obvious that they follow the same trend (though with a larger scatter). The *dotted gray line* is a guide to the eye

progressively higher b values are observed. For most main group cations (exceptions are p block cations in their maximum oxidation state in bonds to chalcogenides) the appropriate b value (in Å) can thus be reasonably approximated from the softness difference $\sigma_X - \sigma_M$ (in eV^{-1}) by the 5th order polynomial fit $b = \sum_{i=0}^{5} a_i(\sigma_X - \sigma_M)^i$ shown in Fig. 8 with the coefficients $a_5 = 2479.6$ Å eV^5, $a_4 = -1384.2$ Å eV^4, $a_3 = 198.75$ Å eV^3, $a_2 = 10.428$ Å eV^2, $a_1 = -2.1316$ Å eV, $a_0 = 0.5009$ Å. For the p block cations in their maximum oxidation state in bonds to chalcogenides a simpler second order polynomial fit with $a_2 = 1.9108$ Å eV^2, $a_1 = 0.8287$ Å eV, $a_0 = 0.2946$ Å (gray broken line in Fig. 8) can be used to predict the systematically lower b values, since the softness difference for all observed cases was $> 0.05\ eV^{-1}$. Analogous polynomial fits (based on the smaller set of reference data available at that time) have been used to derive the systematic b values in the *softBV* parameter set.

The scatter and small number of data points with negative softness differences makes it impossible to decide whether the slopes for the two branches of the correlation between b and the softness difference (anion > cation or anion < cation) differ. The apparent shift of the minimum in the correlation to positive softness differences is most likely just another indication that the rough estimate of the anion softness (by assuming equal softnesses for neutral atoms and monovalent anions) consistently overestimates anion softness.

It should be emphasized that the typical b values found in this way are somewhat larger than the conventionally chosen "universal value" of 0.37 Å. This difference can partly be understood as a self-fulfilling prophesy, since limiting the bond valence contributions to the first coordination shell requires a significantly steeper drop of the bond valence with increasing bond length. As demonstrated in our earlier work [1] for the case of the Li–O bond, for which the freely refined b value as refined from bond distances in Li environments increases systematically with the chosen cutoff radius from ca. 0.45 Å for a cutoff radius corresponding to the boundary of the first coordination shell to the limiting value of 0.515 Å for a sufficiently large cutoff radius. More generally it is advisable to fit the bond valence parameters (especially if both R_0 and b are variable) not only for one more or less arbitrarily chosen cutoff distance R_{cutoff}, but also for a range of R_{cutoff} values, so that it can be judged whether the refined value of b is stable vs. small variations of R_{cutoff}. The cases shown in Fig. 9 exemplify that for R_{cutoff} of the order of the first coordination shell radius no meaningful free refinement of R_0 and b was possible as small changes in the cutoff radius lead to extreme changes in b value (with an inherent tendency to huge b values). When increasing R_{cutoff} typically a minimum of b is observed for values somewhat beyond the first coordination shell, followed – as mentioned above – typically by a systematic increase that (after a slight local maximum for distances in the range of the 2nd coordination shell) leads to a stable value for cutoff distances larger than 5–8 Å depending on the size of the affected ions. This long-range limiting value appears to be the most suitable choice of the b value. It can of course not be ruled out that in some cases inaccurate reference data drive the free simultaneous refinements of R_0 and b more often to too high than to too low b values. Therefore it

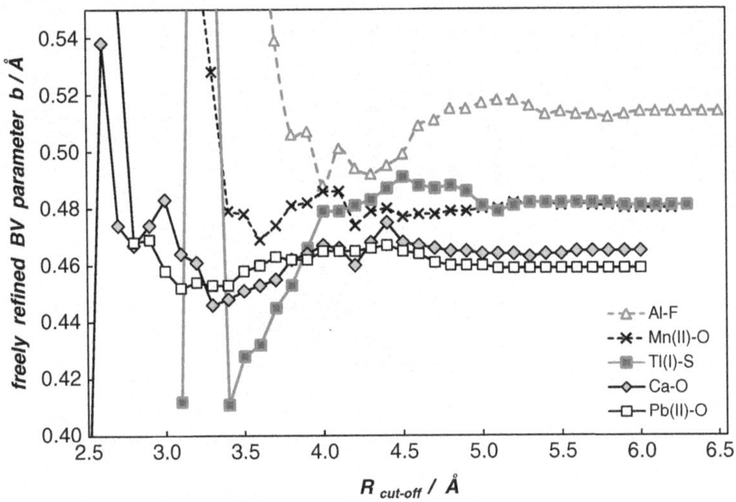

Fig. 9 Selected examples for the correlation of the b values obtained from free refinements of bond valence parameter sets to the cutoff distance R_{cutoff} of the respective cation–anion pair. The set of reference structures for each cation anion pair contained about 100 cation environments and remained unchanged for all cutoff distances

appears advisable to derive b values for cation anion pairs, for which reference data sets would contain only a small number of distinct cation environments from the correlations shown in Fig. 8 rather than from individual refinements. The increased correlation coefficient for the $s(\rho_{BCP})$ correlation when s is calculated from the softness-sensitive *softBV* data that follow this guideline may be taken as a hint that the b values chosen there are – while probably not perfect – at least superior to the alternative of a universal b value.

3.3 Special Considerations for Compounds with Multiple Anions

For compounds that contain more than one type of anions (i.e., different types of elements with negative oxidation states) the effectively employed bond valence parameter b should not just be the tabulated *softBV* value for compounds that contain only this anion but needs to be adapted to account for the mutual equalization of bond softnesses.

From a set of 128 compounds containing both a halide and a chalcogenide anion it could be shown that if that was not accounted for then bond valences of the bonds to softer (harder) anions in the compounds are systematically overestimated (underestimated). In principle a detailed description of this environmental sensitivity

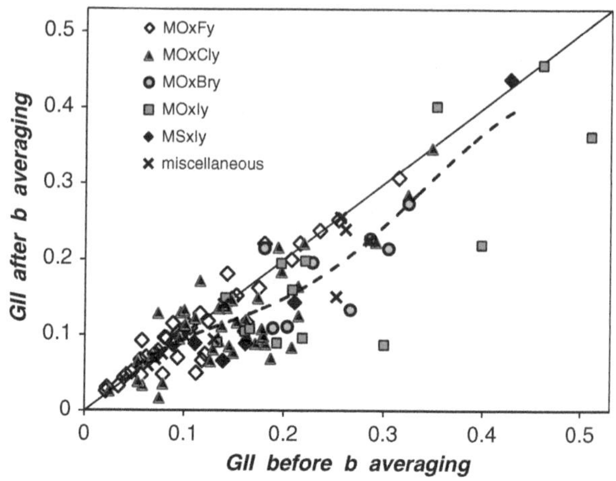

Fig. 10 Influence of *b* averaging on the global instability index G for 128 compounds that contain both a chalcogenide and a halide anion. The *solid line* marks the ratio 1:1, the *broken line* a fourth order polynomial over all data

of bond valence parameters should be treated based on the spatial arrangement of the different anions. A satisfactory agreement can, however, also be reached by a simple average treatment: For each cation first a weighted mean bond valence parameter $b_{\text{average}}(\text{M})$ averaging over all anion types [weighting factor = number of these anions per unit cell × abs(oxidation state)] is determined and then the effective *b* value $b_{\text{effective}}$ for the cation anion pair A–X in the specific multi-anion compound is calculated as

$$b_{\text{effective}}(\text{M},\text{X}) = f\, b_{\text{average}}(\text{M}) + (1-f)\, b\,(\text{M},\text{X}), \tag{32}$$

with an empirical weighting factor $f = 2/3$.

As seen in Fig. 10 this helps to significantly reduce the average *G* for structures in the reference data set from 0.157 (when using unaveraged *b* values) to 0.131 [when using the $b_{\text{effective}}$ values according to (32)].[8] This averaging becomes the more relevant the more the anions in a compound differ in their softnesses. For the MO_xF_y compounds in the reference set *G* was practically unchanged by the *b* averaging, while the relative reductions in *G* become more pronounced with increasing softness difference from 18% for MO_xCl_y to 24% for MO_xBr_y and 27% for MO_xI_y compounds cf. Fig. 10). It should be noted that compounds containing different types of anions are – as mentioned above – excluded from the reference data sets for the determination of *softBV* parameters (while most compounds in the reference data sets contained different types of cations so the anyways smaller softness differences between different cations in the structures are already factored in to some extent).

[8] For conventional bond valence values with fixed $b = 0.37$ Å the average *G*-value for the same set of reference structures was 0.184.

3.4 Cutoff Criteria and Coordination Numbers

In this section we have a closer look at ways to determine the limit of the first coordination shell and thereby also the preferred coordination number that in the energy-scaled version of the bond valence approach becomes an additional empirical parameter required to determine the expected equilibrium bond distance R_{min}. Coordination numbers N_C(M–O) for the oxides of 158 types of cations M were redetermined from 8829 reference cation environments. To this end all cation oxygen distances up to 5–8 Å were calculated for the reference structures and therefrom the running coordination number N_{RCN} calculated as a function of the bond distance $R_{M–O}$, which should yield a plateau in-between coordination shells and correspondingly a minimum in a plot of the differentiated term dN_{RCN}(M–O)/ $dR_{M–O}$ vs. $R_{M–O}$. A synopsis of both plots was then used to identify the preferred coordination number N_C for cations (cf. the example M = B^{3+} in Fig. 11).

For some cations in uncommon oxidation states the number of available data did not allow for a direct determination from the above-mentioned procedure. In 12 of the 158 cases N_C was extrapolated based on the trend for the same element when varying the oxidation states.

While in Brown's earlier work [35] a fixed N_C-independent bond valence threshold value of 0.038 v. u. was used to decide whether a cation–anion pair should be counted as bonded and hence to contribute to N_C, here we assign the coordination numbers devising the above-mentioned geometric approach. This leads to a coordination number-dependent threshold value for the minimum bond valence of bonds that are to be considered as part as contributing to the first coordination shell (see Fig. 12). Selected N_C coordination numbers for oxides determined in this way are tabulated in Table 1. More data are listed in the electronic appendix.

A comparison of our results to the preferred coordination numbers determined by Brown [35] in 1985 cf. Fig. 13) based on the smaller set of reference structures available at that time shows that for high N_C, the values determined by the approach presented here tend to be higher than those reported by Brown. This is mainly an expected result of the different approach in defining the boundary of the first coordination shell. For lower N_C the results of both studies are not significantly different, as the clear gap between the bond lengths of nearest and second-nearest neighboring shells of anions around a cation make the detailed choice of the threshold value less relevant.

In Brown's approach for a monovalent cation[9] with $N_C = 12$ the bond valence in a symmetric arrangement is only slightly more than twice the bond valence for the threshold of the coordination shell, whereas for $N_C = 4$ the ratio between the two bond valence values becomes ca. 6.6. In our approach this ratio remains constant ($\approx$6). As a plausibility check systematic trends of N_C against the atomic number both along periods and down the groups of the periodic table were checked. As

[9] In order to eliminate the effect of the oxidation state of the central cation the bond valence values in Fig. 12 are scaled by the oxidation state V_{id}(M).

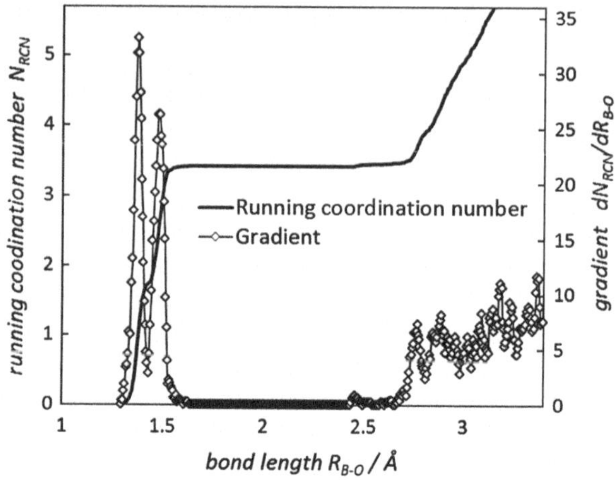

Fig. 11 The preferred coordination number for B^{3+} in oxides is determined to be $N_{RCN} = 3.417$ from (1) the plateau region in the distance-dependent running average coordination number and (2) the minimum in a plot of $dN_{RCN}/dR_{B\text{–}O}$ vs. $R_{B\text{–}O}$. Parameters for other cation-oxygen pairs were determined analogously

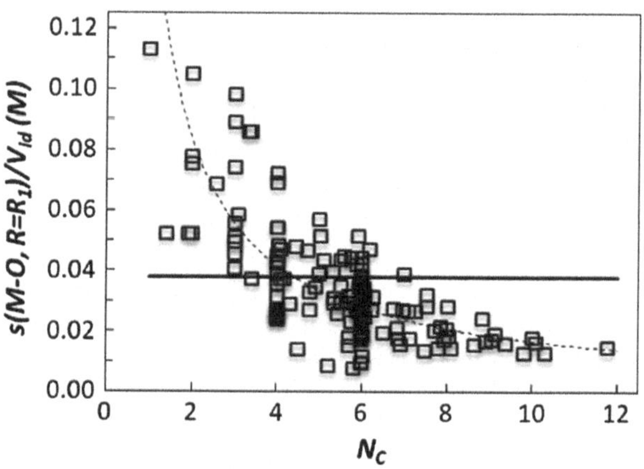

Fig. 12 Variation of dimensionless relative bond valence values $s(M\text{–}O)/V_{id}(M)$ for the bonds of the 158 studied cations M to oxygen at $R(M\text{–}O) = R_1$, i.e. at the boundary of the first coordination shell. The *solid horizontal line* marks the threshold value chosen by Brown; the *broken line* is a power law fit to the data $s/V_{id} = N_C/6$ as a guide to the eye

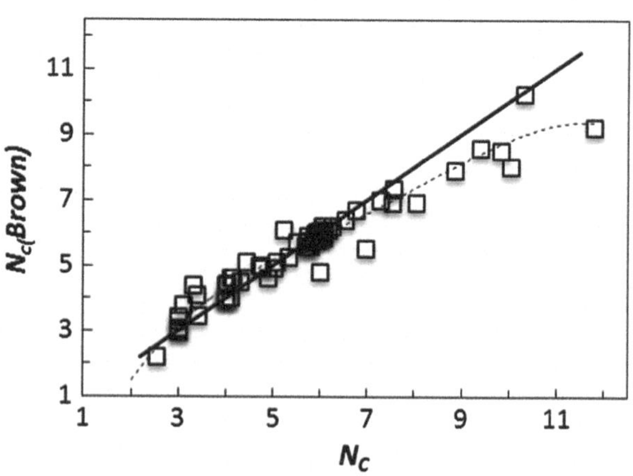

Fig. 13 Comparison of values of preferred coordination numbers N_C(Brown) as reported in Brown's earlier work [35] and N_C values determined in this work according to the method described in the text. This comparison comprises 73 cation types covered in both studies. The *solid line* marks the case $N_C(\text{Brown}) = N_C$; the *broken line* is a polynomial fit to the data as a guide to the eye

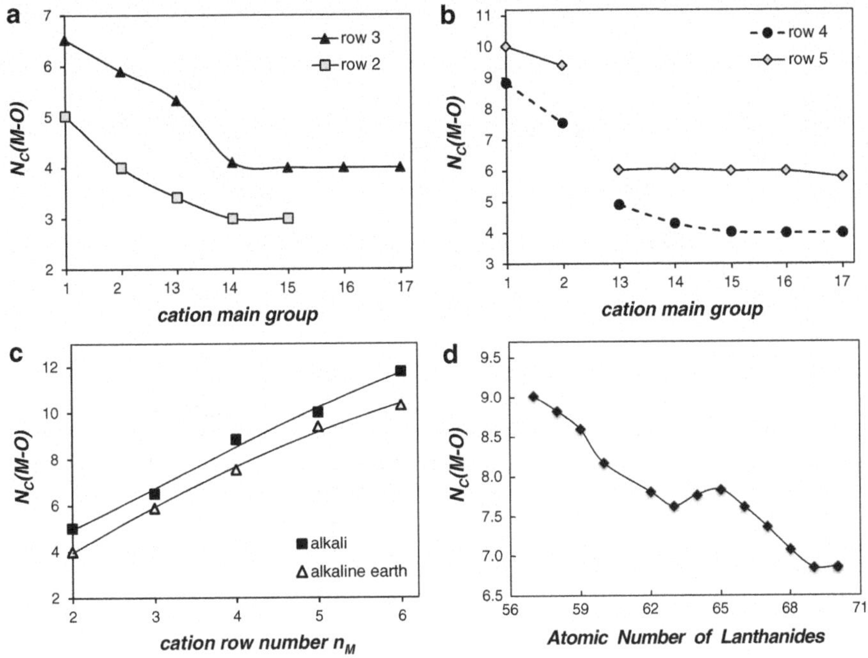

Fig. 14 Variation of the coordination number N_C(M–O) of selected metals in their maximum oxidation state in oxides: (**a**, **b**) variation of N_C with the group number along selected rows of the periodic table, (**c**) with the row number for alkali and alkaline earth cations. (**d**) Variation of N_C in lanthanide cations M^{3+} with the atomic number

exemplified in Fig. 14, N_C reproduces the expected trends: it decreases across a period (increase in the effective nuclear charge) and increases with the row number of the periodic table. In the case of the lanthanide ions M^{3+} the variation of N_C with the atomic number even clearly shows the anomaly for the half-filled f-orbital.

In the same way as for oxides we have also looked into refining parameters for M–X couples where the anion X is a non-oxygen chalcogenides (75 M–S, 29 M–Se, 20 M–Te parameters from 1672, 266, or 110 cation environments) or a halide (70 M–F, 68 M–Cl, 31 M–Br and 25 M–I parameters from 1043, 483, 169, or 167 cation environments, respectively). The reliability of these parameters is often somewhat lower due to the limited number of suitable reference structures. When comparing coordination numbers N_C and threshold radius values of the first coordination shell R_1 for different anions,[10] comparisons between oxides and sulfides are relatively straightforward, yielding a ca. 0.6 Å larger radius of the first coordination shell radius for the sulfides and a reduction of the coordination number N_C(M–S)

[10] Here we limit the comparison to O^{2-}, S^{2-}, F^- due to the considerably lower number of data sets available for the other anions.

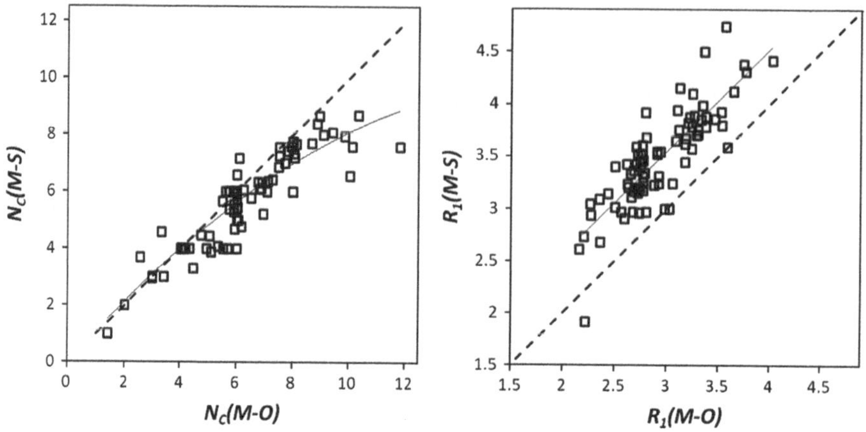

Fig. 15 *Left*: Comparison of coordination number found for sulfides and oxides of the same cation. For small N_C(M–S) values both coordination numbers hardly differ, while for large coordination numbers the larger size of S^{2-} leads to a significantly smaller N_C(M–S). *Right*: Radius R_1 of the first coordination shell of a cation M in a sulfide environment vs. R_1 in an oxide environment. *Broken lines* in both graphs mark a 1:1 ratio, while the *solid lines* are polynomial fits to the data

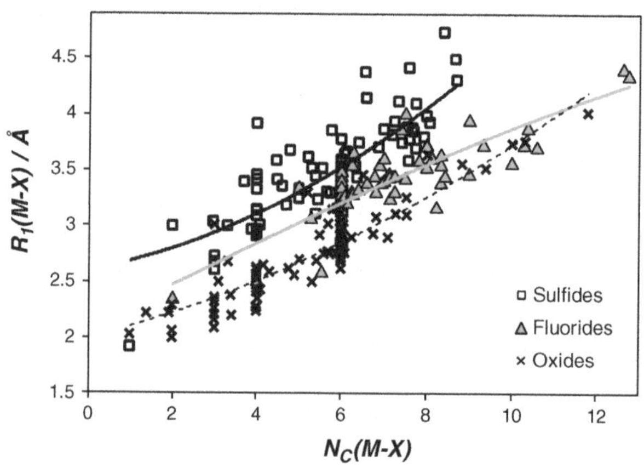

Fig. 16 Variation of the radius of the first coordination shell R_1(M–X) with the coordination number N_C for oxides (*crosses*), sulfides (*squares*), and fluorides (*triangles*). Lines are polynomial fits as a guide to the eye (*black solid line*: sulfides, *broken line*: oxides, *gray solid line*: fluorides)

compared to N_C(M–O) only for high coordination numbers, where the larger size of the S^{2-} becomes a limiting factor (see Fig. 15).

Analogous comparisons between oxides and fluorides are less clear because of the dominance of N_C(M–F) = 6 for the more ionic fluorides. Thus as seen in Fig. 16 the radius of the first coordination shell of fluorides is not found to be smaller than for an oxide of the same coordination number but typically in between the values for sulfides and oxides and the trend with the coordination numbers seems to somewhat deviate from the parallel trends for sulfides and oxides.

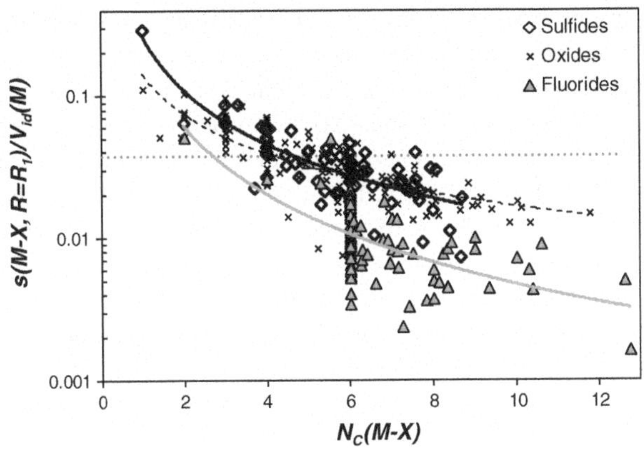

Fig. 17 Logarithmic variation of dimensionless relative bond valence values $s(\text{M–X})/V_{\text{id}}(\text{M})$ for bonds to oxygen, sulfides, and fluorides at the respective boundary for the first coordination shell R_1 with the coordination number. The *dotted horizontal line* marks the threshold value chosen by Brown; the lines are power law fits to the data as a guide to the eye (*black solid line*: sulfides, *broken line*: oxides, *gray solid line*: fluorides)

Comparisons of the scaled bond valence at the limit of the first coordination shell for the three anions (Fig. 17) show again that a universal bond valence threshold treats softer and harder anions differently. The coordination number-dependent values for $s(\text{M–X}, R = R_1)$ values are found to be consistently lower for the harder anions than for the softer sulfides and especially for the fluorides are in almost all cases smaller than Brown's universal threshold value. So also from the point of view of a suitable comparison of coordination polyhedra involving anions of different softnesses it may be useful to employ an N_C and anion softness dependent cutoff criterion, as it automatically results from our geometric approach.

4 Concluding Remarks

The discussions above aim to show that the success of the bond valence concept has a physical basis, as bond valences can be understood as a mass (or principal quantum number *n*, or atomic number, ...) dependent functional of electron density. The particular close correlation with the electron density at the bond critical point in predominantly ionic compounds (or more generally in compounds for which the Laplacian $\nabla^2\rho_{\text{BCP}}$ remains >0) also provides a better understanding why the bond valence concept though in principle applicable to a wide range of bond types and originally building on Pauling's concepts for covalent compounds or metals, is more appropriate for at least partially ionic compounds.

The link to the electron density also implies that concepts that proved essential in density functional theory such as the electronic chemical potential $\mu = (\partial E/\partial)_v = -\chi$ and its derivative, the absolute hardness $\eta = 1/2(\partial\mu/\partial N)_v$ or its inverse the *absolute softness* $\sigma = \eta^{-1}$ and are experimentally accessible via their links to the ionization energy or electron affinity can be advantageously employed in fitting both bond valence parameters that extract as much as possible of this fundamental information contained in the underlying reference crystal structures.

If this is done suitably, then the (squared) bond valences can also – as demonstrated [25, 26] – be linked to the absolute energy scale in static ways such as predicting energy thresholds in solid electrolytes cf. [6], predicting NMR chemical shifts [36] or other spectroscopic properties, but also in dynamic ways as an effective atomistic forcefield.

We should also consider this link between energy and bond valence in the opposite sense. Instead of quantifying bond valence mismatches one could in principle translate them into site energy increases with the help of the equations discussed in this chapter. If we are interested in the chemical plausibility of a structure, then a bond valence-based criterion such as the global instability index *G* [37] (which can also be used for structure fitting [38]) that does not involve the scaling of the bond valence mismatches by the atomic row number index *n* (or other closely correlated quantities) may lead to biased results when comparing compounds in which all atoms have similar masses to compounds containing heavy and light atoms. If for the sake of simplicity the not precisely known exponent for the *n*-scaling is set to agree with (25), then a modified instability index G_n could be defined as

$$G_n = \sqrt{\sum_{A}^{N} \frac{(V(A) - V_{id}(A))^2}{n_A N}}. \quad (33)$$

To be consistent with an energy-related measure of structural stability *N* also has to run over all atoms A in the unit cell (or equivalently in the formula unit), in line with the definition of *G*, rather than – as it is frequently seen – over the unweighted list of distinct atoms, since the absolute influence of a certain bond valence mismatch will increase with the number of symmetry copies inside the unit cell.

Acknowledgments I am grateful to Dr. R. Prasada Rao for his contribution to the development of atomistic forcefields from bond valence parameters (Sect. 2.2), and to undergraduate students Wilson Low Kai Bin and Jenson Tham who contributed to the data collection for Sect. 3.4 in the frame of their Final Year Projects at NUS Singapore. Discussions with Jerry Gibbs (Virginia Tech) on the relationship between bond valence and electron density proved fruitful while elaborating Sect. 2.1. This research is supported by the National Research Foundation Singapore under its Competitive Research Programme (NRF-CRP 8-2011-4).

References

1. Adams S (2001) Relationship between bond valence and bond softness of alkali halides and chalcogenides. Acta Crystallogr B Struct Sci 57:278
2. Adams S, Swenson J (2002) Bond valence analysis of transport pathways in RMC models of fast ion conducting glasses. Phys Chem Chem Phys 4:3179–3184
3. Brown ID (2009) Recent developments in the methods and applications of the bond valence model. Chem Rev 109:6858–6919

4. Adams S (2006) Bond valence analysis of structure – property relationships in solid electrolytes. J Power Sources 159:200–204
5. Adams S (2006) From bond valence maps to energy landscapes for mobile ions in ion-conducting solids. Solid State Ionics 177:1625–1630
6. Adams S, Prasada Rao R (2013) Understanding ionic conduction and energy storage materials with bond valence-based methods. In: Brown ID, Poeppelmeier K (eds) Bond valences. Structure and bonding. Springer, Berlin
7. Morrell MM, Parr RG, Levy M (1975) Calculation of ionization potentials from density matrices and natural functions, and the long-range behavior of natural orbitals and electron density. J Chem Phys 62(2):549–554
8. Bader RFW (1990) Atoms in molecules: a quantum theory. Clarendon, Oxford
9. Weinhold F (2012) Natural bond critical point analysis: quantitative relationships between natural bond orbital-based and QTAIM-based topological descriptors of chemical bonding. J Comput Chem 33:2440–2449
10. Mohri F (2003) Molecular orbital study of the bond-valence sum rule using Lewis-electron pair theory. Acta Crystallogr B 59:190–208
11. Gibbs GV, Finger LW, Boisen MB (1987) Molecular mimicry of the bond length–bond strength variations in oxide crystals. Phys Chem Miner 14:327–331
12. Bartelmehs KL, Gibbs GV, Boisen MB Jr (1989) Bond-length and bonded-radii variations in sulfide molecules and crystals containing main-group elements. Am Mineral 74:620–626
13. Brown ID, Shannon RD (1973) Empirical bond-strength–bond length curves for oxides. Acta Crystallogr A29:266–289
14. Shannon RD (1976) Revised effective ionic radii and systematic studies of interatomic distances in halides and chalcogenides. Acta Crystallogr A 32:751–767
15. Gibbs GV, Rosso KM, Cox DF, Boisen MBP Jr (2003) A physical basis for Pauling's definition of bond strength. Phys Chem Miner 30:317–320
16. Downs RT, Gibbs GV, Boisen MB Jr, Rosso KM (2002) A comparison of procrystal and ab initio model representations of the electron-density distributions of minerals. Phys Chem Miner 29:369–385
17. Gibbs GV, Tamada O, Boisen MB Jr, Hill FC (1999) Laplacian and bond critical point properties of the electron density distributions of sulfide bonds: a comparison with oxide bonds. Am Miner 84:435–446
18. Brown ID (2011) Compilation of bond valence parameters. http://www.iucr.org/resources/data/datasets/bond-valence-parameters
19. Mulliken RS (1934) A new electroaffinity scale; together with data on valence states and on valence ionization potentials and electron affinities. J Chem Phys 2:782–793
20. Duffy JA (1996) Optical basicity: a practical acid–base theory for oxides and oxyanions. J Chem Educ 73:1138
21. Duffy JA, Ingram MD, Fong S (2000) Effect of basicity on chemical bonding of metal ions in glass and its relevance to their stability. Phys Chem Chem Phys 2:1829
22. Eck B, Dronskowski R (2002) Atomistic simulations of solid-state materials based on crystal-chemical potential concepts: basic ideas and implementation. J Alloys Comp 338:136
23. Eck B, Kurtulus Y, Offermans W, Dronskowski R (2002) Atomistic simulations of solid-state materials based on crystal-chemical potential concepts: applications for compounds, metals, alloys, and chemical reactions. J Alloys Comp 338:142
24. Shin Y-H, Cooper VR, Grinberg I, Rappe AM (2005) Development of a bond-valence molecular-dynamics model for complex oxides. Phys Rev B 71:054104
25. Adams S, Prasada Rao R (2009) Transport pathways for mobile ions in disordered solids from the analysis of energy-scaled bond-valence mismatch landscapes. Phys Chem Chem Phys 11:3210–3216
26. Adams S, Prasada Rao R (2011) High power lithium ion battery materials by computational design. Phys Status Solidi A 208(8):1746–1753

27. Rappé AK, Casewit CJ, Colwell KS, Goddard WA III, Skid WM (1992) UFF, a full periodic table force field for molecular mechaniscs and molecular dynamics simulations. J Am Chem Soc 114:10024
28. Brown ID (1992) Chemical and steric constraints in inorganic solids. Acta Crystallogr Sect B Struct Sci 48:553
29. Parr RG, Pearson RG (1983) Absolute hardness: companion parameter to absolute electronegativity. J Am Chem Soc 105:7512–7516
30. Pearson RG (1985) Absolute electronegativity and absolute hardness of Lewis acids and bases. J Am Chem Soc 107:6801–6806
31. Pearson RG (1973) Hard and soft acids and bases. Dowden, Hutchinson and Ross, Stroudenburg
32. Ho TL (1977) Hard and soft acids and bases in organic chemistry. Academic, New York
33. Pauling L (1947) Atomic radii and interatomic distances in metals. J Am Chem Soc 69:542–553
34. Pauling L (1960) The nature of the chemical bond, 3rd edn. Cornell University Press, Ithaca
35. Brown ID (1988) What factors determine cation coordination numbers? Acta Crystallogr B 44:545–553
36. Adams S, Swenson J (2004) Predictability of ion transport properties from the structure of solid electrolytes. Ionics 10:317
37. Salinas-Sanchez A, Garcia-Munoz JL, Rodriguez-Carvajal J, Saez-Puche R, Martinez JL (1992) Structural characterization of R_2BaCuO_5 (R = Y, Lu, Yb, Tm, Er, Ho, Dy, Gd, Eu and Sm) oxides by X-ray and neutron diffraction. J Solid State Chem 100:201–211
38. Adams S, Moretzki O, Canadell E (2004) Global instability index optimizations for the localization of mobile protons. Solid State Ionics 168:281–290

Struct Bond (2014) 158: 129–160
DOI: 10.1007/430_2013_137

Published online: 12 February 2014

Understanding Ionic Conduction and Energy Storage Materials with Bond-Valence-Based Methods

Stefan Adams and R.Prasada Rao

Abstract The analysis and prediction of ion transport in solids from static and dynamic structure models has become an interesting application for the bond valence approach. Specific adaptations of the bond valence approach for this application area are discussed, and the resulting predictions are compared to those from alternative screening approaches. A particular advantage is that the bond-valence-based approach can be applied to both crystalline and glassy solids and that the level of computational effort can be easily adjusted to the level of detail required in the prediction from static pathway models for screening purposes to bond-valence-based molecular dynamics simulations for analyzing the coupling between the migration of the mobile species and rearrangements in the immobile substructure.

Keywords Battery materials · Bond valence site energy model · Ion migration pathways · Solid electrolytes

Contents

S. Adams (✉) and R.P. Rao
Department of Materials Science and Engineering, National University of Singapore, Singapore 117576, Singapore
e-mail: mseasn@nus.edu.sg

Abbreviations

1D	One-dimensional
2D	Two-dimensional
3D	Three-dimensional
σ	Ionic conductivity
$\Delta\rho_{pro}$	Procrystal electron density threshold for a migration pathway
ΔE_{BVSE}	Bond valence site energy threshold for a migration pathway
$\|\Delta V(M)\|$	Absolute bond valence sum mismatch
BVSE(M)	Bond valence sit energy of ion M
DFT	Density functional theory
DFT + U	Density functional theory with augmented description of Coulomb repulsion between localized *d* and *f* electrons by an additional Hubbard-like, localized term U
F	Volume fraction of percolating ion migration pathway clusters
G	Global instability index
K	Force constant
MAE	Mixed alkali effect (mixed mobile ion effect)
MD	Molecular dynamics (simulation)
M_r	Reduced mass of a vibrating system of atoms
p_{A-X}	Penalty function to complement bond valence sum mismatch term by a measure for the deviation from the equal valence rule
RMC	Reverse Monte Carlo (modeling)
R_{min}	Equilibrium distance M − X for a given coordination number
s_{M-X}	Bond valence of interaction between cation M and anion X
s_{min}	Bond valence corresponding to $R = R_{min}$
SOF	Site occupancy factor
V(M)	Bond valence sum of ion M
V_{id}(M)	Oxidation state of ion M

1 Introduction

In this chapter we will discuss to which extent local and long-range ion mobility can be understood from the structural and energetic local environment of the mobile ions. The level of accuracy and detail to be expected will of course depend on the computational effort that we can spend on an individual structure.

For screening purposes such as finding promising structure types for identifying or designing new solid electrolytes or mixed conductors, it may be sufficient to base predictions on static structure models. In principle both crystal structure data from diffraction experiments and energy-minimized computational models from ab initio or atomistic simulations can serve as such static structure models as long as they reflect the actual local structure and to a reasonable approximation the distribution of free volume in the structure. As we will see below, the important role of free volume for ionic motion leads to the conclusion that experimental data measured under ambient conditions often yield more precise predictions than ab initio models, since the latter tend to slightly overestimate the unit cell volume and hence the free volume. This overestimation of free volume will result in a slight underestimation of the activation energy for ionic motion.

For experimental structure data of not too complex crystal structures (and to a lesser extent for simulated structure models), the key issue is typically not so much the precision of the structure determination, but much more how to deal with deviations of the instantaneous local atomic arrangement from the idealized average crystal structure. Static structure models at the relevant temperatures are time averages over atoms in more or less strongly correlated motion. A static pathway prediction model will ignore the coupling between these motions, most notably the relaxation of the immobile substructure during the transport step. This will lead to an inherent tendency to overestimate energy barriers that has to be accounted for by scaling. For the same reason the predictions can be expected to be the more reliable the weaker the coupling between the motion of the mobile atom and the immobile substructure will be. Therefore, it may not be too surprising that the first successful examples of such pathway predictions dealt with heavy monovalent cations (such as Ag^+), where both the low charge and the high mass of the moving particle contribute to the degree of independence in its motion. An additional though not completely independent factor facilitating predictions of Ag^+ ion conductors is the ability of this transition metal ion to adapt to various coordination numbers down to two [1]. In contrast, pathway models for the motion of strongly coupled small, light cations (in the extreme case, protons) will be more precise, if energetic effects on the immobile substructure can be incorporated, e.g., as proposed in our earlier work by mapping the variation of the global instability index as a function of the position of a probe ion of the mobile species [2].

Moreover, the compounds of interest for ion transport pathway prediction systematically exhibit some degree of static disorder, and pathway predictions need to be based on representative local structure models. Even in the extreme case of a complete absence of long-range order, i.e., for ion conducting glasses, relevant local structure models may be derived from diffraction data, but additional constraints based on density, chemical plausibility, or spectroscopic information are required to generate static snapshot-type models that capture essential features of the local structure by reverse Monte Carlo (RMC) fitting [3–7]. Such RMC models are quantitative in the sense that they are in the ideal case fully compatible with any experimental information provided in the fitting process, but they will not generate unique structure solutions, and their interpretation should therefore focus on statistical analysis methods that extract characteristic features of the observed pathway network rather than on individual migration barriers.

Alternatively, the relation between structure and conductivity in both ordered and disordered compounds can be investigated using molecular dynamics (MD) simulations. In principle, MD simulations will lead to comprehensive structural and dynamic information within the limitations imposed by the system size, the simulated period, and the agreement of the employed interaction potential parameters with reality. Both diffraction data (crystal structures for crystalline compounds, RMC fits for glasses) and MD approaches are valuable tools to obtain insight into the conduction mechanism and its correlation to the atomic structure, though in the case of MD simulations it has to be verified that the force field chosen for the simulations leads to structure models that are consistent with experimental information.

2 Bond Valence Models of Transport Pathways from Static Structure Models

2.1 *Bond Valence Mismatch Pathways Versus Bond Valence Site Energy Pathways*

For screening purposes, bond valence or bond valence site energy (BVSE) pathway models derived from static structure models appear to be the most straightforward approach. In a range of earlier studies, it has been discussed how the bond valence method can be used to analyze ion transport pathways statistically, yielding predictions of ionic conductivity from crystal structure data and RMC- or MD-generated structure models [4, 8–10].

For the bond valence mismatch pathways, regions of low bond valence sum mismatch (and hence implicitly low site energy) for an ion M^+ are modeled as regions in the structure, where the bond valence sum $V(M)$

$$|\Delta V(\mathrm{M})| = \left|\sum_{\mathrm{X}} s_{\mathrm{M-X}} - V_{\mathrm{id}}(\mathrm{M})\right| + \sum_{\mathrm{X}} p_{\mathrm{M-X}} \tag{1}$$

over bond valences $s_{\mathrm{M-X}}$ from all adjacent counterions X approaches the oxidation state $V_{\mathrm{id}}(\mathrm{M})$. To enhance the chemical plausibility of "BV mismatch landscapes," the bond valence sum mismatch term $|\Delta V(\mathrm{M})|$ needs to be complemented by penalty functions $p_{\mathrm{A-X}}$ that (a) discriminate against sites where a matching $V(\mathrm{M})$ is achieved by unfavorable strongly asymmetric coordinations [11] and (b) exclude sites too close to other cation types. The cation–cation or anion–anion penalty functions may simply take the form of exclusion radii, but truncated Coulomb repulsions yield a more physical description.

This approach has been first successfully applied to understand ionic conduction in solids by Garrett et al. [1] who demonstrated the close agreement between a (sharpened) bond valence mismatch map of α-AgI and the disordered Ag probability density map from neutron diffraction studies. Here, "sharpened" means that when comparing to a set of equally spaced isolines of constant silver density, the best match was not achieved when comparing to equally spaced isolines of bond valence mismatch but when using a high power (or an exponential function) of the bond valence mismatch. The simple AgI is an ideal candidate for such comparisons, since it contains only a single type of cations, eliminating the need for a penalty function to model repulsions between mobile and immobile cations and more importantly a scaling factor between the bond valence and penalty term.

The use of bond valence maps itself had been originally proposed by Waltersson in 1978 [12] as a tool to locate the light atom Li in nine complex oxide structures, finding already that the correct position for dense structures typically corresponds to a minimum of the bond valence sum. Along the same lines, localization of ions in complex superstructures with the help of bond valence maps was, e.g., studied by Withers et al. [13], van Smaalen [14], and Adams et al. [15]. While Waltersson's approach was focusing on locating the low-energy (equilibrium and interstitial) sites only, Garrett et al. [1] then introduced a first attempt to link the variation of the bond valence mismatch to the variation in the atomic probability density not only at cation sites but also for the paths in-between these sites. From an empirical comparison to the experimental probability density map, it was proposed that for α-AgI the atomic probability density should be proportional to V^{-16} and the same approach was then used to derive ion transport paths in $Ag_{16}I_{12}P_2O_7$.

In continuation of this approach, the bond valence mismatch in a large variety of silver and alkali ion conductors has been analyzed based on the bond valence formalism. A range of examples is discussed below. As briefly mentioned elsewhere in this volume [16], the intended use of our bond valence parameters in bond valence maps was a major reason for us to deviate from the nearest neighbor convention otherwise commonly applied in bond valence calculation. When calculating bond valence sums strictly from interactions to the nearest neighbors only, the extremely asymmetric nature and sudden changes of the first coordination shells of a mobile ion along its path could otherwise lead to artifacts: When the mobile ion passes the boundary of its own coordination shell, an unphysical steplike change of the coordination shell and hence of the bond valence sum would result as exemplified in Fig. 1. Therefore, it appears to be more straightforward to use a shell of a sufficiently large radius that moves with the mobile ion. For dynamic applications of the method, it is also important that even a minor steplike change in the bond valence mismatch or BVSE is eliminated to avoid infinite gradients that would translate into infinite accelerations of the moving atom. Thus, at least for any applications of the approach in molecular dynamics simulations, it is advisable to subtract the (necessarily minute) BVSE for the cutoff distance from all site energies forcing the interaction to reach exactly zero at the cutoff distance.

In most cases stable solid electrolytes are nearly densely packed so that sites with the matching valence will form well-defined minima in these bond valence maps. This unfortunately led some authors to the misconception that ions would generally

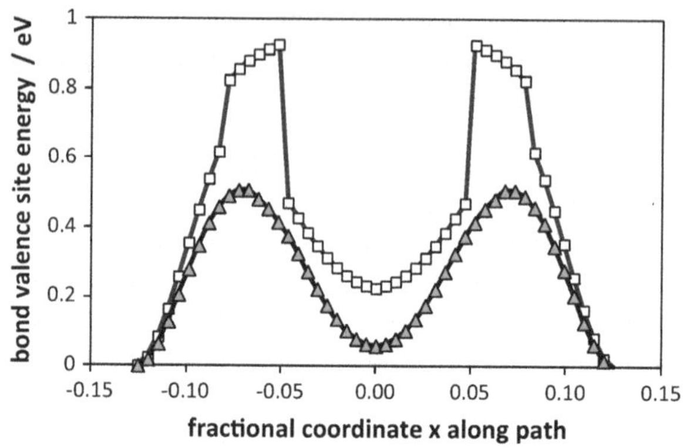

Fig. 1 Variation of BVSE along the transport path of a model cation conductor (equilibrium sites at $x = \pm 0.125$; interstitial site at $x = 0$) when calculating the BVSE of the mobile ion along the path based on the *softBV* parameter and a cutoff radius of 8 Å around the respective position of the mobile ion (*dark triangles*) compared to the site energies that result from the formal application of the same conversion algorithm to conventional bond valence parameters with universally fixed value of the bond valence parameter b in combination with the first coordination shell cutoff criterion (*open squares*). The different choice of the bond valence parameter b translates into a difference in the predicted activation energy. More importantly, for the latter approach, jumps in site energy (or of bond valence mismatch) occur along the path when the mobile ion crosses the boundaries between coordination shells (here at $x \approx \pm 0.04$ and $x \approx \pm 0.08$). Such artifacts when using standard bond valence conventions would render the approach less practicable

tend to migrate in the direction of lowest bond valence sum rather than in the direction of lowest bond valence sum mismatch. As discussed more in detail in Sect. 3.3, the difference becomes most obvious when we consider the surface of a solid electrolyte, where a driving force to minimize the bond valence sum would cause mobile ions to escape from the solid into the free space.

Moreover, it has to be kept in mind that bond valence maps (or more precisely, bond valence sum mismatch maps) use bond valence units scale rather than an energy scale, and attempts to link energy or probability density to the bond valence mismatch were difficult to achieve in a general transferable way as the calculation of effective bond valence mismatches according to Eq. 1 requires scaling between bond valence terms and cation–cation (or anion–anion) repulsions or (for open structures) to discriminate between sites of identical bond valence sum based on the degree to which the equal valence rule is fulfilled.

In order to overcome these challenges, we have more recently [9, 17] proposed to convert the bond valence mismatch first into an absolute contribution to the ion's (bond valence) site energy BVSE(M) so that the penalty functions for cation–cation repulsions, $E_{\mathrm{repulsion}}$, or deviations from the equal valence rule, ΔE_{BVR}, can be incorporated in a natural way according to

$$\mathrm{BVSE(M)} = D_0|\Delta V(\mathrm{M})|^2 + \Delta E_{\mathrm{EVR}} + E_{\mathrm{repulsion}}. \qquad (2)$$

An implementation of this approach exploiting the formal analogy between the squared bond valence sum mismatch and the Morse-type potential is discussed in detail in a previous chapter of this book [16].

Transport pathways for various compounds with both cation (Ag^+, Cu^+, Li^+, Na^+, K^+, etc.) and anion (O^{2-}, F^-, etc.) mobile systems were investigated by several groups including our group. Here, we briefly review outcomes of a few such studies. It should be noted that the methods applied vary in detail, which can only be sketched briefly, so that the reader is referred to the cited original publications for the exact procedure in which the transport pathways are derived in the respective study.

2.2 *Ion Transport Pathways from Bond Valence Maps in Crystalline Cation Conductors*

As mentioned above, Garrett et al. [1] first demonstrated the equivalence of the bond valence sum mismatch pathways in α-AgI with pathways determined experimentally from an anharmonic atomic displacement refinement of neutron diffraction data as a justification for predicting transport pathways for their then new fast Ag^+ ion conductor $Ag_{16}I_{12}P_2O_7$ using the same approach. In our earliest work on pathway models for similar Ag^+ ion conducting oxyhalide systems [3, 18], we used a parameter set by Radaev et al. [19],

$$s_{\mathrm{Ag-O}} = \exp\left[\frac{1.89\text{Å} - R_{\mathrm{Ag-O}}}{0.33\ \text{Å}}\right], \quad s_{\mathrm{Ag-I}} = \exp\left[\frac{2.08\text{Å} - R_{\mathrm{Ag-I}}}{0.53\text{Å}}\right], \tag{3}$$

as their parameters – like our later systematic *softBV* parameter set – included the influence of higher coordination shells and thus facilitated the modeling of pathways by avoiding cutoff effects at the boundaries of coordination shells or the need for rescaling the bond valence sums if conventional bond valence parameter sets are used (bond valence parameter sets derived assuming the first coordination shell convention will otherwise obviously yield overestimations of the bond valence sum if contributions from more distant counterions are included to avoid the cutoff effects). Moreover, it was realized that the adjustment of the bond valence parameter b allows accounting for differences in ion polarizability.

Three-dimensional bond valence maps of Ag^+ ion conductors were constructed by summing up bond valence contributions to all anions up to a distance of 8 Å for any point of a three-dimensional grid. It is also advisable to choose the number of grid points across a unit cell as a multiple of 12 or 24 so that common special positions at fractional coordinates become explicit grid points. Initially, the bond valence sum mismatch term was complemented by hard cutoff criteria to account for the repulsion between mobile and immobile cations. Thus, grid points at a distance to other cations M smaller than the sum of radii of Ag^+ and M were treated as inaccessible. It should however be noticed that the choice of these minimum

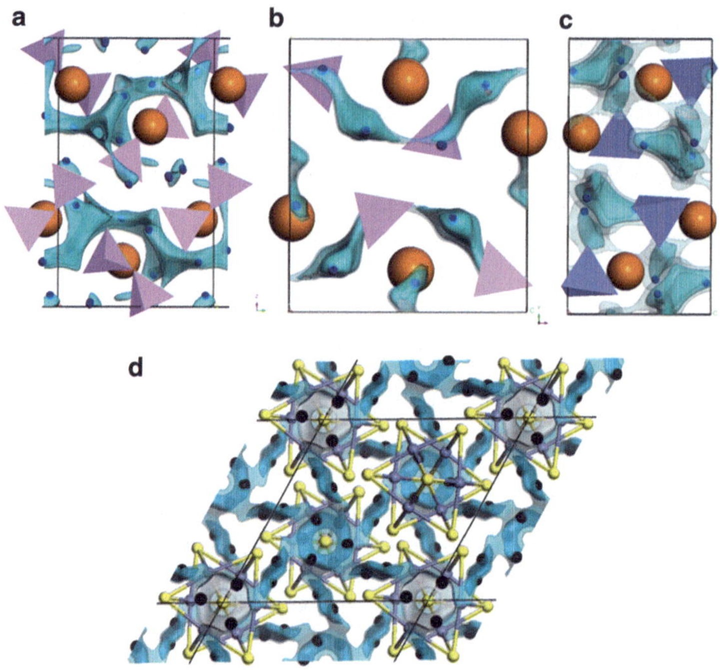

Fig. 2 Models of ion transport pathways determined as regions of low bond valence sum mismatch in Ag^+ ion conducting solids (**a**) $Ag_5IP_2O_7$ [22], (**b**) Ag_4IPO_4, (**c**) Ag_3IMoO_4 (*large spheres*: I; *small spheres*: Ag; oxyacid polyanions shown as polyhedra), and (**d**) Li^+ ion conduction pathways in the mixed conducting Chevrel phase $Li_4Mo_6S_8$. In (**d**), the *dark* (*light*) isosurfaces correspond to local (long-range) pathways for Li^+

acceptable distances between mobile and immobile cations may strongly influence the resulting bond valence pathway maps. Thus, a universal choice of minimum cation–cation distances (as it seems to be, e.g., implemented in the software 3DBVSMAPPER [20]) may result in artifacts compromising the reliability of pathway predictions. Such artifacts from insufficient consideration of cation–cation repulsions may also be the reasons why a detailed bond valence sum mismatch-based discussion of alternative paths in Chevrel phases $A_xMo_6S_8$ by Levi et al. [21] contradicts a BVSE model that (at least for $Li_4Mo_6S_8$) just finds the expected local pathway hexagons $(Li(1))_6$ interconnected by pairs of Li(2) sites to a 3D network (Fig. 2d).

A more physical description of the repulsion effects is obviously achieved when such hard cutoff radii are replaced by a Coulomb repulsion term so that with decreasing cation–cation distance pathway regions close to immobile cations become gradually less favorable for the mobile cation. On the other hand, this approach requires a scaling between a bond valence sum mismatch description of attraction terms and an energy-scaled Coulomb repulsion. Despite these shortcomings, ion transport in a variety of crystalline Ag^+ ion conducting systems such as

$Ag_8I_4V_2O_7$ (*P-62 m*) [23], $Ag_5IP_2O_7$ [22], $Ag_8I_{18}W_4O_{16}$ (*C2*), $Ag_{16}I_{12}P_2O_7$ (*P6/mcc*), $Ag_4P_2O_7$ ($P3_221$), α-Ag_2HgI_4, β-Ag_2HgI_4, or various Ag argyrodites [24] has in the subsequent years been investigated by bond valence techniques [25, 26]. Furthermore, the bond valence mismatch pathway analysis was applied to other alkali ion conductors as well as to selected fluoride and oxide ion conductors [11]. Three-dimensional graphical representations of isosurfaces with constant bond valence mismatch like the ones in Fig. 2 provide a vivid image of the pathway topology. Still, additional information, e.g., on the occupancy of the cation sites and jump distances, is generally required to derive a transport mechanism from such a bond valence pathway model, since a low-energy migration pathway may be of limited use if it is impossible for the considered mobile species to reach it from actually occupied sites.

The same approach was also applied in a first comparison of bond valence and DFT results on antifluorite-type $Li_{14}Cr_2N_8O$, where both methods accordingly came to the conclusion that (in contrast to analogous Mn compounds) there are no low-energy pathways for Li^+ in $Li_{14}Cr_2N_8O$ [27]. A more recent comparison of DFT + U and bond valence analysis of Li^+ motion in the modulated structure of $LiFeBO_3$ was undertaken by Janssen et al. [28]. In DFT + U calculations, the description of Coulomb repulsion between localized *d* electrons (and if present *f* electrons) is augmented by an additional Hubbard-like, localized term U. Again, both approaches agreed not only in the identification of the subset of Li sites that are crucial to Li motion and those Li sites not involved in the transport but also in the overall transport pathways that is also consistent with our earlier description of the path in the unmodulated structure [17]. Slight differences in the evaluation of Li^+ sites close to the immobile Fe and B cations were ascribed by the authors to their simple hard-sphere exclusion approach in treating cation–cation repulsions.

The limited computational demand makes the bond valence approach particularly suited to study large sets of data. Recently, Avdeev et al. [10, 20] presented an analysis of 9,701 compounds containing the potentially mobile monovalent cations Li^+, Na^+, K^+, Ag^+, or Cu^+ with respect to the existence of infinite networks of pathways for the respective mobile ion. Alkali ion pathways for four compounds discussed as examples in their work are shown in Fig. 3.

Besides the test for the existence of infinite pathways at a given bond valence mismatch threshold, the authors also propose to link the absolute conductivity to the fractional accessible volume (*F*, cf. Fig. 4), i.e., the fraction of the unit cell volume that for a given bond valence mismatch threshold belongs to the infinite pathway. While this approach (that had been used before for ion conducting glasses, see below) appears oversimplifying for crystalline compounds, as it does not directly account for the strong influences of other factors such as the degree of disorder and the concentration of mobile charge carriers on the absolute ionic conductivity, it was found that the values of the fractional accessible volume per ion of the mobile species in Li^+, Na^+, and K^+ oxides commonly occur in known ion conducting solids such as $Li_{3x}La_{2/3-x}TiO_3$ perovskites, P2- and β-$NaFeO_2$-type oxides, as well as birnessite-type compounds.

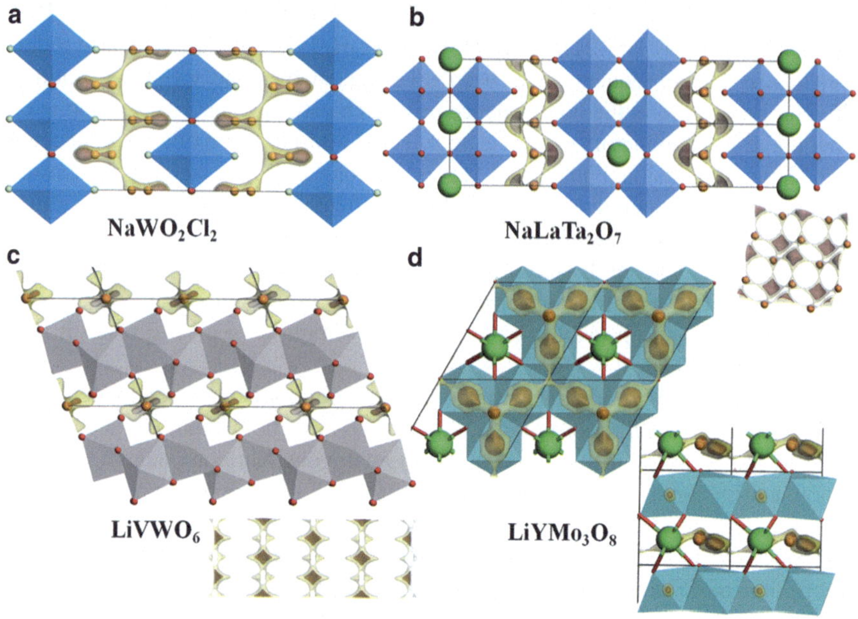

Fig. 3 Bond-valence-based models of ion migration pathway in selected alkali ion conductors: (**a**) $NaWO_2Cl_2$, (**b**) $NaLaTa_2O_7$, (**c**) $LiVWO_6$, (**d**) $LiYMo_3O_8$. *Dark isosurfaces* correspond to regions of low-energy alkali ion sites, while the *light isosurfaces* correspond to continuous pathways. *Insets* in graphs (**b**) and (**c**) show the 1- or 2-dimensional pathways within a single interlayer region of the respective compound

2.3 Ion Transport Pathways from Bond Valence Maps of Ion Conducting Glasses

The simplicity of the bond valence approach keeps comparatively large systems manageable so that the bond valence method is well suited to investigate ionic conduction in complex systems and hence also in glassy ion conductors, provided that the local coordination for the mobile ions is known. The structures for glassy systems were developed by combining molecular dynamics simulations or RMC modeling with the bond valence approach. Glasses in which Ag^+ or alkali ions with mixed oxide–halide coordination are the mobile species are discussed as an example. The bond valence approach was found to be particularly suited to provide new insight about this issue.

$AgI_{0.75}$-$(Ag_2MoO_4)_{0.25}$ and $(AgI)_{0.6}$-$(Ag_2O$-$2B_2O_3)_{0.4}$ are discussed here as representative model systems for molecular and network glasses, respectively [18]. Slices through the bond valence isosurfaces for the two glassy systems are shown in Fig. 5a and b, respectively. At first sight the complex isosurfaces may look rather puzzling, since there is no more translational order except for the periodic boundary conditions. What is already seen from these graphs is that while the

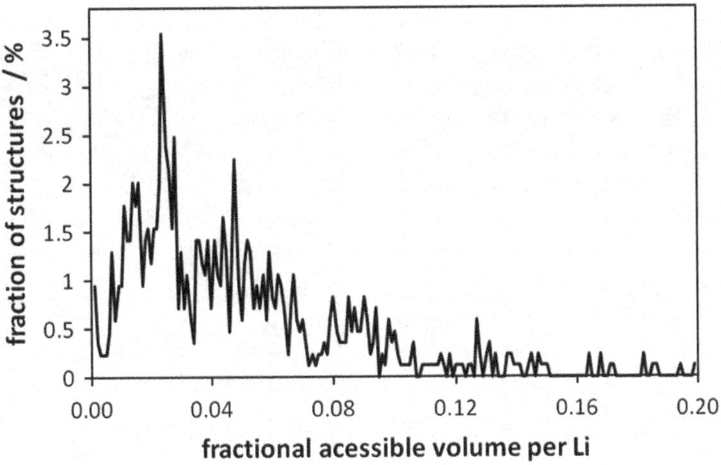

Fig. 4 Distribution of 1,312 lithium oxides extracted from the ICSD database with respect to fractional accessible volume of the crystal structure space with bond valence mismatch of less than 0.2 valence units. The histogram bin size is 0.001. According to Avdeev et al. [10], structures with high values of the fractional accessible volume per ion of the mobile species may be expected to have high ionic conductivity

long-range mobility on glasses is isotropic and homogeneous, locally ions have to follow pathways of restricted dimensionality.

For each model system, the mean fraction of the valence sum that originates from bonds to oxide ions is in accordance to the respective overall halide/oxide ratio both for an averaging over all grid points and over all Ag^+ positions. A more detailed inspection reveals that in contrast to an earlier literature model, the *dc* ionic conductivity in both crystalline and amorphous silver oxyhalide compounds cannot rely on transport in halide-rich regions alone, as such regions form too small a volume fraction of the oxyhalide structure (see Fig. 5c) and do not form a continuous network. On the other hand, the local mobility of Ag^+ ions is still somewhat larger in halide-rich regions. A comparison of borate glasses with various alkali and silver halide dopants [29] shows that pathways exhibit a similar change in ordering upon salt doping as the glass network itself and that the higher degree of ordering in AgI-doped glasses and hence in the pathways of these systems is not an important factor for their high ionic conductivity. The same study shows that the major effect of the halide doping is the increase in the minimum of the local pathway dimensionality. In other words, the bottlenecks for ionic transports become wider, and the mobile ions have more directions into which transport can proceed.

For the fast ion conducting glasses (where migration barriers within the path are not too large compared to the thermal energy of the mobile ions), the cation migration can be reasonably approximated even by a simple random walk that is restricted to regions of low bond valence mismatch [25, 26].

For a given choice of the valence interval $V \pm \Delta V$ that is defining the pathway, the ratio between the volume fractions of the maximum pathway clusters in

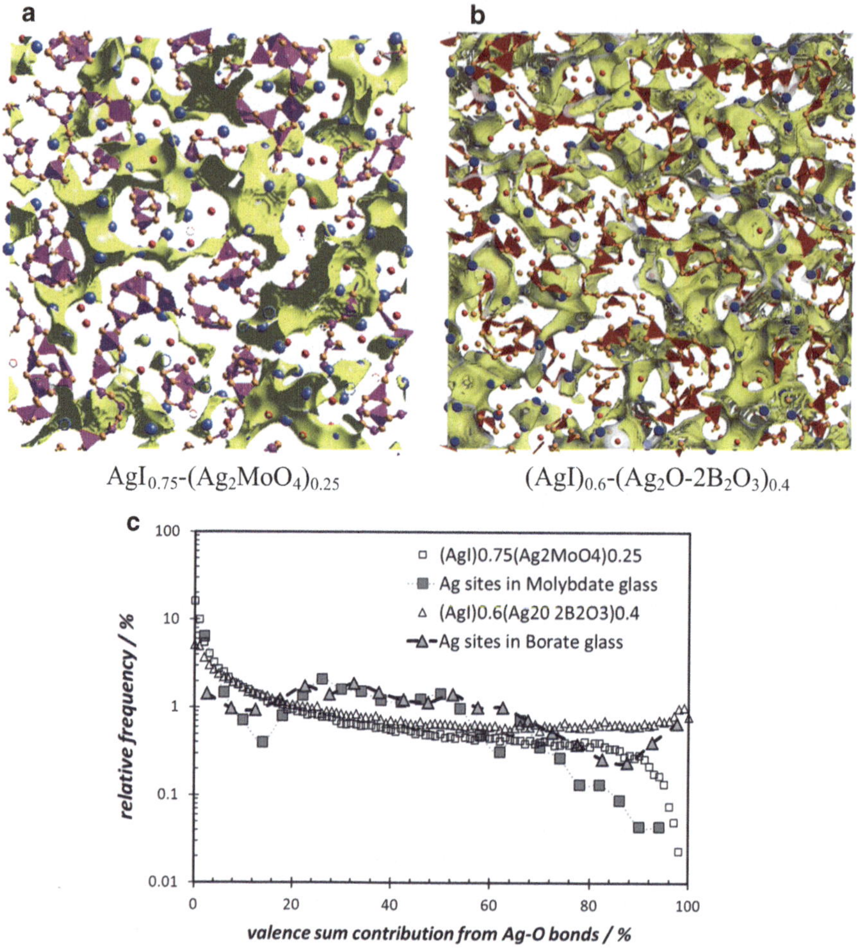

Fig. 5 10-Å-thick slices through bond valence isosurfaces for RMC models of the glassy phases (**a**) $(AgI)_{0.75}$–$(Ag_2MoO_4)_{0.25}$ and (**b**) $(AgI)_{0.6}$–$(Ag_2O$–$2B_2O_3)_{0.4}$ (Ag: *large spheres*; I: *small spheres*, polyanions as polyhedral) and (**c**) the occurrence of sites with a certain oxygen contribution to the total Ag bond valence sum: (*open symbols*: for all pathway volume elements with $V(Ag) = 1.00 \pm 0.05$ valence units; *filled symbols*: for RMC-modeled Ag sites with $V(Ag) = 1.00 \pm 0.05$). In both glasses, Ag^+ ions are slightly enriched in pathway regions with mixed oxide/halide coordination

different ion conducting glasses provides a convenient tool for estimating the activation energies E_A and absolute room temperature conductivities in a wide range of silver ion conducting systems: the cube root of the volume fraction F of the percolating pathway cluster (i.e., the accessible volume fraction for the mobile ion with an activation energy corresponding to the bond valence mismatch ΔV) is linearly related to the activation energy (see Fig. 6) [3]. When comparing glasses

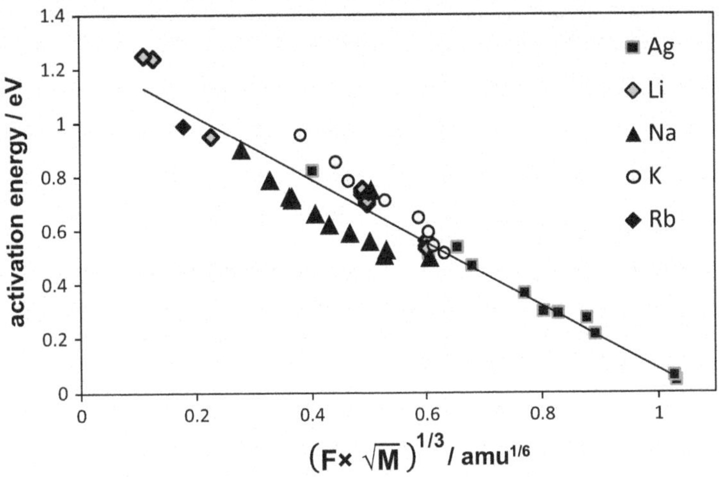

Fig. 6 Correlation between experimental ionic conductivities and volume fraction F of the infinite Ag conduction pathway clusters for RMC models of disordered Ag^+ ion conductors (oxyhalide glasses $(AgI)_{0.75}$–$(Ag_2MoO_4)_{0.25}$, $(AgI)_{0.6}$–$(Ag_2O$–$2B_2O_3)_{0.4}$, $(AgI)_{0.6}$–$(Ag_2O$–$2B_2O_3)_{0.4}$, $(AgI)_{0.75}$–$(Ag_2WO_4)_{0.25}$, oxide glasses Ag_2O–$2B_2O_3$ and Ag_2O–$4B_2O_3$, as well as AgI at 525 K and at 740 K) and of various alkali ion conducting borate, phosphate, or silicate glasses with or without chloride doping. Data redrawn after Adams and Swenson [4] including a more recent series of molecular dynamics simulation models of sodium silicate glasses [30]

with different types of mobile ions, the linear trends for each ion can be unified to a single correlation, if the value of F is scaled by the square root of the mass M of the mobile ion [4]:

$$E_A \approx A \times \sqrt[3]{F\sqrt{M}} + B. \tag{4}$$

From the wide range of oxyacid glass systems plotted in Fig. 6, approximate values for the empirical constants are $A \approx -1.17$ eV/amu$^{1/6}$ and $B \approx 1.26$ eV. While the lack of an accepted theory on the origin of the activation energy for ion transport in amorphous systems makes it difficult to link this empirically observed scaling to a theoretically expected correlation, it should be noted that this empirical mass dependence will, as discussed before [16], at least partially be a mass dependence of the correlation between bond valence mismatch and the energy scale. An empirical distinction whether the fundamental scaling parameter is the mass or the principal quantum number or the cation electronegativity remains ambiguous due to the strong correlation between these quantities [4].

It appears however straightforward that a scaling between the pathway volume fraction $\boldsymbol{F}$ and the absolute (room temperature) conductivity should contain an additional mass-dependent scaling factor to compensate for the mass dependence of the pre-exponential factor. The application of the Nernst–Einstein equation to a description of ion motion as a random walk consisting of isolated jumps yields for the temperature dependence of the conductivity

$$\sigma T = \alpha \frac{nq^2 d^2 \upsilon_0}{k_B} \exp\left[-\frac{E_A}{k_B T}\right]. \quad (5)$$

Here, n represents the concentration of charge carriers, q their charge, d a typical jump distance, ν_0 the vibration frequency of the ion within the potential well, and α a geometrical constant describing the success probability of the attempted jump. The vibration frequency itself depends on the inverse square root of the reduced mass M_r of the oscillator (as well as on the force constant K of the interaction, which in general is related to the cation charge):

$$\upsilon_0 = \sqrt{\frac{K}{M_r}}, \text{ where } M_r = \frac{M \times M_{network}}{M + M_{network}} \approx M \quad (6)$$

As the cation in fast ion conducting glasses vibrates against a comparatively heavy rigid anion network, the reduced mass M_r of the vibrating particles may be approximated by the mass of the mobile cation. This proportionality of cation motion frequencies to the inverse square root of the cation mass has effectively been observed for alkali-metal ions in diborate glasses by far-infrared spectroscopy [31]. Consequently, the increase of the pre-exponential factor with decreasing cation mass can be compensated in a comparison of glass systems with different monovalent mobile ions by an additional $M^{1/2}$ scaling, so that a correlation analogous to Eq. 4 for the conductivity should take the form

$$\ln\left(\sigma T \sqrt{M}\right) \approx A' \times \sqrt[3]{F\sqrt{M}} + B'. \quad (7)$$

Here, the empirical constants for the same set of reference systems adopt the values $A' \approx -42.1$ amu$^{-1/6}$ and $B' \approx -33.9$.

The mixed alkali effect (MAE, or mixed mobile ion effect), i.e., the strongly nonlinear variation of transport properties with composition in glasses that contain more than one type of mobile ions, was only poorly understood, partly due to the difficulty to determine the conduction pathways for the mobile ions. Since the drastic drop in conductivity (compared to the corresponding single alkali glasses) for an intermediate composition tends to increase with increasing size difference between the two types of mobile ions, we studied both the glass system $Li_xRb_{1-x}PO_3$ ($x = 0, 0.25, 0.5, 0.75$, and 1) [5], where a large MAE is to be expected, and the system $Ag_xNa_{1-x}PO_3$ ($x = 0, 0.25, 0.5, 0.75$, and 1) [32] where the MAE should be less pronounced. In each glass, the transport pathway volume fractions F were determined for the two potentially mobile ions in the same way as described above for the single mobile ion glasses and then the ionic conductivities as well as their activation energies were estimated from the correlations in Eqs. 4 and 7.

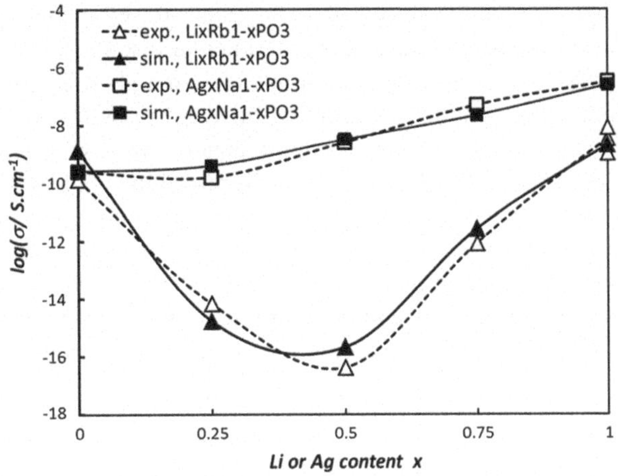

Fig. 7 Room temperature dc ionic conductivity σ versus composition for the glass systems $Li_xRb_{1-x}PO_3$ and $Ag_xNa_{1-x}PO_3$ (*open symbols*: experimental data, *filled symbols*: values predicted from the pathway volume fractions F of the RMC structural models employing Eq. 7). The lines are guides to the eye

Figure 7 shows how in the $Li_xRb_{1-x}PO_3$ system the ionic conductivities of the single alkali glasses decrease by 6–8 orders of magnitude at the intermediate compositions ($x = 0.5$). This dramatic drop in conductivity cannot be explained by any major structural alteration upon the mixing of alkali ions. Since the local structural environments of Li^+ and Rb^+ ions are distinctly different, there is a large energy mismatch for Li^+ ion jumps to Rb^+ sites and vice versa. This fact in combination with the low dimensionality of the pathways and the non-statistic cation distribution causes the strong blocking effect in the mixed alkali glasses. This reduces the possibility for, e.g., the Li^+ ions to perform energetically favorable ionic jumps, giving an on average higher activation energy and a lower ionic conductivity.

In the same way, the analysis of conduction pathways and the extraordinarily weak MAE in $Ag_xNa_{1-x}PO_3$ is reproduced by bond valence analysis of reverse Monte Carlo produced structural models (cf. Fig. 7). In this case the MAE is suppressed due to the joint motion of Ag^+ and Na^+ via a common cooperative hopping process in heavily overlapping pathways, while Li^+ and Rb^+ in $Li_xRb_{1-x}PO_3$ move on distinctly different conduction pathways. Thus, the bond valence analysis shows in the simple way that the MAE is a natural consequence whenever the mobile ions have sufficiently different sizes and/or polarizabilities (bond softnesses).

The fact that the MAE can be reproduced and understood from static structural models (i.e., without including any kind of structural relaxation, except the hopping motion of the alkali ions) further implies that the MAE had been too weak if we allowed the local structure to relax on a time scale similar to the inverse hopping rate of the alkali ions. Thus, at room temperature such site relaxation must be considerably slower and therefore not of importance for the MAE. However, at higher temperatures sufficiently close to the respective glass transition temperature, the time scale of site relaxation will approach the inverse hopping rate of the alkali ions, and this makes it possible for an A ion to move to a previous B site and vice versa. This will considerably reduce the efficiency of the blocking, and thereby the MAE, in accordance to experimental results.

While mixing different mobile ions in a glass is as discussed detrimental to the total ionic conductivity, mixing different glass formers in the immobile matrix is often found to enhance the ionic conductivity, which is sometimes referred to as "mixed glass former effect." As discussed in detail by Tho et al. [33], the variation of activation energy and thereby of the dc conductivity can be directly predicted by the bond valence pathway analysis from (MD-simulated) structural models of the borophosphate glass series $0.45Li_2O$-$(0.55 - x)P_2O_5$-xB_2O_3 $(0 \leq x \leq 0.55)$, and the role of the concentration variation of P–O–B units in the glass matrix for the conductivity enhancement in this mixed glass former system could be quantified.

2.4 Static and Dynamic BVSE Models

In complex structures with mobile and immobile cations, the existence of continuous migration pathways for the mobile species in bond valence models may critically depend on the (in the previously discussed version of the bond valence approach to some extent arbitrary) scaling between the attractive terms in bond valence units and the cation–cation repulsion terms, which can be treated straightforwardly as soft energy-scaled Coulomb repulsions. To eliminate this arbitrariness and to translate results of the bond valence analysis directly to the energy scale, we have, as elaborated elsewhere in this book [16], transformed the concept of ion migration following paths of minimum bond valence mismatch into paths of minimum BVSE. In brief, the BVSE of a mobile cation M at a given point in the structure is determined as the sum over bond valence terms for the interactions with each of the N_X adjacent anions X_j and (screened) cation–cation repulsions $E_{Coulomb}(M - M_i)$ for the interaction with other immobile cations M_i.

$$\mathrm{BVSE(M)} = D_0\left[\sum_{j=1}^{N_X}\frac{\left(s_{\mathrm{M-X}_j} - s_{\min}\right)^2}{s_{\min}^2} - N\right] + \sum_{i=1}^{N_M} E_{\mathrm{Coulomb}}(\mathrm{M} - \mathrm{M_i}). \quad (7)$$

When calculating pathways within the *BVSE* landscape for a mobile ion, it is advisable not to include the repulsions among the ions of the mobile species. In this way we can see all occupied or vacant sites that the ions can reach with a given activation energy (be it by individual hops or collective transport modes such as interstitialcy mechanisms) as a continuous path, whereas low-energy sites that can only be reached via high energy barriers will remain isolated.

While it has to be acknowledged that the calculation of these *BVSE* landscapes requires a little bit more computational effort than the calculation of bond valence mismatch landscapes, the fact that the same parameters used to calculate these static pathway models are also suitable as a highly transferable force field for molecular dynamics simulations of a wide range of oxides appears to be a major

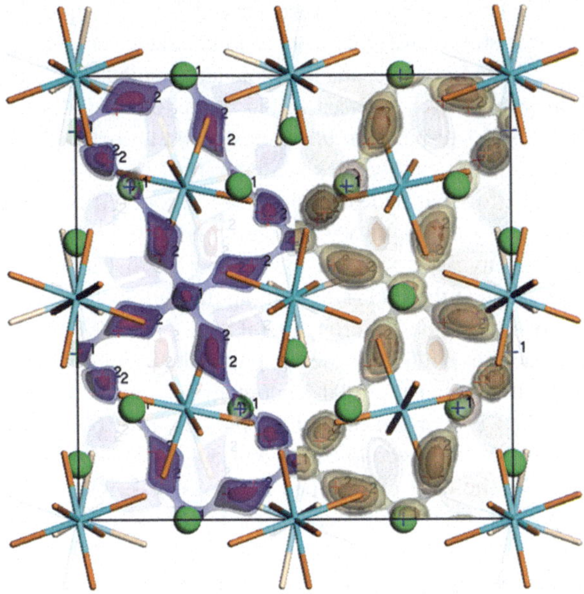

Fig. 8 Li^+ pathways in the cubic high-temperature phase of $Li_7La_3Zr_2O_{12}$ superimposed on the crystal structure. In the *left-hand side* part of the graph, the pathways are shown as derived from the BVSE analysis of the static structure model, while in the *right-hand side* part, the paths are based on the Li distribution density averaged over 500 ps of a constant volume molecular dynamics simulation run at $T = 1{,}000$ K. On both sides, pathways in the back half of the unit cell are shown in *pale colors* to facilitate distinction from the paths in the front half of the unit cells. *Numbers* indicate the two types of Li sites

advantage: it both underlines the physical plausibility of the parameter set and opens the possibility to employ the *BVSE* analysis on MD simulations calculated with a consistent parameter set.

Here, we first discuss a few recent studies of ion migration models for Li^+ ions in fast ion conducting solids using this *BVSE* approach. For the understanding of transport in Li^+ ion conductors, it is a major advantage that the method requires detailed structural information only for the immobile substructure. Although, e.g., a direct determination of lithium occupancies from XRD data of complex disordered compounds can hardly be conclusive, the position of the other atoms can often be determined with sufficient precision to map the potential Li sites and their connection by bond valence or *BVSE* calculations.

The *BVSE* analysis for the Li^+ disordered cubic high-temperature phase of the garnet-related high stability fast ion conductor $Li_7La_3Zr_2O_{12}$ (see Fig. 8) from the Li density distribution in high-temperature molecular dynamics simulations quantitatively agrees with the predictions of the *BVSE* model. Within this pathway network, neighboring sites cannot be occupied simultaneously, as this would lead to unphysically short Li(1)–Li(2) distances of 1.6 Å and Li(2)–Li(2) distances $<$1 Å. As the *BVSE* analysis shows that there are no further interstitial sites, the limits of possible Li distributions can be derived: Each occupied Li(1) blocks the occupation of four distinct Li(2) sites. In contrast, each occupied Li(2) site only blocks one Li(2) and one Li(1) site, and the same unoccupied Li(1) site can have four occupied Li(2) neighbors [34]. As seen in Fig. 8 the tetrahedrally coordinated 24*d* site Li(1) and the Li(2) site pair (on 96 *h* sites round an octahedrally coordinated 48 *h* site) are the only low-energy sites available for Li^+. No additional interstitial sites exist in contrast to what had previously been proposed in the literature [35]. A 3D network of pathways of lowest activation energy for Li^+ ion migration in $Li_7La_3Zr_2O_{12}$ (LLZ) involves both site types: four local Li(1)–Li(2)–Li(2)–Li(1) paths are interconnected at the Li(1) site forming the pathway network. As also seen in Fig. 8, a pathway network derived for general $Li_{15-2y}La_3(M^y)_2O_{12}$ phases, where M^y represents a cation in oxidation state *y*+, puts a the more stringent constraint on the Li distribution, the higher the total Li content per unit cell (and the lower the oxidation state of M^y) is. For the theoretical maximum of 60 Li per unit cell, all 48 octahedral voids would be occupied leaving only 12 of the 24 tetrahedral sites accessible for Li. For the case of undoped $Li_7La_3Zr_2O_{12}$, where 56 Li have to be distributed, the possible range site occupancy factor, SOF, is thereby limited to $0.333 \leq \text{SOF(Li(1))} \leq 0.555$ and $0.444 \leq \text{SOF(Li(2))} \leq 0.5$. Analogously, the SOF values for the doped samples with $x = 0.25$ are restricted to the ranges $0.25 \leq \text{SOF(Li(1))} \leq 0.583$ and $0.417 \leq \text{SOF(Li(2))} \leq 0.5$, respectively.

A nearly full occupancy of tetrahedral sites that had previously been suggested for cubic LLZ from X-ray diffraction results can thus be clearly ruled out [36]. More recent neutron diffraction data and the Li distribution found in our molecular dynamics simulations are consistent with the blocking model [34]. The MD simulations using the bond-valence-based parameters moreover allow for a quantitative prediction of the structural phase transition in undoped LLZ, its suppression in aliovalently doped LLZ and the ionic conductivity of the materials [34].

The transport mechanism of $Li_{10}GeP_2S_{12}$ (LGPS) [37], a recently identified compound with ultrahigh Li ion conductivity, was investigated by combining the BVSE approach with molecular dynamics (see Fig. 9) [38]. Molecular dynamics simulations for LGPS using the BV-based parameters reveal the dynamic lithium distribution, interstitial sites, and the structural prerequisites for its extraordinarily high bulk Li^+ ion conductivity. An additional Li site missing in the original structure determination based on powder diffraction data was found by a bond valence analysis and has in the meanwhile been confirmed by a single crystal diffraction study [41]. Besides an arrangement of Li sites with nearly equivalent energy and high vacancy concentration in equidistant chains, the cross-linking of 1D channels for fast ion conduction to a 3D network by paths with moderate activation energies and compositional and rotational disorder in the immobile sublattice containing polyatomic anions of slightly different size are identified as crucial factors for a systematic design of fast ion conductors. High symmetry structures with a 3D network of cation transport pathways and similarly high density of vacancies would almost inevitably be structurally less stable. At the same time a typical problem of 1D ion conductors, the easy blocking of transport

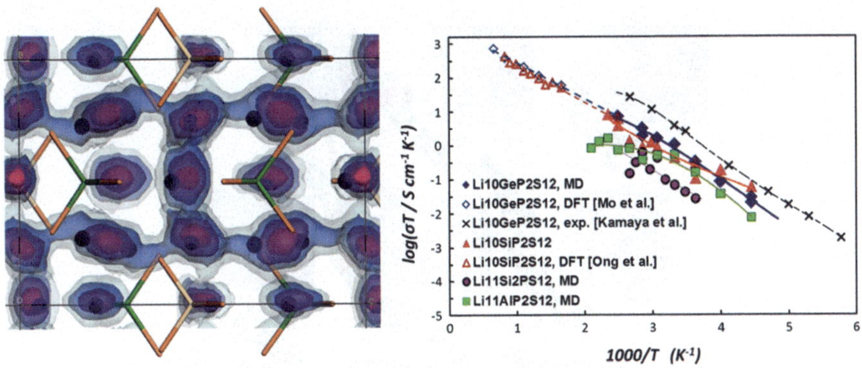

Fig. 9 *Left-hand side*: Li distribution from a 10 ns MD simulation of a 3 × 3 × 2 supercell at $T = 300$ K projected into a single unit cell shown along [100] (redrawn after Adams and Rao [38]). Regions of highest Li density (*darkest isosurface*) coincide with the 4 Li sites, the easiest path for transport (*lighter isosurface*) corresponds to the Li(3)–Li(1) channels along [001]. The *lightest isosurface* reveals a significant probability for hops between channels via Li(2) and Li(4) establishing a 3D network of pathways. *Right-hand side*: Arrhenius plots of temperature-dependent ionic conductivity in LGPS and several structurally related thiophosphates from bond-valence-based MD simulations, literature DFT simulations at high temperatures [39, 40], and experimental data [37]

channels by structural defects is effectively mitigated in LGPS by the existence of two types of cross-linking Li^+ pathways perpendicular to the channel direction. The existence of a redundant system of pathways perpendicular to the main Li^+ channels not only ensures that the high conductivity is robust with respect to high defect concentrations. The rotational mobility of the slightly smaller PS_4^{3-} in LGPS (or the respective smaller MS_4^{n-} group in analogous compounds $Li_{24-(3-z)\cdot x-z\cdot y}X^x{}_{3-z}Y^y{}_zS_{12}$ (X,Y = P, Si, Ge, Sn,..., where superscripts x and y refer to the valences of cations X and Y, respectively) and the preferential simultaneous occurrence of anion rotation and cation diffusion steps imply that Li^+ migration in LGPS is dynamically coupled to anion reorientations.

Framework materials based on phosphate or sulfate polyanion building blocks are increasingly regarded as favorable replacements for conventional oxide-based cathode materials in lithium-ion battery applications. The Li insertion phases Li_xFePO_4 [42] and $Li_xV_2(PO_4)_3$ (for which Li paths are shown later in Fig. 11) were the first of such materials identified and characterized. Recently, a few groups have described the insertion properties of the lithium vanadium fluorophosphates, "tavorite-type" compounds $LiVPO_4F$ [43] and $LiMSO_4F$ (M = Fe, Co, Ni) [44, 45], which both exhibit a relatively high operating potential versus Li. From projections of the framework structure, Recham et al. [44] originally suggested three tunnels with large cross sections (along [100], [010], and [101]) as pathways for a presumed 3D Li migration in $LiFeSO_4F$ (for a brief discussion of such geometric pathway predictions, see Sect. 3.1). For clarifying the characteristics of Li^+ ion migration pathways in $LiVPO_4F$ and $LiFeSO_4F$, we have chosen

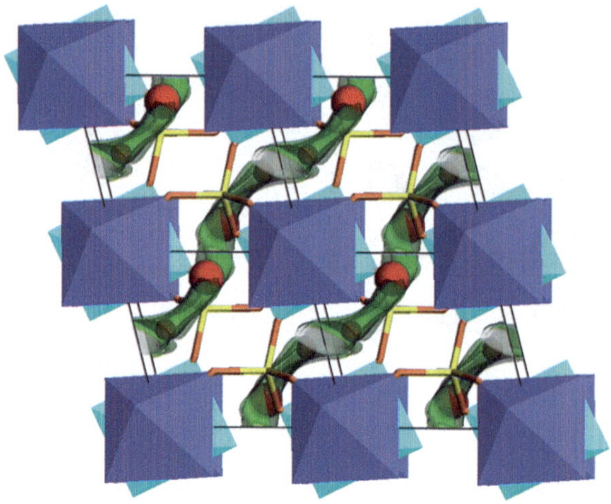

Fig. 10 *Left*: Structure of $LiFeSO_4F$ from MD simulations projected along c (Fe(1)O5F light, Fe (2)O5F dark octahedra; SO_4 ball and stick; Li black crosses (occupancy 0.6 for Li(1): *large spheres*, 0.4 for Li(2):*small spheres*). Isosurfaces of constant Li density (*dark*) from MD simulations ($T = 600$ K) and BV pathway model (*light*) that are superimposed on the structure model accordingly propose [111] as the direction of lowest activation energy for Li^+ migration

BVSE-based MD simulations. The relaxed MD-simulated supercell at 300 K (Fig. 10) closely resembled the published preliminary XRD data, except for the more pronounced disorder along channels extending along [111] with two weak Li density maxima: Li(1) at 0.15, 0.57, and 0.63 (site occupancy factor s.o.f. $\approx$ 0.6) and Li(2) at 0.45, 0.85, and 0.85 (s.o.f. $\approx$ 0.4) that form Li(2)–Li(1)–Li(1)–Li(2) channels along [111] with distances of about 2 Å that are favorable for Li transport, while migration in other directions – e.g., the previously proposed tunnels – requires hop distances of 3.7 Å. Static BV models for the Rietveld-refined and our MD-simulated structure models accordingly suggest zigzag-shaped 1D paths along [111] involving both Li sites as pathways of lowest migration energy barrier (0.22 eV), while a migration energy of ca. 0.97 eV is required to connect the channels in the [010] direction, and an only slightly higher activation energy of 1.1 eV leads to a 3D network of Li paths. Since low-energy pathways [111] connect partially occupied Li sites, a defect formation is not required for migration along these channels. The experimentally observed activation energy of 0.99 eV is much higher than the one predicted for migration along [111] channels, but closely resembles the activation energy for the formation of a 2D pathway network (paths along [111] and [010]).

Molecular dynamics simulations of the isostructural LiVPO4F using our Morse-type softBV force field in accordance to static pathway models show that the Li^+ mobility is only slightly higher along zigzag *z*-axis channels (that correspond to the [111] direction in $LiFeSO_4F$) and the activation energy (ca. 0.5 eV) is nearly isotropic due to interconnections along [011] with nearly the same activation energy [17]. This pronounced difference in the degree of anisotropy may serve as an example that ionic motion in isostructural compounds does not necessarily follow the same pathways and especially the relative contribution of alternative paths to the overall conductivity may vary substantially with the size and charge of ions in the immobile substructure.

3 Comparison to Alternative Approaches

In this chapter BVSE pathway models and bond valence mismatch approach are compared to alternative pathway modeling approaches starting from a geometric, Voronoi–Dirichlet space partition approach, via electron density-based approaches to models based on the absolute bond valence discussing both similarities and the underlying reasons for differences in the predictions. As exemplified in Fig. 11, a particular advantage of the BVSE models is that besides the vivid visualization of low-energy pathways, it also yields both the location of interstitial sites irrespective of their occupancy and a quantitative prediction of activation energy barriers.

3.1 *Geometric Analysis of Channels and Cages*

To compare the results of the BVSE approach with even simpler methods, we reanalyzed Li^+ migration paths discussed in a study by Anurova et al. [46] who used a geometric, Voronoi–Dirichlet partition approach to investigate cages and channels in crystalline lithium oxides. Differences are particularly pronounced for the 33 types of ternary oxides listed by these authors as containing one-dimensional Li pathways: in the BVSE models, 1D migration channels with low to moderate activation energies are observed for 19 of these structures only, while three exhibit 2D pathways (LT–$LiPO_3$, $Li_2W_2O_7$, Li_2TeO_3), six even 3D pathways (α-Li_3BO_3, Li_4GeO_4, Li_2SeO_4, $Li_2T_2O_5$ (T = Si, Ge), Li_4TeO_5, $Li_4Mo_5O_{17}$), and in further six cases, the literature structure models employed in their study were implausible or do not yield any paths.

A main reason for the significant deviations is the commonly complex curved nature of ion migration paths (cf. Fig. 11), which are difficult to identify from a geometric approach emphasizing straight channels. Unsurprisingly, a closer agreement is found for structures that Anurova et al. suggested to be 2D or 3D conductors. The main difference besides avoiding hard exclusion radii is however that the BVSE pathway analysis yields not only a model for the pathway geometry but also approximate energy thresholds along the pathways and hence allows a direct estimate of activation energies.

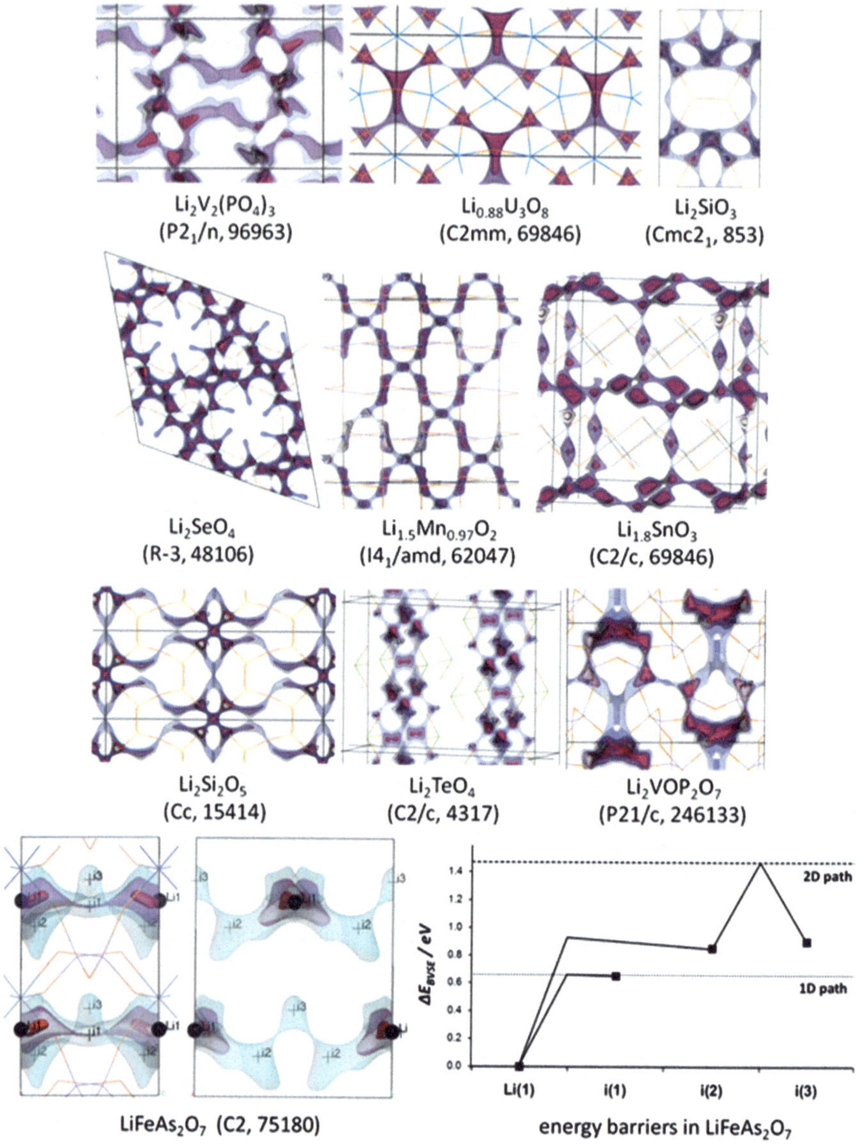

Fig. 11 *Rows 1–3*: Isosurfaces of constant $\Delta E_{BVSE}(Li)$ as models of Li^+ ion migration pathways in selected oxides. In each graph three isosurfaces corresponding to increasing ΔE_{BVSE} are superimposed (*red*, *magenta*, *light blue*). Li atoms are indicated as crosses and labeled; other atom positions are indicated by line or stick models. *Labels* below the graphs indicate the respective compound name, space group, and the ICSD database code of the underlying structure data. *Row 4*: Projections ǁx and ǁz of the $\Delta E_{BVSE}(Li)$ pathway models for $LiFeAs_2O_7$ indicating the location of the three types of interstitial sites (local minima of ΔE_{BVSE}) as *crosses* and the equilibrium Li site as a *black sphere*. The *right-hand side* diagram specifies the energy barriers for Li^+ migration in $LiFeAs_2O_7$ derived from the ΔE_{BVSE} analysis. *Horizontal dotted lines* mark the resulting migration barriers 0.66 eV for transport along the 1D path Li(1)–i(1)–Li(1), and 1.47 eV for 2D paths Li(1)–i(2)–i(3)–(i2)–Li(1)

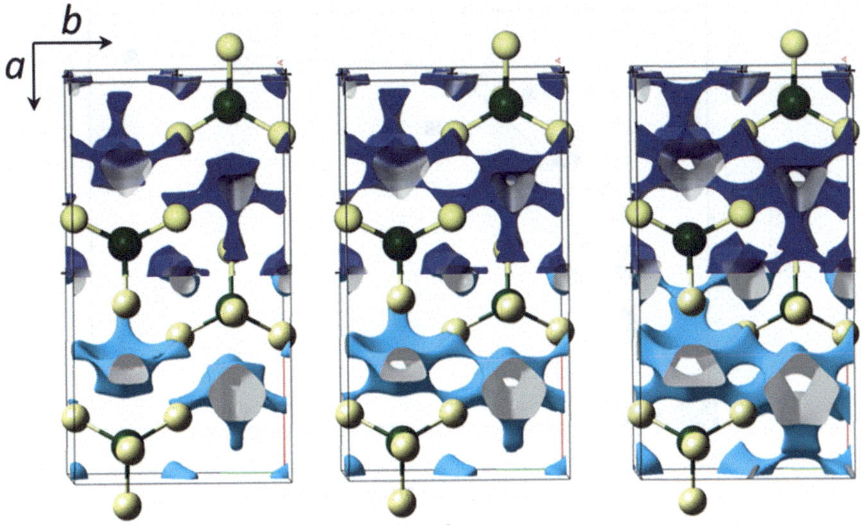

Fig. 12 Regions in the Li_4GeS_4 structure accessible to moving Li^+ ions according to BVSE energy calculations (*top half* of the unit cell) for ΔE_{BVSE} of (**a**) 0.95 eV, (**b**) 1.1 eV and (**c**) 1.35 eV and the procrystal analysis (*bottom half* of the unit cell) showing paths with electron density isovalues of 0.0016 au, 0.0018 au, and 0.0024 au, respectively

3.2 Electron Density-Based Approaches

Given the correlation between bond valence and electron density, it appears tempting to compare also what electron density maps and maps of the BVSE predict as ion transport pathways. Hirshfeld surface analysis has been explored to characterize intermolecular interactions in molecular crystals [47, 48]. This analysis is based on the procrystal, which is obtained from superposition of spherical atomic electron densities placed at the crystal structure positions, a quantity that can readily be calculated from the structure using software tools such as CrystalExplorer [49]. The approach was also explored as a tool to map out voids in porous crystals such as metal organic framework materials and zeolites [50].

Starting from the simplifying assumption that ions migrate through the crystal following a path of lowest electron density, the procrystal method was recently also tested by Filsø et al. [51] as a tool to visualize migration pathways of a static atomic arrangement in dense, inorganic crystals, and the findings were compared to those of various other approaches including the BVSE approach. A qualitative comparison between BVSE-based results (ΔE_{BVSE}) and procrystal calculations is shown for the example of Li_4GeS_4 in Fig. 12. Essentially, both models suggest the most accessible path to run along *b* (perpendicular to the plane of view in Fig. 12) involving two types of Li sites, while the third Li site is isolated. Mapping a slightly higher energy/electron density isovalue then renders a transport within the *bc* plane

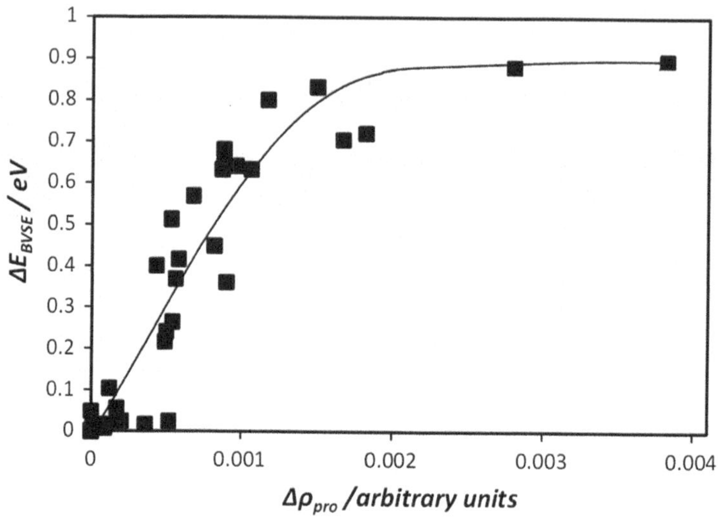

Fig. 13 Correlation between ΔE_{BVSE} and the difference in procrystal electron density $\Delta\rho_{\mathrm{pro}}$ for interstitial sites and pathway bottlenecks in various lithium-ion conducting oxides relative to the respective minimum of the same quantity in the respective structure (equilibrium sites). Reinterpretation of data from the procrystal study by Filsø et al. [51]

possible, whereas an even higher energy/isovalue is necessary to achieve three-dimensional transport. In such relatively dense structures, the morphologies of the migration paths are strikingly similar despite the different assumptions on which they are based.

As seen from Fig. 13, a comparison of ΔE_{BVSE} and procrystal electron density thresholds $\Delta\rho_{\mathrm{pro}}$ for a wider range of Li ion conducting or mixed conducting oxides yields an approximate correlation between the ΔE_{BVSE} and $\Delta\rho_{\mathrm{pro}}$ observed for interstitial sites and paths. The correlation with bond valence is more pronounced for the region of lowest ΔE_{BVSE} and $\Delta\rho_{\mathrm{pro}}$, where $\Delta\rho_{\mathrm{pro}}$ is essentially determined by the valence electron density, while the influence of core electrons on higher $\Delta\rho_{\mathrm{pro}}$ is not reflected in higher ΔE_{BVSE} values. In both cases the reference points for the differences are the lowest BVSE or ρ_{pro} value observed in the same structure model (typically at the location of the equilibrium sites).

For stable compounds not containing major voids, BVSE and procrystal methods (as well as DFT studies) predict similar migration pathways. This can be understood as for nearly densely packed compounds equilibrium sites form local minima of both the procrystal electron density and the bond valence (mismatch). Thus while the procrystal analysis appears to be a graphically appealing computationally straightforward way to get a rapid overview of possible migration pathways in novel materials, it can only be applied (1) to relatively densely packed structures and (2) only $\Delta\rho_{\mathrm{pro}}$ values corresponding to valence electron densities (<0.002 a.u.) should be considered as the correlation breaks down for higher $\Delta\rho_{\mathrm{pro}}$ values.

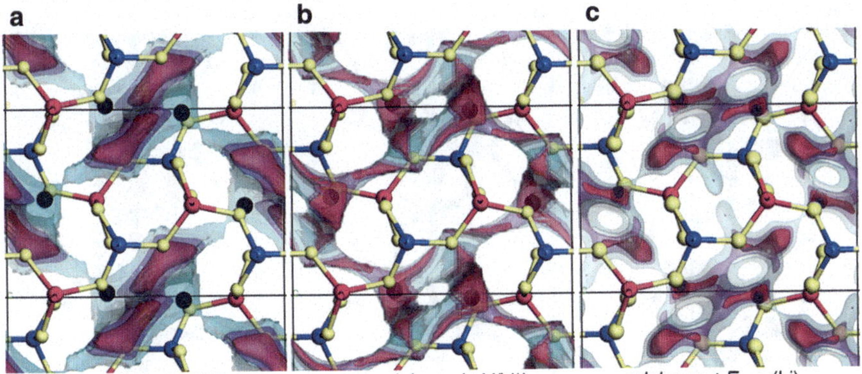

Fig. 14 Comparison of different Li^+ ion migration pathway models for $LiAlSiO_4$ (dehydrated zeolite LiABW, ICSD dataset 97909) showing (**a**) isosurfaces of minimum absolute bond valence sum V(Li), (**b**) paths of minimal bond valence sum mismatch $|\Delta V(\mathrm{Li})|$ and (**c**) isosurfaces of minimal BVSE ΔE_{BVSE}(Li). (**a**) and (**b**) use the same arbitrary scaling for Coulombic cation–cation repulsions and no penalty function for deviations from the equal valence rule. In each graph (**a**)–(**c**), three isosurfaces (*red*, *magenta*, *light blue*) are superimposed on the structure model (Li *black spheres*; Si (Al): *blue* (*magenta*) *large spheres*, O: *small spheres*) to distinguish regions that the Li^+ ions can reach with different activation energies

3.3 Pathways of Minimum Absolute Bond Valence Sums

Analogous caveats as for the procrystal analysis also apply to the approach based on the (generally incorrect) postulate that a mobile ions would prefer to move into regions where its bond valence sums is as low as possible [52–54]. This assumption is only true in densely packed structures as long as the regions of lowest bond valence sum coincide with regions of lowest bond valence sum mismatch. In open structures, such as the zeolite LiABW-type structure of $LiAlSiO_4$ shown in Fig. 14 as an example, the postulate will erroneously lead to paths in the central region of any existing large voids. Only for valence mismatch model (b) and the BVSE model (c) of Fig. 14 the equilibrium Li sites are located within the innermost isosurface (i.e., the region of lowest "site energy" for a Li according to the respective model), while the absolute bond valence sum model (a) suggests (vacant) migration paths as regions of lowest bond valence sum (down to $V(\mathrm{Li}) = 0.73$ valence units) in the open channels of the zeolite structure. Moreover, it should be noted that the BVSE model in detail shows some deviations from both the bond valence sum model (a) and the bond valence sum mismatch model (b). Deficiencies of (a) and (b) originate from both the nonideal scaling of the Coulomb repulsions with respect to the bond-valence-based attraction terms and the lack of a penalty term for deviations from the equal valence rule in (a) and (b).

That said, since many stable structures including solid electrolytes are not too far away from a dense packing and thus equilibrium sites for the mobile ions are

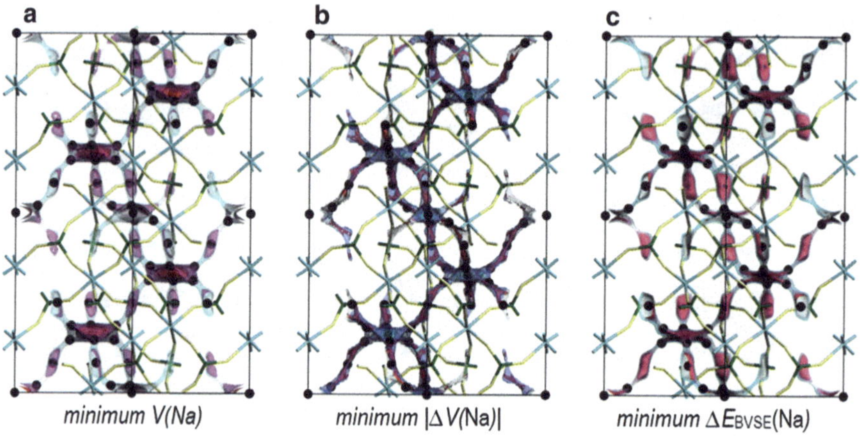

Fig. 15 Comparison of different Na^+ ion migration pathway models for NASICON-type $Na_{3.05}Zr_2Si_{2.05}P_{0.95}O_{12}$ (ICSD dataset 62383) showing (**a**) isosurfaces of minimum absolute bond valence sum V(Na), (**b**) paths of minimal bond valence sum mismatch $|\Delta V(\text{Na})|$, and (**c**) isosurfaces of minimal BVSE ΔE_{BVSE}(Na). In each graph (**a**)–(**c**), three isosurfaces (*red, magenta, light blue*) are superimposed on the structure model (Na *black spheres*; immobile ions as stick model) to distinguish regions that the Li^+ ions can reach with different activation energies

minima of the bond valence sum as well as of the bond valence sum mismatch, the fundamentally flawed approach of identifying regions of lowest absolute bond valence sum with pathways can yield the same correct pathway topology as the (much safer approach) of identifying regions of minimum bond valence sum mismatch or of lowest BVSE. Figure 15 exemplifies that for NASICON-type $Na_{3.05}Zr_2Si_{2.05}P_{0.95}O_{12}$, where the same pathway topology is predicted from all three approaches and differences occur only in minor details (e.g., that again the isosurface of lowest absolute bond valence sum does not include the equilibrium positions of the Na^+). For both the bond valence sum and the (as seen for fundamental reasons more advisable) bond valence sum mismatch calculations, it is in the end often the suitable weighting of the bond valence terms, the repulsions between immobile and mobile ions, and a suitable penalty function for pronounced deviations from equal valence configurations that is more decisive for an accurate pathway prediction. This degree of arbitrariness in scaling between different contributions can be avoided by the BVSE approach.

In the extreme case of surfaces, where the bond valence sum falls to zero, the minimum valence sum postulate would predict that mobile ions should easily escape from their host solid, which is obviously not true. In contrast, both the bond valence BVSE and the bond valence sum mismatch approach correctly yield a large energy threshold for mobile ions at surfaces (see Fig. 16), and the BVSE approach can also provide a plausible estimate for the (significantly smaller) variation of the Li site energies across interfaces between ionic conductors.

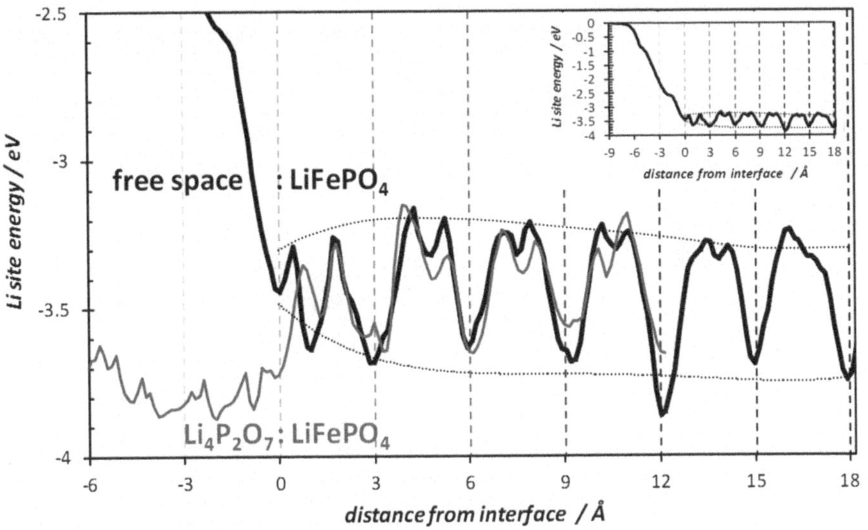

Fig. 16 Variation of the Li site energy E_{BVSE} as a function of the distance from the surface of a $LiFePO_4$ crystal (*black solid line*). It is assumed that the surface is perpendicular to the lithium channel direction. Energy values are calculated based on a snapshot from an MD-simulated structure model of a reconstructed $LiFePO_4$ surface. Negative distance values refer to free space, positive distance values to the $LiFePO_4$ region. *Dotted lines* mark the time averaged minima and maxima of the site energy along the Li migration path. The *inset* emphasizes that the BVSE of the Li reaches the free space limit of 0 eV at a distance of ca. 6 Å from the surface. The *grey line* indicates the qualitatively different variation of the Li (bond valence) site energy as a function of the distance from the interface between glassy $Li_4P_2O_7$ and $LiFePO_4$. Again, positive distances refer to the $LiFePO_4$ phase, while negative distances now refer to the $Li_4P_2O_7$ glass phase

4 Concluding Remarks: Scaling, Statistical Accuracy, and Limitations of Static Bond Valence Models

The BVSE model essentially attempts to predict the energetic environment of mobile ions from a static structure model. Thereby, the approach cannot be expected to yield activation energies with high precision, but should still yield useful semiquantitative information. The static nature of the approach inherently leads to an overestimation of migration barriers, as the approach essentially calculates the energy that would be required (in the frame of the assumed force field) to move the mobile ion without taking into account relaxations of the atomic positions in the environment. To mitigate this overestimation, the resulting energies should be scaled by a factor that depends on how rigid the network is. For a wide range of nearly dense framework structures, a factor of 0.8 works reasonably well, but considerably lower scaling factors (ca. 0.4) would be appropriate to take into account low-energy structural relaxations in structures held together by weak van der Waals forces only, such as the layered compounds (e.g., the classical cathode materials Li_xMO_2). The low number of compounds for which it is known that the

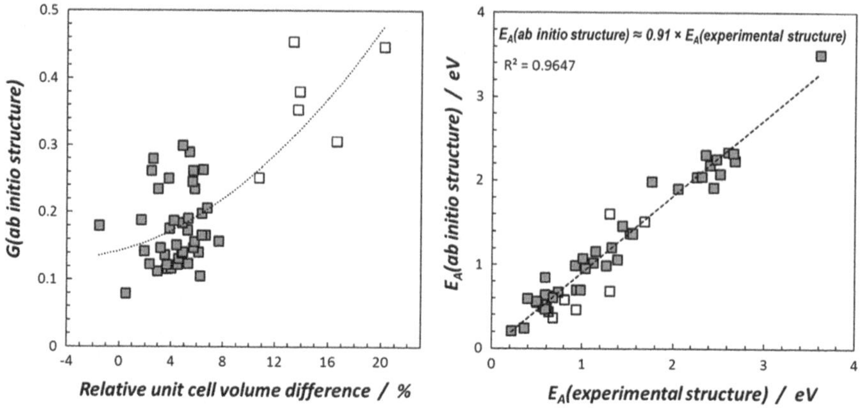

Fig. 17 *Left-hand side*: variation of global instability index G of 48 ab initio structure models of Li^+ ion conducting oxides versus the relative difference between the volume of the ab initio structure models and the experimentally determined unit cell volume for the same phase. *Right-hand side*: linear correlation between the (*unscaled*) activation energy barriers derived by the BVSE method for experimental and for ab initio calculated structure models. In both graphs the six cases for which the ab initio model overestimates the unit cell volume by more than 8% are marked as open symbols. These less reliable data points are excluded from the calculation of the r.h.s. linear correlation

ion migration mechanism involves complex reorientations in the immobile substructure in the sense of the paddle-wheel mechanism makes it difficult to validate the applicability of the approach in these rare circumstances and for such structures only dynamic structure models (series of snapshots from a MD simulation run) should be used. Then, the appropriate scaling may be intermediate between the abovementioned limiting values.

If the (static) structure models originate from ab initio simulations, it should also be noted that the common tendency of ab initio approaches to slightly overestimate the unit cell volume will affect the appropriate scaling factor. As the bond valence parameters are derived from fits to experimental reference crystal structure, they tend to yield somewhat lower than ideal bond valence sums when applied to ab initio models. A way to quantify this source of error is the global instability index G specified for each structure. Figure 17 summarizes the comparison of Li^+ ion pathway models that we calculated for 48 oxide structures based on both their experimentally determined structure and ab initio structure models: in 47 out of 48 studied cases the unit cell volume is overestimated by the ab initio simulations and G tends to increase with the relative overestimation of the volume. The activation energies for the formation of continuous Li^+ ion migration pathways predicted from ab initio structure models are proportional to those derived from experimental structure data yet systematically underestimate the activation energy barrier. Hence, for estimating activation energy barriers from ab initio structure models, the recommended scaling factors are 0.9 for network structures or 0.45 for interlayer paths in layered structures.

Finally, the bond valence parameters can also be used to both generate and analyze dynamic structure models by providing a highly transferable force field for molecular dynamics simulations of inorganic solids. This of course requires a higher computational effort, but as demonstrated above, e.g., in the case of $Li_{10}GeP_2S_{12}$, allows for more detailed findings on the transport processes including correlations between ion diffusion and rearrangements in the immobile substructure. Thus, it is a major advantage of the bond valence methods the approach can be gradually refined to provide the required level of details while utilizing consistently the wealth of crystal chemical information stored in the bond valence parameters.

References

1. Garrett JD, Greedan JE, Faggiani R, Carbotte S, Brown ID (1982) Single-crystal growth and structure determination of $Ag_{16}I_{12}P_2O_7$. J Solid State Chem 42:183–190
2. Adams S, Moretzki O, Canadell E (2004) Global instability index optimizations for the localization of mobile protons. Solid State Ion 168:281–290
3. Adams S, Swenson J (2000) Determining ionic conductivity from structural models of fast ionic conductors. Phys Rev Lett 84:4144–4147
4. Adams S, Swenson J (2002) Bond valence analysis of transport pathways in RMC models of fast ion conducting glasses. Phys Chem Chem Phys 4:3179–3184
5. Swenson J, Adams S (2003) Mixed alkali effect in glasses. Phys Rev Lett 90:155507
6. Adams S, Swenson J (2005) Bond valence analysis of reverse Monte Carlo produced structural models: a way to understand ion conduction in glasses. J Phys Condens Matter 17:S87
7. Hall A, Adams S, Swenson J (2006) Comparative study of ion conducting pathways in borate glasses. Phys Rev B 74:174205
8. Müller C, Zienicke E, Adams S, Habasaki J, Maass P (2006) Comparison of ion sites and diffusion paths in glasses obtained by molecular dynamics simulations and bond valence analysis. Phys Rev B 75:014203
9. Adams S, Prasada Rao R (2009) Transport pathways for mobile ions in disordered solids from the analysis of energy-scaled bond-valence mismatch landscapes. Phys Chem Chem Phys 11: 3210–3216
10. Avdeev M, Sale M, Adams S, Prasada Rao R (2012) Screening of the alkali-metal ion containing materials from the Inorganic Crystal Structure Database (ICSD) for high ionic conductivity pathways using the bond valence method. Solid State Ion 225:43–46
11. Adams S (2006) From bond valence maps to energy landscapes for mobile ions in ion-conducting solids. Solid State Ion 177:1625–1630
12. Waltersson K (1978) A method, based upon "Bond-Strength" calculations, for finding probable lithium sites in crystal structures. Acta Crystallogr A34:901–905
13. Withers RL, Schmid S, Thompson JG (1998) Compositionally and/or displacively flexible systems and their underlying crystal chemistry. Prog Solid State Ch 26:1–96
14. Van Smaalen S (1999) Atomic valences in aperiodic crystals studied by the bond valence method. In: Jelsi DA, George TF (eds) Computational studies of new materials. World Scientific, Singapore, pp 273–294
15. Adams S, Ehses KH, Spilker J (1993) Proton ordering in the Peierls distorted hydrogen molybdenum bronze $H_{0.33}MoO_3$: structure and physical properties. Acta Crystallogr B 49: 958–967
16. Adams S (2013) Practical considerations in determining bond valence parameters. Struct Bond. doi:10.1007/430_2013_96

17. Adams S, Prasada Rao R (2011) High power lithium ion battery materials by computational design. Phys Status Solidi A 208:1746–1753
18. Adams S, Swenson J (2000) Migration pathways in Ag-based superionic glasses and crystals investigated by the bond valence method. Phys Rev B 63:054201
19. Radaev SF, Fink L, Trömel M (1994) Bond valences for halides and oxides based on geometrical coordination. Z Kristallogr Suppl 8:628
20. Sale M, Avdeev M (2012) 3DBVSMAPPER: a program for automatically generating bond-valence sum landscapes. J Appl Crystallogr 45:1054–1056
21. Levi E, Gershinski G, Aurbach D, Isnard O, Ceder G (2009) New insight on the unusually high ionic mobility in Chevrel phases. Chem Mater 21:1390–1399
22. Adams S, Preusser A (1999) Silver ion conduction pathways in $Ag_5IP_2O_7$. Acta Crystallogr C 55:1741–1743
23. Adams S (1996) Crystal structure and Ag^+ conductivity of the solid electrolyte $Ag_8I_4V_2O_7$. Z Kristallogr 211:770–776
24. Wada H, Sato A, Onoda M, Adams S, Tansho M, Ishii M (2002) Phase transition and crystal structure of silver-ion conductor Ag12-nM + nS6 (M = Ti, Nb, Ta). Solid State Ion 154: 723–727
25. Adams S (2000) Modelling ion conduction pathways by bond valence pseudopotential maps. Solid State Ion 136(137):1351–1361
26. Adams S, Hariharan K, Maier J (1996) Investigations on silver iodide oxysalt glass ceramics. J Solid State Ionics 86–88:503
27. Cabana J, Ling CD, Oró-Solé J, Gautier D, Tobias G, Adams S, Canadell E, Palacin MR (2004) Antifluorite-type lithium chromium oxide nitrides: synthesis, structure, order, and electrochemical properties. Inorg Chem 43:7050–7060
28. Janssen Y, Middlemiss DS, Bo S-H, Grey CP, Khalifah PG (2012) Structural modulation in the high capacity battery cathode material $LiFeBO_3$. J Am Chem Soc 134:12516–12527
29. Hall A, Adams S, Swenson J (2006) Local dimensionality and intermediate range ordering of conduction pathways in borate glasses. J Non-Cryst Solids 352:5164–5169
30. Prasada Rao R, Tho TD, Adams S (2011) Ion transport pathways in molecular dynamics simulated alkali silicate glassy electrolytes. Solid State Ion 192:25–29
31. Kamitsos EI, Chryssikos GD (1998) Alkali sites in glass. Solid State Ion 105:75–85
32. Hall A, Swenson J, Adams S, Meneghini C (2008) Mixed mobile Ion effect and cooperative motions in silver-sodium phosphate glasses. Phys Rev Lett 101:195901
33. Tho TD, Prasada Rao R, Adams S (2012) Structure property correlation in lithium borophosphate glasses. Eur Phys J E 35:8
34. Adams S, Prasada Rao R (2012) Ion transport and phase transition in Li7-xLa3(Zr2-xMx)O12 (M = Ta5+, Nb5+, x = 0, 0.25). J Mater Chem 22:1426–1434
35. Geiger CA, Alekseev E, Lazic B, Fisch M, Armbruster T, Langner R, Fechtelkord M, Kim N, Pettke T, Weppner W (2011) Inorg Chem 50:1089–1097
36. Awaka J, Takashima A, Kataoka K, Kijima N, Idemoto Y, Akimoto J (2011) Crystal structure of fast lithium-ion-conducting cubic $Li_7La_3Zr_2O_{12}$. Chem Lett 40:60–62
37. Kamaya N, Homma K, Yamakawa Y (2011) A lithium superionic conductor. Nat Mater 10: 682–686
38. Adams S, Prasada Rao R (2012) Structural requirements for fast lithium migration in $Li_{10}GeP_2S_{12}$. J Mater Chem 22:7687–7691
39. Mo Y, Ong SP, Ceder G (2012) First principles study of the $Li_{10}GeP_2S_{12}$ Lithium super ionic conductor material. Chem Mater 24:15–17
40. Ong SP, Shyue P, Mo YF, Richards WD, Miara L, Lee HS, Ceder G (2013) Phase stability, electrochemical stability and ionic conductivity of the Li10 +/− 1MP2X12 (M = Ge, Si, Sn, Al or P, and X = O, S or Se) family of superionic conductors. Energ Environ Sci 6:148–156
41. Kuhn A, Köhler J, Lotsch BV (2013) Single-crystal X-ray structure analysis of the superionic conductor $Li_{10}GeP_2S_{12}$. Phys Chem Chem Phys 15:11620–11622

42. Adams S (2012) Ultrafast lithium migration in surface modified $LiFePO_4$. Appl Energ 90: 323–328
43. Barker J, Gover RKB, Burns P, Bryan AJ (2005) Performance evaluation of lithium vanadium fluorophosphate in lithium metal and lithium-ion cells. J Electrochem Soc 152:A1776
44. Recham N, Chotard J-N, Dupont L, Delacourt C, Walker W, Armand M, Tarascon J-M (2010) A 3.6 V lithium-based fluorosulphate insertion positive electrode for lithium ion batteries. Nat Mater 9:68–74
45. Barpanda P, Recham N, Chotard JN, Djellab K, Walker W, Armand M, Tarascon J-M (2010) Structure and electrochemical properties of novel mixed Li(Fe1-xMx)SO4F (M = Co, Ni, Mn) phases fabricated by low temperature ionothermal synthesis. J Mater Chem 20: 1659–1668
46. Anurova NA, Blatov VA, Ilyushin GD, Blatova OA, Ivanov-Schitz AK, Dem'yanets LN (2008) Migration maps of Li^+ cations in oxygen-containing compounds. Solid State Ion 179: 2248–2254
47. Spackman MA, McKinnon JJ (2002) Fingerprinting molecular interactions in molecular crystals. Cryst Eng Comm 4:378–392
48. McKinnon JJ, Spackman MA, Mitchell AS (2004) Novel tools for visualizing and exploring intermolecular interactions in molecular crystals. Acta Crystallogr B 60:627–668
49. Wolff SK, Grimwood DJ, McKinnon JJ, Turner MJ, Jayatilaka D, Spackman MA (2010) CrystalExplorer (version 3.0). University of Western Australia, Australia
50. Turner MJ, McKinnon JJ, Jayatilaka D, Spackman MA (2011) Visualisation and characterization of voids in crystalline materials. Cryst Eng Comm 13:1804–1813
51. Filsø MØ, Turner MJ, Gibbs GV, Adams S, Spackman MA, Iversen BB (2013) Visualizing Lithium-ion migration pathways in battery materials. Chemistry-A European Journal 19: 15535–15544
52. Mazza D, Ronchetti S, Bohnké O, Duroy H, Fourquet JL (2002) Modeling Li-ion conductivity in fast ionic conductor $La_{2/3-x}Li_{3x}TiO_3$. Solid State Ion 149:81–88
53. Ouerfelli N, Guesmi A, Mazza D, Zid MF, Driss A (2008) Arsenate Na3Fe2(AsO4)3: structural study of the low temperature forms and property simulation of alkaline cation conduction. Acta Crystallogr C 64:I41–I44
54. Mounir F, Karima H-N, Ben Saad K, Férid M (2012) Modeling Li-ion conductivity in LiLa (PO3)4 powder. Physica B 407:2593–2600

Struct Bond (2014) 158: 161–190
DOI: 10.1007/430_2013_91

Published online: 19 April 2013

Crystallization and Dissolution in Aqueous Solution: A Bond-Valence Approach

Frank C. Hawthorne and Michael Schindler

Abstract In many groups of minerals, structural diversity occurs by polymerization of a small number of clusters (or fundamental building blocks). Where these minerals crystallize from aqueous or hydrothermal solutions, the fundamental building blocks occur as aqueous species in solution, and it seems reasonable to conclude that crystallization of these minerals occurs by condensation of these clusters in solution. The variation in Lewis acidity of these clusters is a function of the pH of the aqueous solution in which they occur, in accord with the different structures crystallizing from similar aqueous solutions at different pH. Strongly bonded polyhedron chains (equivalent to periodic bond-chains) control the morphology of crystals. Anions at the surface of a mineral (i.e., exposed to an ambient aqueous solution) are called *terminations*, and the residual valence at a termination controls its reactivity (i.e., is the driving force for reaction with the aqueous solution). The residual valence of a polyhedron chain controls the growth or dissolution rate at the crystal face associated with that chain and may be calculated as the net residual valence of the terminations per repeat of the polyhedron chain. Edges involving polyhedron chains with low normalized residual valence will grow slowly, whereas edges involving polyhedron chains with high normalized residual valence will grow rapidly, and the relative morphology of crystals will be controlled by the relative magnitudes of the residual valence of polyhedron chains parallel to specific faces. The observed morphology of selected uranyl-oxide hydroxyl-hydrate and borate minerals is in reasonable accord with this approach.

Keywords Aqueous speciation · Bond valence · Borate minerals · Crystallization · Dissolution · Morphology · Surface structure · Uranyl minerals

F.C. Hawthorne (✉)
Department of Geological Sciences, University of Manitoba, Winnipeg, MB, Canada R3T 2N2
e-mail: frank_hawthorne@umanitob.ca

M. Schindler
Department of Earth Sciences, Laurentian University, Sudbury, ON, Canada P3E 2C6

Contents

Abbreviation

v.u Valence units

1 Introduction

A basic axiom of bond-valence theory is the valence-sum rule [1]: *The sum of the bond valences at each atom is equal to the magnitude of the atomic valence*. This rule has been shown to hold (within a few percent) for a large number of crystal structures. For most of the structures in which we are interested (minerals), the maximum valences of the cations (4^+ to 6^+) generally exceed the maximum valences of the anions (2^-). The result is that we can identify strongly bonded oxyanions, e.g., $(SiO_4)^{4-}$, $(PO_4)^{3-}$, $(SO_4)^{2-}$, in these structures, and it is these oxyanions that dominate their structural characteristics and geochemical behavior [2, 3]. A primary interest in mineralogy is the behavior of minerals in geological processes, in particular crystallization and dissolution, as the structural and chemical characteristics of minerals can carry a lot of information on the conditions under which they form. Hence a mechanistic understanding of the processes of crystallization and dissolution of minerals is of considerable interest. Many minerals crystallize from aqueous solution or magma, both of which have some regularity in their structure (i.e., a nonrandom arrangement of their constituent atoms). Hawthorne [4] proposed the *reaction principle*: *During a chemical reaction, atoms move relative to each other such that they continually minimize local deviations from the valence-sum rule.* This rule implies that the atomic arrangements in the reactants affect the atomic arrangements in the products of any chemical reaction. If this is the case in geological processes, we need to focus not just only on the atomic arrangements in the resulting minerals but also on the atomic arrangements in the precursor phases, i.e., aqueous solutions and magmas, and the atomic arrangements of any intermediate phases, e.g., the interfaces between the reactant phase(s) and the product phase(s).

1.1 Aqueous Solutions

1.1.1 Borates

Hawthorne [5–7] showed that structural diversity in oxysalt minerals (specifically phosphates) occurs by polymerization of a small number of clusters (fundamental building blocks) to form chains, sheets, and frameworks and suggested that structural units in many low-temperature minerals form by condensation of fundamental building blocks that occur as aqueous complexes in hydrothermal or aqueous solutions. With regard to borates, Ingri and coworkers (reviewed in [8]) showed that the following borate species occur in highly concentrated aqueous borate solutions with decreasing pH: $[B(OH)_3]^0$, $[B_5O_6(OH)_4]^-$, $[B_3O_3(OH)_4]^-$, $[B_3O_3(OH)_5]^{2-}$, $[B_4O_5(OH)_4]^{2-}$ and $[B(OH)_4]^-$ and noted that "polyanions of the kind found in crystals exist in solution.....and are readily available for the building of crystals." Furthermore, ^{11}B-NMR spectroscopy [9–11] and Raman spectroscopy [12, 13] have confirmed the occurrence of all these aqueous species except $[B_3O_3(OH)_5]^2$. With regard to borate minerals, Christ et al. [14] used the data summarized in [8] to illustrate the variation in aqueous borate species and their variation in abundance in solution as a function of pH (Fig. 1). $[B(OH)_3]^0$ is the stable species at low pH and $[B(OH)_4]^-$ is the stable species at high pH, while around a pH of 8, the more complicated species $[B_5O_6(OH)_4]^-$, $[B_3O_3(OH)_4]^-$, $[B_3O_3(OH)_5]^{2-}$, and $[B_4O_5(OH)_4]^{2-}$ occur in solution and show their maximum concentrations at slightly different values of pH. From Fig. 1, Schindler and Hawthorne [2, 15] calculated the percentage of $^{[4]}$B in aqueous solution and showed that it is a smooth function of pH (Fig. 2). This suggests that the stability (i.e., existence) of each cluster and the relative amounts of each cluster as a function of pH are controlled by bond-valence matching between the clusters and their host aqueous solution.

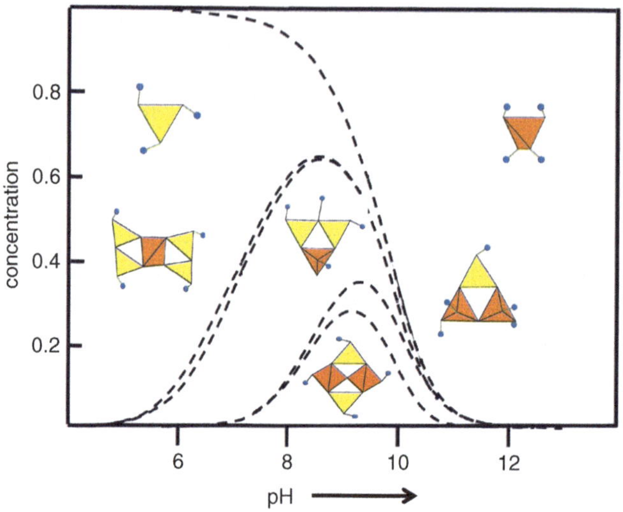

Fig. 1 The distribution of B species in aqueous solution of 0.40 M on total $B(OH)_3$; ($B\varphi_3$): *yellow/pale-gray*; ($B\varphi_4$): *orange/medium-gray*; after [14] from the data of [8]

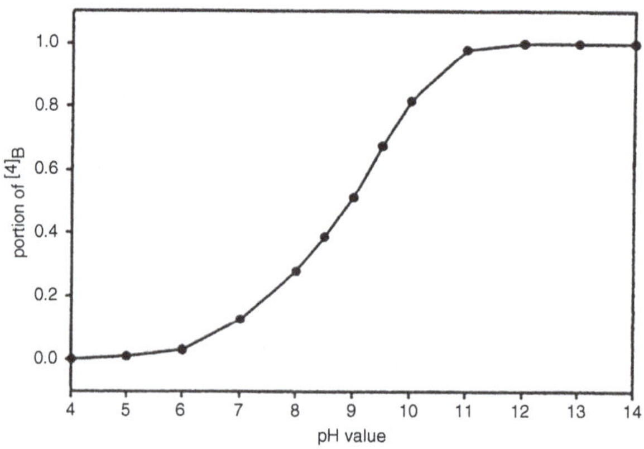

Fig. 2 The fraction of $^{[4]}B$ in the aqueous species in solution as a function of pH in the aqueous solution shown in Fig. 1

1.1.2 Uranyl Oxide-Hydroxy-Hydrates

The hydrolysis of dioxouranium(VI) can be formulated as follows:

$$m\mathrm{UO_2}^{2+} + n\mathrm{H_2O(l)} \rightarrow (\mathrm{UO_2})_m(\mathrm{OH})_{2m-n}$$

Grenthe et al. [16] reviewed the thermodynamic data for $(UO_2)_m(OH)_{2m-n}$ species and listed eighteen different species. However, spectroscopic evidence exists only for a few species, and Grenthe et al. [16] indicated that the thermodynamic data for some of these 18 species are questionable. Tsushima et al. [17]

showed by EXAFS, FTIR, and UV–vis spectroscopy and DFT calculations that in aqueous solution, $(U^{6+}O_2)^{2+}$ always has five peripheral anions, the monomer, dimer and trimer are present, and there is edge-sharing between uranyl polyhedra. Although we do not know the exact distribution of all $(UO_2)_m(OH)_{2m-n}$ species as a function of pH, the information given in [17] is sufficient to allow us to consider crystallization and dissolution mechanisms for uranyl-oxide hydroxyl-hydrate minerals.

1.2 The Effect of pH on Aqueous Complexes and Crystal Structure

1.2.1 Borates

The relative amount of $^{[4]}B$ in solution is a function of the pH of that solution, and it seems reasonable to propose that aqueous borate complexes adjust to varying pH by varying the relative amounts of $^{[3]}B$ and $^{[4]}B$. In the crystal structures of the sedimentary borate minerals, these clusters in aqueous solution are the fundamental building blocks of all the borate minerals [18]. This implies that the structural units in borate minerals form by condensation of these complexes in hydrothermal or aqueous solutions, and their relation with pH gives us a direct relation between the pH of the nascent solution and the chemical compositions and structures of the crystallizing minerals.

1.2.2 Uranyl Oxide-Hydroxy-Hydrates

The situation for the uranyl oxide-hydroxy-hydrates is somewhat different from that of the borate minerals. The coordination number of U^{6+} does not change as a function of pH; it is [7] across the whole range. What will change as a function of pH is the polymerization and the ligancy in the clusters. If minerals form by condensation of these complexes in aqueous solution, the pH at which they crystallize should relate to the stability of the clusters that constitute the fundamental building blocks of the structure. There is general consensus on the occurrence of the aqueous species $(UO_2)^{2+}$ and $(UO_2)_3(OH)_5{}^+$, which occur in strongly and weakly acidic to neutral aqueous solutions, respectively [19]. Schoepite, $(UO_2)_8O(OH)_{12}(H_2O)_{12}$, is a uranyl-oxide hydroxy-hydrate mineral which contains sheets of polymerized trimers and dimers of (edge-sharing) $(U^{6+}O_2\varphi_5)$ polyhedra [$\varphi = (OH), (H_2O)$]. Schindler and Putnis [20] synthesized well-crystaline schoepite under weakly acidic conditions (pH = 5.5–6.5) which overlap with the general occurrence of the $(UO_2)_3(OH)_5{}^+$ species in solution, suggesting that the crystallization of schoepite is controlled by the polymerization of aqueous species present in solution (see below).

2 Crystal Faces

Hartman and Perdok [21–23] proposed that where an atom or cluster of atoms attaches to a growing surface of a crystal, the probability of subsequent detachment is inversely proportional to the number of strong bonds between the atom or cluster and the crystal surface. The focus is on uninterrupted chains of strong bonds between building units, called periodic bond-chains. Periodic bond-chains define the major growth direction(s) of a crystal. There are three types of faces: F (or flat) faces with two or more types of periodic bond-chains parallel to the face; S (or stepped) faces with one type of periodic bond-chain parallel to the face, and K (or kinked) faces with no periodic bond-chains parallel to the face. The morphology of a crystal is controlled primarily by the occurrence of F faces and secondarily by the occurrence of S faces. Complicated oxide and oxysalt structures are generally represented as arrangements of polyhedra, where each polyhedron consists of a central cation and its coordinating anions. This type of representation leads to major simplification in representing and understanding the topological aspects of the arrangements of chemical bonds, and the linkage of such polyhedra is used as a basis for hierarchical classification of crystal structures (e.g., [18, 24–26]). We will follow this polyhedron approach in considering surfaces of crystals and will consider periodic bond-chains as *polyhedron chains*.

Molecular modeling may be used to examine the morphology of crystals by calculating surface energies or step energies, provided good interaction potentials are available for the constituent species. This is usually not the case for hydroxy-hydrated oxysalt minerals which contain unusual coordination geometries and both (OH) and (H_2O) groups, e.g., althupite, $AlTh[(UO_2)\{(UO_2)_3(PO_4)_2(OH)O\}_2](OH)_3(H_2O)_{15}$. As many such minerals are important phases from an environmental perspective, and such minerals constitute the bulk of the mineral kingdom, we need an approach that is tractable for such complicated materials. Bond-valence theory is a key part of such an approach.

3 Interaction of a Surface with an Aqueous Solution: A Bond-Valence Perspective

The valence-sum rule (Eq. 2 in chapter "Bond Valence Theory") requires that the sum of the bond-valences incident at any cation or anion must be equal to its valence. It is useful to regard bond valences as directed, as this keeps track automatically of the nature (cation or anion) of the ion occupying any site in a structure. Here, we adhere to the convention that bond valences from a cation to an anion are positive, and bond valences from an anion to a cation are negative.

For a surface, we may identify two situations: (1) the surface of the crystal is adjacent to a vacuum and (2) the surface of the crystal is adjacent to a liquid (or gas). In the first situation, the surface ions must have coordinations different from those in the bulk crystal, and the surface structure responds to these differences by lengthening or shortening specific bonds (relaxation), and there may be a reorganization of the topology of the chemical bonds at the surface (cf. in Chapter "Structure and Acidity in Aqueous Solutions and Oxide–Water Interfaces"). In the second situation, the bond-valence requirements of the surface atoms are partly met by adjacent atoms in the coexisting liquid (or gas), and surface relaxation will be much less than where the surface is exposed to a vacuum. For a crystal surface in equilibrium with an aqueous solution, the surface is partly or fully hydrated, depending on the pH of the solution, and aqueous species in the solution bond to anions or cations on the surface. The type of anion or cation on the surface and the conditions in the coexisting solution will affect the degree of hydration and type of attachment. The atoms of the liquid will tend to arrange themselves such that surface relaxation is minimized, and one may well be able to consider local atom interactions as the average of what occurs at the surface over a longer timescale. In turn, this suggests that we may be able to approximate bond valences of near-surface bonds as equal to the bond valences of the analogous bonds in the bulk structure.

The conditions in the coexisting aqueous solution and the anions or cations at the surface will affect the degree of hydration and the type of reactions that can occur at the surface. The degree of hydration will be a function of the bond-valence requirements of the anions at the surface and the pH of the solution. The bond-valence requirements of the surface anions can be predicted using the MUSIC ("MUltiSIte Complexation") model of Hiemstra et al. [27] using the equation

$$\mathrm{p}K_{\mathrm{a}} = -A\left(\sum s_j + V\right) \tag{1}$$

where $\mathrm{p}K_{\mathrm{a}}$ is the intrinsic acidity constant, A equals 19.8, V is the valence of the surface O atom (−2), and $\sum s_j$ is the bond-valence sum at the surface O atom and is defined as

$$\sum s_j = \{s_{\mathrm{M}} + m s_{\mathrm{H}} + n(1 - s_{\mathrm{H}})\} \tag{2}$$

where s_{M} is the valence of the M–O bond, s_{H} is the bond valence of the H–O bond to the surface oxygen if the base is an (OH) group (assumed to be 0.80 v.u. [1]), $(1 - s_{\mathrm{H}})$ is the valence of weak hydrogen bonds from aqueous species to surface anions, and m and n are the numbers of stronger O–H and weaker O···H bonds, respectively. Hiemstra et al. [27] used bond valences from unrelaxed bulk structures. Conversely, for the calculation of intrinsic $\mathrm{p}K_{\mathrm{a}}$ values, Bickmore et al. [28, 29] considered unevenly distributed charge densities and relaxed metal–oxygen bonds on surfaces.

The intrinsic acidity constant is a measure of the strength of an acid (HA) in an acid–base equation, $HA + H_2O \leftrightarrow A^- + H_3O^+$, and is closely related to the ability of the conjunctive base (H_2O is a Lewis base) to donate electrons to the acid (H^+,

polyhedron chain, ΔpK_{PC}, is a function of the types and numbers of different anion-terminations along the chain, and may be written as $\sum \Delta pK_{PC}$, i.e., the sum of the pK_a values of acid–base reactions at the corresponding anion-terminations:

$$\sum \Delta pK_{PC} = [b \times pK_{a1} + c \times pK_{a2}]/(b + c) \tag{5}$$

$$\sum \Delta pK_{PC} = [b \times \Delta s_1 + c \times \Delta s_2]/(b + c) \tag{6}$$

The right-hand side of Eq. (6) is the O-atom residual valence for a polyhedron chain and correlates with the average pK_a-value and the free energy of the acid–base reactions along a polyhedron chain, indicating the affinity of the constituent O-atoms for hydrogen bonds or O–H bonds [31].

4.2 Point of Zero Charge and Net Proton Charge: A Bond-Valence Perspective

The point of zero charge is the point where the total net surface charge is zero [33]. The total net surface charge involves (1) the permanent structural charge caused by isomorphic substitution(s); (2) the net proton charge (i.e., the charge due to binding of H or OH at the surface); (3) the charge of the inner-sphere-complex; (4) the charge of the outer-sphere-complex. An inner-sphere complex occurs where a cation or anion in solution bonds directly to terminations on the surface, whereas an outer-sphere complex occurs if a cation or anion in the solution bonds via (H_2O) groups to terminations on the surface; both types of complex change the net proton charge of the surface. We may simplify this issue from a bond-valence perspective by factoring inner- and outer-sphere complexes into two components: (1) surface ions and (2) aqueous complexes in solution, i.e., we treat inner- and outer-sphere complexes as part of the aqueous solution, considering only the difference in interaction between edges with different net proton charge and the aqueous solution. The net proton charge may be considered as the difference between the incident and exident bond-valences between the terminations at the surface and the species in aqueous solution. A termination that accepts bonds is a *Lewis base* and a termination that donates bonds is a *Lewis acid* [31]. At zero net proton charge, the net strength of the Lewis bases and Lewis acids is zero. Where a surface has zero net proton charge, the pH of the coexisting solution is called the *point of zero net proton charge*, which we will designate as the *point of zero charge*. Weak Lewis bases and Lewis acids occur on a surface at the point of zero charge, and (depending on the intrinsic acidity constant of the acid–base reaction), strong Lewis bases and Lewis acids occur at pH values that differ considerably from the point of zero charge. Thus we may define the point of zero charge of a surface from a bond-valence perspective: *At the point of zero charge of a surface, there is a minimum in the number of strong Lewis acids and Lewis bases* (i.e., *highly charged terminations) on*

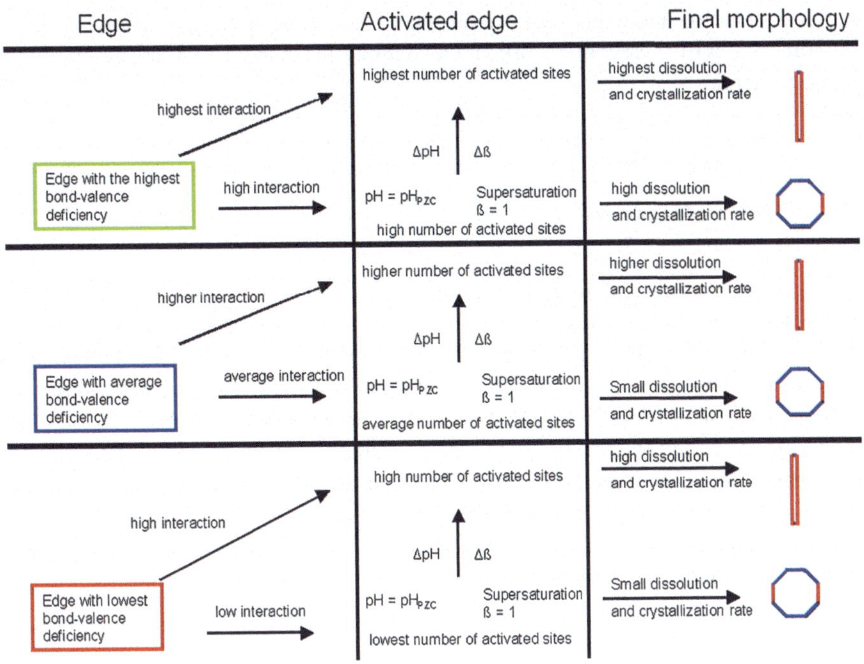

Fig. 4 Summary of the relations between an edge, the corresponding activated edge and the type of crystallization or dissolution. The left column gives the possible types of edges with low, average, and high residual valence, respectively. The central column indicates schematically the increase in activated sites with (**a**) the initial residual valence of an edge, (**b**) ΔpH, the difference between the pH of the solution and the pH_{pzc} of the edge, and (**c**) $\Delta\beta$, the difference between the supersaturation, β, of a solution and the supersaturation at equilibrium (where $\beta = 1$). Increasing ΔpH and $\Delta\beta$ is indicated by an *arrow*. The right column lists the corresponding rates of crystal growth and dissolution, and the final morphologies of a theoretical (0 0 1) face; from [31]

the surface, which results in low bond-valence transfer between surface acceptors and donators, and aqueous species. Thus the lowest interaction between a face and the ambient aqueous solution occurs where the pH of the solution is equal to the point of zero charge of that face, and hence the crystal has very low growth and dissolution perpendicular to that face. The possible types of edges, ambient conditions, and growth rates of faces of different type are summarized in Fig. 4.

5 Application to Minerals

Schindler and Hawthorne [34] and Hawthorne and Schindler [3] showed that many structural and chemical aspects of the uranyl-oxysalt minerals may be understood in terms of bond-valence theory, and Schindler et al. [31, 35] applied the ideas outlined above to understand aspects of the surface morphology of uranyl-oxysalt minerals. We will review this work and also apply these ideas to hydroxy-hydrated borate minerals.

5.1 *Uranyl Minerals*

The crystal structures of uranyl minerals are dominated by the structure of the complex uranyl cation: $(UO_2)^{2+}$. The central U^{6+} is coordinated by two O^{2-} anions that form an approximately linear O^{2-}–U^{6+}–O^{2-} group with U^{6+}–O^{2-} bond-valences of ~1.6–2.0 v.u., and the coordination of this complex cation is completed by four to six equatorial anions which receive bond valences of ~0.30–0.70 v.u. The bond-valence requirements of the equatorial anions are most easily satisfied by polymerization of the uranyl polyhedra in the plane of the equatorial anions, φ (=O, OH), and hence the structures of these minerals are dominated by sheets of edge-sharing uranyl polyhedra that are linked in the third dimension by weak alkali-metal–oxygen bonds and hydrogen bonds.

5.1.1 Polyhedron Chains

The $[(UO_2)_8O_2(OH)_{12}]$ sheet is a characteristic structural unit in uranyl minerals and occurs in the crystal structures of schoepite, $[(UO_2)_8O_2(OH)_{12}](H_2O)_{12}$ [36, 37], and fourmarierite, $Pb_2^{2+}[(UO_2)_8O_6(OH)_8](H_2O)_8$ [38]. The sheets are linked by weak Pb–O and/or hydrogen bonds, and the former mineral is of interest as the primary phase formed by alteration of nuclear-fuel rods by aqueous solution. A view of the $[(UO_2)_8O_2(OH)_{12}]$ sheet is shown in Fig. 5, together with the linear polyhedron chains parallel to [1 0 0], [0 1 0], [1 2 0], [2 1 0], and [1 1 0] in the plane of the sheet. The apical uranyl O-atoms project outward from the (0 0 1) surface and determine the reactivity of the (0 0 1) surface itself. These O-atoms receive an average of 1.6–1.7 v.u. from the central U^{6+} cation and can accept only bonds of less than 0.3–0.4 v.u. Hence the uranyl O-atoms can accept only hydrogen bonds and weak bonds from alkali or alkaline-earth cations; they cannot be protonated and are not involved in acid–base reactions at the surface. The situation is very different for surface features such as terraces or steps, as here the equatorial O-atoms in the polyhedron sheet are exposed at the surface. Within the sheet, these equatorial O-atoms bond to two or three U^{6+}-atoms (Fig. 5) and their incident individual bond-valences vary between approximately 0.2 and 0.8 v.u. Where exposed at terraces or edges, these equatorial O-atoms deviate significantly from the valence-sum rule and can protonate and deprotonate in acid–base reactions with the adjacent aqueous solution. Thus edges and terraces are much more reactive than the basal surface, as is the case in dissolution of phyllosilicates which is controlled by acid–base reactions on the corresponding edges (e.g., [39]).

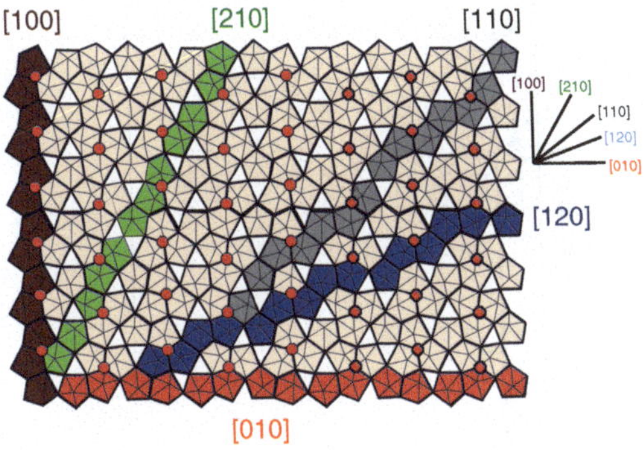

Fig. 5 Polyhedron representation of the uranyl-oxide hydroxy-hydrate sheet in schoepite, $[(UO_2)_8O_2(OH)_{12}](H_2O)_{12}$, showing polyhedron chains parallel to [1 0 0], [0 1 0], [1 2 0], [1 1 0] and [2 1 0], respectively; equatorial O^{2-} anions of the uranyl-polyhedra are shown as *orange/gray circles*, equatorial edges are shown as *heavy black lines*; from [31]

5.1.2 Calculation of pK_a and Lewis Basicity for Different U–O Anion-Terminations

Average equatorial U–φ bond-valences for [6]-, [7]- and [8]-coordinated U^{6+} are 0.64, 0.54, and 0.45 v.u., respectively ([25], calculated with the parameters of [40]). Individual equatorial $^{[a]}$U–φ bond-valences vary over a larger range; for example, the $^{[7]}$U–φ bond-valences in schoepite vary between 0.27 and 0.73 v.u. [36], giving rise to a range of intrinsic acidity constants for one type of anion termination. The types of anion termination in uranyl minerals are limited by the occurrence of [6]-, [7]- and [8]-coordinated U^{6+}: e.g., [6]- and [8]-coordinated U^{6+} never occur together, and always occur with [7]-coordinated U^{6+}. The conformation of an anion termination can be denoted by the symbol $^{[a]}$U–φ–$n^{[b]}$U, where φ is an anion that bonds to one U atom in [*a*]-coordination and *n* U atoms in [*b*]-coordination. In uranyl-oxide minerals, the anion terminations that can occur on edges are as follows: $^{[8]}$U–φ, $^{[7]}$U–φ, $^{[6]}$U–φ, $^{[8]}$U–φ–$^{[8]}$U, $^{[8]}$U–φ–$^{[7]}$U, $^{[7]}$U–φ–$^{[7]}$U, $^{[7]}$U–φ–$^{[6]}$U, $^{[8]}$U–φ–2$^{[7]}$U, $^{[7]}$U–φ–2$^{[7]}$U, and $^{[6]}$U–φ–2$^{[7]}$U (Table 1). The Lewis base-strength of an anion termination is the bond valence required by the termination (to satisfy the valence-sum rule) divided by the number of bonds accepted by the anions of that termination. Thus for the anion-termination $^{[7]}$U–(OH), the bond-valence required is $2 - (0.54 + 0.80) = 0.66$ v.u., the number of bonds required by the anion (assuming a coordination number of [4]) is $4 - 2 = 2$, and the resultant Lewis basicity is $0.66/2 = 0.33$ v.u. For the anion termination $^{[a]}$U–(H_2O), the constituent O-atom has an incident bond-valence sum greater than 2 v.u. and is actually a Lewis acid rather than a Lewis base. The (H_2O) group transforms the bond-valence (*v* v.u.) of the $^{[a]}$U–O bond into two weaker hydrogen bonds of bond-valence *v*/2 [41, 42] and the Lewis acidity of the termination $^{[7]}$U–(H_2O) is $0.54/2 = 0.27$ v.u. Lewis basicities and Lewis acidities of the above anion terminations are given in Table 1.

Table 1 Intrinsic acidity constants, Lewis acidities, and Lewis basicities of equatorial anions terminating the edges on uranyl-sheet minerals

Code	Acids and bases of the anion-termination	pK_{a1}	pK_{a2}	pK_{a3}	Lewis acidity (v.u.)	Lewis basicity (v.u.)
$^{[8]}$U–φ	$^{[8]}$U–OH$_3$ ↔ $^{[8]}$U–OH$_2$^{a} ↔ $^{[8]}$U–OH^{b} ↔ $^{[8]}$U–O^{c}	−5	6.9	18.8	+0.225^{a}	−0.375^{b}, −0.52^{c}
$^{[7]}$U–φ	$^{[7]}$U–OH$_3$ ↔ $^{[7]}$U–OH$_2$^{a} ↔ $^{[7]}$U–OH^{b} ↔ $^{[7]}$U–O^{c}	−6.7	5.1	17	+0.27^{a}	−0.33^{b}, −0.49^{c}
$^{[6]}$U–φ	$^{[6]}$U–OH$_3$ ↔ $^{[6]}$U–OH$_2$^{a} ↔ $^{[6]}$U–OH^{b} ↔ $^{[6]}$U–O^{c}	−8.7	3.1	15	+0.32^{a}	−0.27^{b}, −0.45^{c}
$^{[8]}$U–φ–$^{[8]}$U	$^{[8]}$U–OH$_2$–$^{[8]}$U ↔ $^{[8]}$U–OH–$^{[8]}$U^{a} ↔ $^{[8]}$U–O–$^{[8]}$U^{b}	2	13.8			−0.30^{a}, −0.55^{b}
$^{[8]}$U–φ–$^{[7]}$U	$^{8]}$U–OH$_2$–$^{[7]}$U ↔ $^{[8]}$U–OH–$^{[7]}$U^{a} ↔ $^{[8]}$U–O–$^{[7]}$U^{b}	0.2	12			−0.21^{a}, −0.51^{b}
$^{[7]}$U–φ–$^{[7]}$U	$^{[7]}$U–OH$_2$–$^{[7]}$U ↔ $^{[7]}$U–OH–$^{[7]}$U^{a} ↔ $^{[7]}$U–O–$^{[7]}$U^{b}	−1.6	10.3			−0.12^{a}, −0.46^{b}
$^{[7]}$U–φ–$^{[6]}$U	$^{[7]}$U–OH$_2$–$^{[6]}$U ↔ $^{[7]}$U–OH–$^{[6]}$U^{a} ↔ $^{[7]}$U–O–$^{[6]}$U^{b}	−3.5	8.3			−0.02^{a}, −0.41^{b}
$^{[8]}$U–φ–2$^{[7]}$U	$^{[8]}$U–OH–2$^{[7]}$U ↔ $^{[8]}$U–O–2$^{[7]}$U^{a}	5.3				−0.47^{a}
$^{[7]}$U–φ–2$^{[7]}$U	$^{[7]}$U–OH–2$^{[7]}$U ↔ $^{[7]}$U–O–2$^{[7]}$U^{a}	3.5				−0.38^{a}
$^{[6]}$U–φ–2$^{[7]}$U	$^{[6]}$U–OH–2$^{[7]}$U ↔ $^{[6]}$U–O–2$^{[7]}$U^{a}	2.5				−0.28^{a}

The superscripts a, b, and c denote the Lewis acidities and Lewis basicities and their corresponding terminations

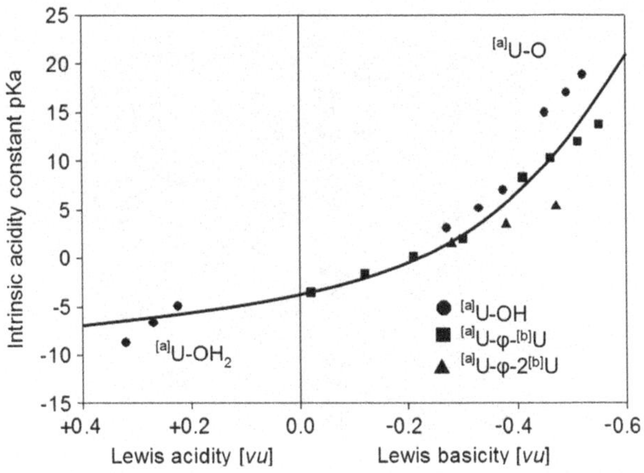

Fig. 6 Lewis basicity as a function of intrinsic acidity constant, pK_a, of anion terminations on the edges of uranyl sheets; from [31]

On the (0 0 1) face of schoepite, there are three different equatorial-anion terminations: $^{[7]}$U–(OH)–2$^{[7]}$U, $^{[7]}$U–(OH)–$^{[7]}$U, and $^{[7]}$U–O–$^{[7]}$U. The characteristic bond-valence for the equatorial $^{[7]}$U–φ bond in schoepite is 0.47 v.u., and as discussed above, we use a coordination number of [4] for O at the surface of the mineral. The acid–base reactions and pK_a values are

[1]. $^{[7]}\mathrm{U}-(\mathrm{OH})-2^{[7]}\mathrm{U} \leftrightarrow {}^{[7]}\mathrm{U}-\mathrm{O}-2^{[7]}\mathrm{U}+\mathrm{H}^+$ p$K_a = 7.7$
[2]. $^{[7]}\mathrm{U}-(\mathrm{OH})-{}^{[7]}\mathrm{U} \leftrightarrow {}^{[7]}\mathrm{U}-\mathrm{O}-{}^{[7]}\mathrm{U}+\mathrm{H}^+$ p$K_a = 13.1$
[3]. $^{[7]}\mathrm{U}-(\mathrm{OH_2})-{}^{[7]}\mathrm{U} \leftrightarrow {}^{[7]}\mathrm{U}-(\mathrm{OH})-{}^{[7]}\mathrm{U}+\mathrm{H}^+$ p$K_a = 1.2$

The pK_a value is calculated using the mean incident bond-valence at O in the anion termination on the LHS of each reaction. Thus in [1], the O atom receives 0.47 × 3 + 0.20 (from a hydrogen bond) which equals 1.61 v.u. Using this value in Eq. (1) gives p$K_a = -19.8(1.61 - 2) = 7.7$. The values in [2] and [3] were calculated in the same way. It should be noted that the calculated pK_a-value of 7.7 for Eq. (1) is in good agreement with an experimentally determined pK_a-value of ~7, which was extrapolated on the basis of titrations of dehydrated schoepite in NaCl solutions of different concentration [31]. The correlation between pK_a and the Lewis basicities and Lewis acidities of the terminations of Table 1 is shown in Fig. 6. These values were calculated with characteristic bond-valences for $^{[6]}$U–O, $^{[7]}$U–O, and $^{[8]}$U–O given in [25]. The values in Table 1 indicate the correlation between the acid- and base-strengths of an anion termination and the magnitude of the residual valence on its anion terminations. Exact pK_a-values for anion-terminations on the basal surfaces of any uranyl mineral must be calculated using the average U–O bond-valence in the corresponding structure (as shown above for the basal surface of schoepite).

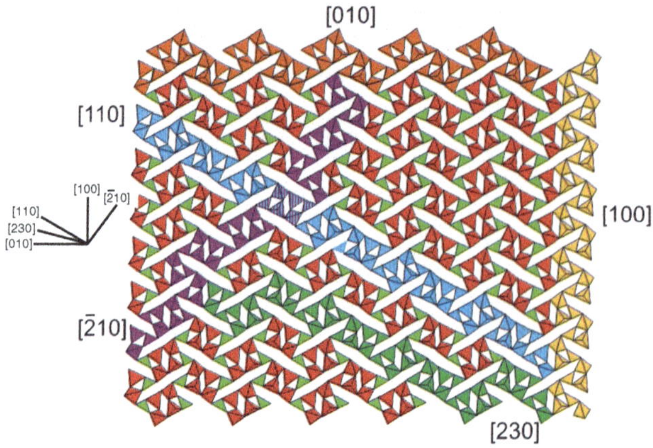

Fig. 7 Polyhedron representation of the borate sheet in fabianite, $Ca_2[B_6O_{10}(OH)_2]$; polyhedron chains are shown parallel to [1 0 0], [0 1 0], [1 1 0], [1 2 0], and [2 3 0], respectively, the directions of which are indicated by the *labeled lines*

5.2 Borate Minerals

The crystal structures of borate minerals are dominated by the presence of both $(B\varphi_3)$ triangles and $(B\varphi_4)$ tetrahedra with B–O bond-valences of ~1.00 and 0.75 v.u., respectively. The valence-sum rule allows polymerization of these two groups and they form structures with isolated polyhedra, clusters, chains, sheets, and frameworks of $(B\varphi_3)$ triangles and $(B\varphi_4)$ tetrahedral [18, 43, 44].

5.2.1 Polyhedron Chains

The crystal structure of fabianite, $Ca_2[B_6O_{10}(OH)_2]$ [45], consists of sheets of composition $[B_6O_{10}(OH)_2]$, parallel to (0 0 1) and linked in the third dimension by Ca. Two $(B\varphi_4)$ tetrahedra each share a corner with a $(B\varphi_3)$ triangle to form a three-membered ring that is very common in borate structures. Two rings share two corners between $(B\varphi_4)$ tetrahedra to form a $[B_6\varphi_{14}]$ group, and these groups share corners to form a $[B_6\varphi_{12}]$ sheet. Figure 7 shows a view of this sheet, together with the polyhedron chains parallel to [1 0 0], [0 1 0], [1 1 0], [2 1 0], and [2 3 0] in the plane of the sheet. The situation with regard to anion terminations is more complicated than in the case of uranyl-oxide minerals. Figure 8 shows the [0 1 0] polyhedron chain on the (0 0 1) surface (cf. Fig. 3) with the incident bond-valence required by the surface anions for adherence to the valence-sum rule. Most anions require an incident hydrogen bond for adherence to the valence-sum rule, and only a small number of surface anions are (OH) groups (and hence available for (de-) protonation reactions with aqueous species in solution). Figure 9 shows the edge of a [0 1 0] polyhedron chain on the (0 0 1) surface (cf. Fig. 3), again with the

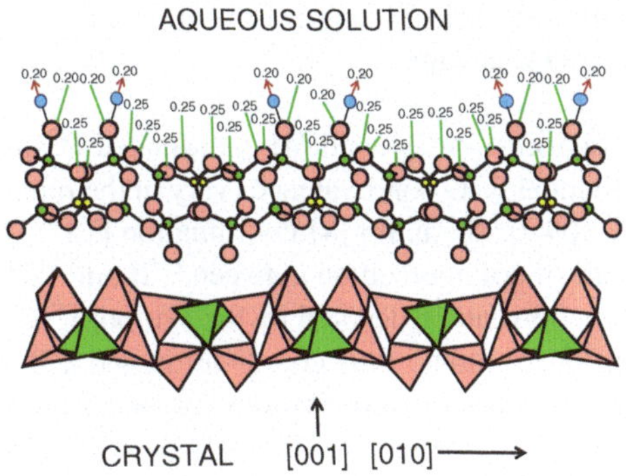

Fig. 8 Polyhedron and ball-and-stick representations of the [0 1 0] polyhedron chain exposed at the (0 0 1) surface in fabianite. The bonds (and corresponding bond-valences required) involving surface anions and constituents of the aqueous solution are shown as *lines* [bond valence donated from the (aquated) cations of the aqueous solution] and as *arrows* (bond valence donated from H atoms of protonated surface anions to anions of the aqueous solution). Note that there are not many protonated anions that can participate in redox reactions with the aqueous solution

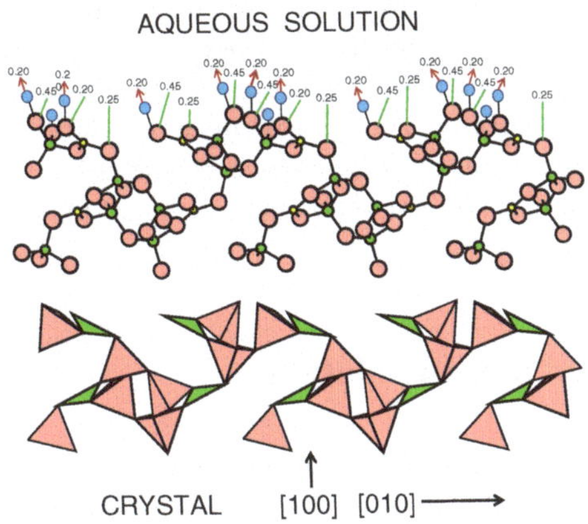

Fig. 9 Polyhedron and ball-and-stick representations of the [0 1 0] polyhedron chain exposed at a (1 0 0) edge on the (0 0 1) surface in fabianite. The bonds (and corresponding bond-valence required) involving surface anions and constituents of the aqueous solution are shown as *lines* (bond valence donated from the aqueous solution) and as *arrows* (bond valence donated from H atoms of protonated surface anions). Note that most of the anions exposed on the edge are protonated and hence can participate in redox reactions with the aqueous solution

incident bond-valence required by the surface anions for adherence to the valence-sum rule. Many anions are now protonated for adherence to the valence-sum rule and hence are available for (de-)protonation reactions with aqueous species in solution. Thus edges (and terraces) exposed on the surface of fabianite are far more reactive, i.e., can act as a nucleus for crystallization or dissolution, than the surface itself.

5.2.2 Lewis Basicity and Acid Strength for Different B–O Anion-Terminations

Average B–φ bond-valences for [3]- and [4]-coordinated B are 1.00 and 0.75 v.u., respectively, and individual B–O bond-valences vary in the ranges 0.86–1.19 v.u. for [3]-coordination and 0.69–0.95 v.u. for [4]-coordination [18]. The valence-sum rule allows all combinations of polymerization between $^{[3]}$B and $^{[4]}$B, and the conformation of an anion termination can be denoted by the symbol $^{[a]}$B–φ–$n^{[b]}$B, where φ is an anion that bonds to one B atom in [*a*]-coordination and *n* B atoms in [*b*]-coordination. In borate minerals, the anion terminations can occur on edges are as follows: $^{[3]}$B–φ, $^{[4]}$B–φ, $^{[3]}$B–φ–$^{[3]}$B, $^{[3]}$B–φ–$^{[4]}$B, $^{[4]}$B–φ–$^{[4]}$B, and $^{[4]}$B–φ–2$^{[4]}$B.

On the [0 1 0] chain on the (0 0 1) face of fabianite, there are three different anion terminations: $^{[4]}$B–φ, $^{[3]}$B–φ–$^{[4]}$B, and $^{[4]}$B–φ–$^{[4]}$B. On the edge of the [0 1 0] chain on the (0 0 1) face of fabianite, there are three different anion terminations on the (0 1 0) chain: $^{[3]}$B–φ, $^{[4]}$B–φ, and $^{[3]}$B–φ–$^{[4]}$B that are involved in acid–base reactions. Oxygen-atoms on the $^{[3]}$B–φ, $^{[4]}$B–φ terminations are highly undersaturated and are most likely protonated ($^{[3]}$B–OH, $^{[3]}$B–OH_2, $^{[4]}$B–OH, $^{[4]}$B–OH_2) in natural waters with pH <12. There is also the possibility that acid–base reactions on the surface of borate minerals involve a change in coordination number of these types of terminations as is observed in aqueous solution: $B(OH)_3{}^0 + H_2O \leftrightarrow B(OH)_4{}^- + H^+$, $pK_a = 9.15$. However, changes in coordination number may occur only at kink sites in polyhedron chains, where there is a higher degree of freedom to allow structural change associated with a transformation of a triangle into a tetrahedron or vice versa (i.e., breaking of bonds and rotation of the polyhedron).

6 Crystallization and Dissolution

For minerals crystallizing from low-temperature aqueous solutions, the primary controls on their stability should be (a) the activity of the species in solution and (b) protonation reactions between solid and solution at the edges of polyhedron chains: deprotonation of edge anions promotes attachment of aqueous cation species (i.e., crystallization), whereas protonation of edge anions weakens their bonds to the bulk structure and promotes dissolution. With regard to crystallization, the character and activity of the aqueous species is of interest as these provide groups of atoms that may attach to the solid during crystallization.

6.1 *Uranyl Minerals*

We have seen that the uranyl cation, $(U^{6+}O_2)^{2+}$, occurs as monomers, dimers, and trimers in aqueous solution. We also discussed that well-crystallized schoepite with

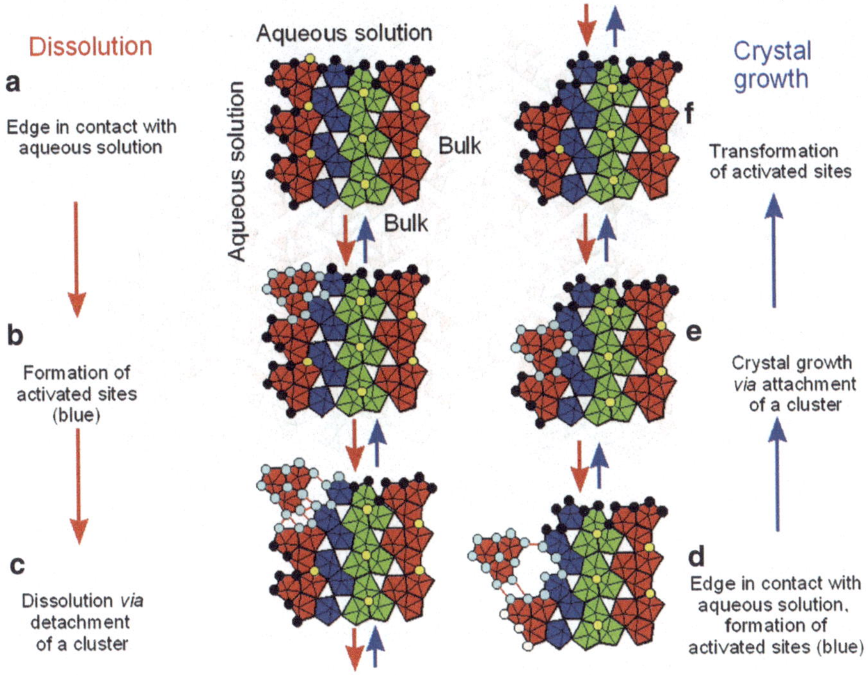

Fig. 10 A schematic of dissolution and growth mechanisms at an edge of a schoepite sheet. The sequence of dissolution is indicated with *downward-directed arrows* and the sequence of crystal growth by *upward-directed arrows*. The schoepite sheet is built of clusters of three (*red/dark-gray* and *green/pale-gray*) and two (*blue/dark-gray*) pentagonal bipyramids which are structurally identical to the principal aqueous species $[(UO_2)_3(OH)_5(H_2O)_5]^+$ and $[(UO_2)_2(OH)_2(H_2O)_6]^{2+}$ in weak acidic solutions (Fig. 4); O^{2-} ligands in the sheet are indicated as *circles*, and ligands which have interacted with the aqueous solutions are indicated in *light blue/white*; (**a**, **f**) activated sites occur only on anion terminations, and activated sites in the layer are transformed to sites; (**b**, **e**) formation of activated sites during dissolution via detachment of a cluster; ligands adjacent to potential detached clusters interact with the solution and are highlighted as *light-blue/white circles*; attachment of a cluster at a kink site occurs via release of one (H_2O) per common ligand between cluster and kink site; (**c**, **d**) breaking (formation) of the U–φ–U bonds and detachment (or approach) of a cluster from (to) an activated kink site on a schoepite layer; from [31]

sheets composed of edge-sharing trimers and dimers of $((U^{6+}O_2)^{2+}\varphi_5)$ polyhedra can be synthesized in a pH range where the trimer $(UO_2)_3(OH_5)^+$ predominates (e.g., [17, 46]). In this regard, Schindler et al. [31] proposed a model for dissolution and crystallization of schoepite that involves the attachment and detachment of trimers, as illustrated in Fig. 10.

Dissolution requires breaking of U–φ–U bonds and subsequent detachment of clusters. The U–φ bonds holding the cluster to the crystal (Fig. 10a) weaken through interaction between the ligands and the adjacent aqueous solution (Fig. 10b), producing what Schindler et al. [31] called an *activated site*, where bonds between ions at activated sites and aqueous species catalyze dissolution or crystal growth at

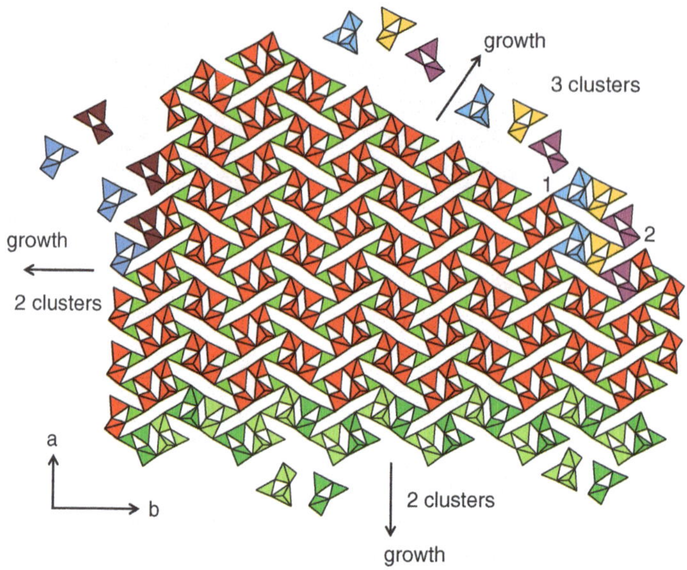

Fig. 11 Schematic of growth at the edges of the (0 0 1) face of fabianite. Growth or dissolution may proceed by attachment or removal of the $[B_3O_3(OH)_5]^{2-}$ cluster that is a major species in aqueous borate solutions

an edge. Here, the U–φ bonds break and the cluster detaches from the crystal, is protonated, and forms an aqueous species (Fig. 10c). Thus an activated site participating in dissolution involves the terminations around a polyhedron cluster where protonation or strong bonds between ligands and aqueous species result in weakening of U–φ bonds.

Crystallization occurs via attachment of clusters at activated sites at anion terminations on the edge of polyhedron chains. Attachment produces one additional (H_2O) or (OH) group per common corner between cluster and anion termination (Figs. 10d,e). Thus at an activated site involved in crystal growth, there are strong hydrogen bonds from anion terminations to a polyhedron cluster in solution. The other anions of the cluster and the (former) activated site stay activated until the anions do not require additional bond-valence from aqueous species (Fig. 10f).

6.2 *Borate Minerals*

As mentioned above, the species in aqueous borate solutions can also be seen embedded in all borate structures, suggesting that crystallization proceeds by condensation of these species to form the solid. This process is illustrated for the structure of fabianite in Fig. 11. The cluster $[B_3O_3(OH)_5]^{2-}$, which shows maximum stability in aqueous solution at a pH of ~9.4 (Fig. 1), consists of a ring of one

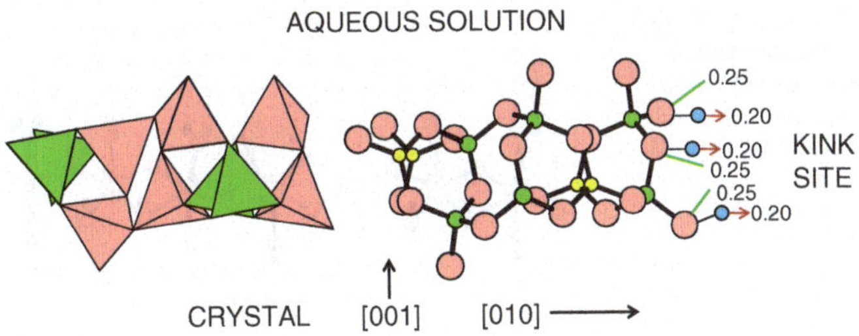

Fig. 12 Polyhedron and ball-and-stick representations of the [1 0 0] polyhedron chain exposed at the (0 0 1) surface in fabianite (as in Fig. 8) but with the polyhedron chain terminated by a kink site. The bonds (and corresponding bond-valence required) involving kink-site anions and constituents of the aqueous solution are shown by *lines* (bond valence donated from the aqueous solution) and *arrows* (bond valence donated from H atoms of protonated surface anions). Note that there are more protonated anions (that can participate in redox reactions with the aqueous solution) at the kink site that there are exposed to the aqueous solution on the (0 0 1) surface (Fig. 7)

$(B\varphi_3)$ group and two $(B\varphi_4)$ groups that is the fundamental building block of the fabianite structure. As indicated in Fig. 11, growth can proceed in any direction on the surface by condensation of $[B_3O_3(OH)_5]^{2-}$ groups in various combinations.

7 Quantitative Aspects of Crystallization and Dissolution

So far, we have put a qualitative mechanism in place for crystallization and dissolution. Now we will make this mechanism more quantitative.

7.1 *Residual Valence, Kink Sites, and O^{2-} Ligands*

Above, we have considered edges (and terraces) in terms of continuous polyhedron chains. However, during growth, such polyhedron chains will not be continuous. A polyhedron chain forming an edge will terminate at one (or more) place where the atoms of the chain have not yet attached themselves, forming a kink site (Fig. 3). At a kink site, the edge anions are protonated and form an activated site (Fig. 12). Hence the residual valence of an edge increases with its number of kink sites. A larger number of kink sites on an edge (i.e., a high residual valence) favors attachment of polyhedra from the aqueous solution because an attached polyhedron can share more common ligands with the polyhedra of the existing structure than it can on an edge with less kink sites. Hence one expects a correlation between the residual valence of an edge and the growth rate of that edge. Moreover, the number of activated sites on an edge during dissolution or crystal growth should correlate with the difference between the pH of the solution and the point of zero charge of the edge.

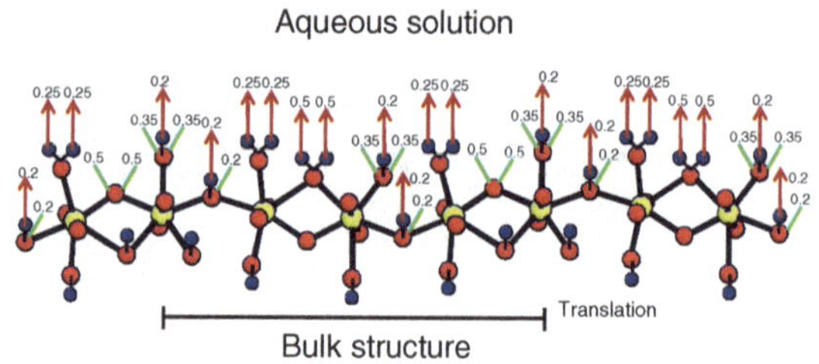

Fig. 13 Ball-and-stick model of a possible activated edge on (0 0 1) parallel to [0 1 0] in schoepite; legend as in Fig. 8; modified from [31]

7.2 *Uranyl Minerals*

7.2.1 Activated Sites and Edges in Schoepite

Consider a polyhedron chain in schoepite parallel to the [0 1 0] edge (Fig. 5). In the repeat period of the chain, there are one $^{[7]}$U–O–$^{[7]}$U, one $^{[7]}$U–(H_2O)–$^{[7]}$U, two $^{[7]}$U–(OH), two $^{[7]}$U–(H_2O), and two $^{[7]}$U–(OH)–$^{[7]}$U terminations that interact with the aqueous solution (Fig. 13). For arithmetic simplicity, we assign an average bond-valence of 0.50 v.u. to a $^{[7]}$U–φ bond (rather than the grand mean value of 0.47 v.u., see above). The O-atom of a $^{[7]}$U–(H_2O) group receives 0.50 v.u. from $^{[7]}$U and requires an additional 2 × 0.75 v.u. from the two H atoms of the (H_2O) group in order to satisfy the valence-sum rule. Accordingly, each H-atom forms a hydrogen bond of 0.25 v.u. with the aqueous species. The O-atom of a $^{[7]}$U–(OH) group receives 0.50 v.u. from $^{[7]}$U and 0.80 v.u. from the H atom of the (OH) group (which in turn forms a hydrogen bond of 0.20 v.u. with the aqueous species), and the O atom requires a further 0.70 v.u. from cations in solution. The central O–atom of an $^{[7]}$U–(OH)–$^{[7]}$U group receives 2 × 0.50 v.u. from two $^{[7]}$U–O bonds and 0.80 v.u. from the H-atom of the (OH) group and requires an additional 0.20 v.u. from a bond (or bonds) from an aqueous species. The central O-atom of an $^{[7]}$U–O–$^{[7]}$U group requires an additional 1.0 v.u. from bonds involving the aqueous species. The O-atom of a $^{[7]}$U–(H_2O)–$^{[7]}$U group accepts 2 × 0.50 v.u. from two $^{[7]}$U–O bonds and requires an additional 2 × 0.50 v.u. from two O–H bonds; therefore, the $^{[7]}$U–(H_2O) group donates two hydrogen bonds with bond-valence of 0.50 v.u. The acid–base equilibria between the Lewis acids and bases are listed in Table 1. The pK_a values indicate that strong Lewis bases (such as $^{[7]}$U–O–$^{[7]}$U) and Lewis acids (such as $^{[7]}$U–(H_2O)–$^{[7]}$U) occur only at high and low pH, respectively: the number of strong Lewis bases and acids on the [0 1 0] edge is very small in weak acidic, neutral, and weak basic solutions [31].

7.2.2 Left Terminations and Right Terminations

The surface of a crystal grows primarily by attaching species (usually polyhedra) to the dominant edges on that surface. The orientations of these edges are controlled by the orientation of the strongly bonded polyhedron chains on that surface. There are two important factors to be considered here: (1) polyhedron chains that

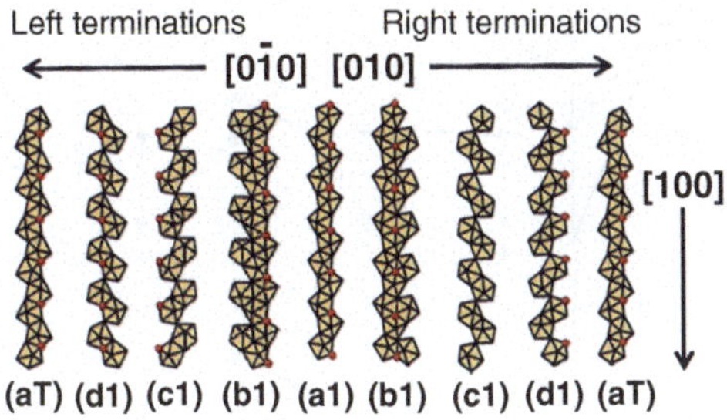

Fig. 14 The left and right chain terminations for different conformations of the polyhedron chain parallel to the [1 0 0] edge in the uranyl sheet of schoepite; the positions of O^{2-} anions are indicated by *circles*. For left terminations, the bulk structure continues to the right, and the surface occurs to the left, and vice versa

are not bilaterally symmetric will expose different atomic arrangements on edges facing in opposite directions, e.g., the (0 1 0) edge of a [1 0 0] chain (as shown in Fig. 14) and the (0 $\bar{1}$ 0) edge of a [1 0 0] chain; thus, the chain shown in the center of Fig. 14 (labeled a1) has different atomic arrangements on the left and right sides (referred to relative to its length), and these we designate as *left terminations* and *right terminations*. (2) As an edge grows by accretion of polyhedra, the atomic arrangement at that edge changes to give a series of arrangements, until the original arrangement occurs again after the accretion of one unit-cell in the direction of growth. When considering the conformation of an edge exposed to an aqueous solution, a series of atomic arrangements must therefore be considered.

This is illustrated in Fig. 14 for chains parallel to [1 0 0] in schoepite. These different terminations are exposed to the left or right of the length of the chain. We designate the different parallel chains as a1, b1, etc. (Fig. 14). There are two $^{[7]}$U–(OH) terminations and four $^{[7]}$U–(OH)–$^{[7]}$U terminations in chain a1. For $^{[7]}U^{6+}$–O = 0.47 and O–H = 0.80 v.u., the O atoms involved in the two $^{[7]}$U–(OH) terminations receive $(2 \times 1 \times 0.47 + 2 \times 0.8) = 2.54$ v.u., and the O atoms involved in the four $^{[7]}$U–(OH)–$^{[7]}$U terminations receive $(4 \times 2 \times 0.47 + 4 \times 0.8) = 6.96$ v.u. The aggregate residual valence of the O atoms involved in the terminations along this chain is $4 - 2.54 + 8 - 6.96 = 2.50$ v.u. Obviously the residual valence along a polyhedron chain is a function of the number of highly charged terminations (e.g., U–(OH): −0.5; U–O–U: −1.0; U–O: −1.5 v.u.).

Figure 15 shows the normalized residual valence of the left- and right-terminations of the chains parallel to [1 0 0], [2 1 0], [1 1 0], [1 2 0] and [0 1 0] in schoepite. Edges involving polyhedron chains with low normalized residual valence will grow slowly, whereas edges involving polyhedron chains with high normalized residual valence will grow rapidly (as high residual valence at anion terminations promotes acid–base reactions with the solution and corresponding crystallization or dissolution). The relative morphology of crystals of schoepite will be controlled in the (1 0 0) plane by the relative magnitudes of the residual valence of polyhedron chains of the form (*h k* 0) (i.e., parallel to the F face): sheets should elongate in the direction of chains with high residual valence and the dominant edges should be controlled by chains with low residual valence. Of the chains shown in Fig. 14, some show major

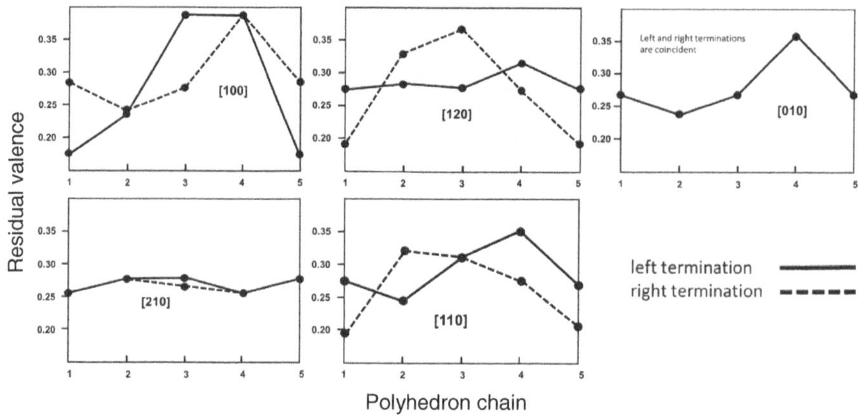

Fig. 15 The calculated residual valence per unit length (v.u./Å) of the polyhedron chains parallel to the [1 0 0], [0 1 0], [1 2 0], [1 1 0], and [2 1 0] edges on the (0 0 1) face for both left and right terminations; the numbers along the abscissa denote the different conformations of each chain, and the corresponding values of the residual valence per repeat are shown on the ordinate

differences in their normalized residual valence with the different types of termination (e.g., the [1 0 0] polyhedron chain), whereas others show little difference with the different types of termination (e.g., the [2 1 0] polyhedron chain). Let us consider the [1 0 0] polyhedron chain. Conformations with high normalized residual valence (e.g., right terminations a3 and a4, Fig. 14) will protonate and deprotonate rapidly in acid–base reactions with the adjacent aqueous solution and change quickly to other conformations. Other conformations with low normalized residual valence (e.g., right terminations a1, a2 and a5, Fig. 14) will protonate and deprotonate far more slowly and the growth rate of the polyhedron chain will be controlled by these conformations with low normalized residual valence. Thus from Fig. 15, we predict the following dominance of edges for the (0 0 1) face of schoepite: [1 0 0] > [1 1 0] ≈ [1 2 0] > [0 1 0] > [2 1 0].

Figure 16a shows the morphology of a crystal of synthetic schoepite grown on the (1 0 4) face of calcite [20, 31]. The edges [1 2 0], [1 0 0], [1 1 0], and [0 1 0] are present (Fig. 16a, right) and [1 2 0] is absent, as predicted above. Figure 16b shows a schoepite crystal from Katanga, Democratic Republic of Congo [47], with a prominent (0 0 1) face slightly elongate along [0 1 0]. The edges defining the (0 0 1) face have indices [1 0 0], [1 1 0], [1 2 0], and [0 1 0] (Fig. 15b, right).

7.3 *Borate Minerals*

Figures 17a,b show the crystal morphology of nobleite, $Ca[B_6O_9(OH)_2](H_2O)_3$ [48], and tunellite, $Sr[B_6O_9(OH)_2](H_2O)_3$ [49], which contain sheets of polymerized borate polyhedra (Fig. 17c). The morphology of these crystals is defined by

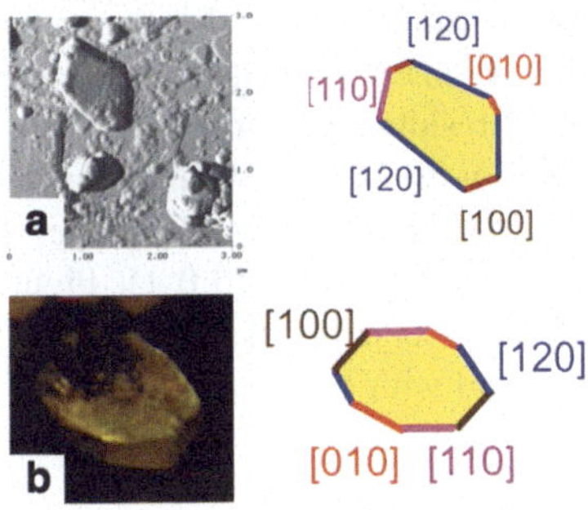

Fig. 16 (**a**) AFM image of a synthetic schoepite crystal grown in a weak acidic solution on the calcite (1 0 4) surface (from [20]), plus a sketch of the crystal which shows the [1 2 0], [1 1 0], [1 0 0], and [0 1 0] edges; (**b**) schoepite crystals (from [47]) plus a sketch of crystal showing a prominent (0 0 1) face that is outlined by the [1 2 0], [1 1 0], [1 0 0], and [0 1 0] edges

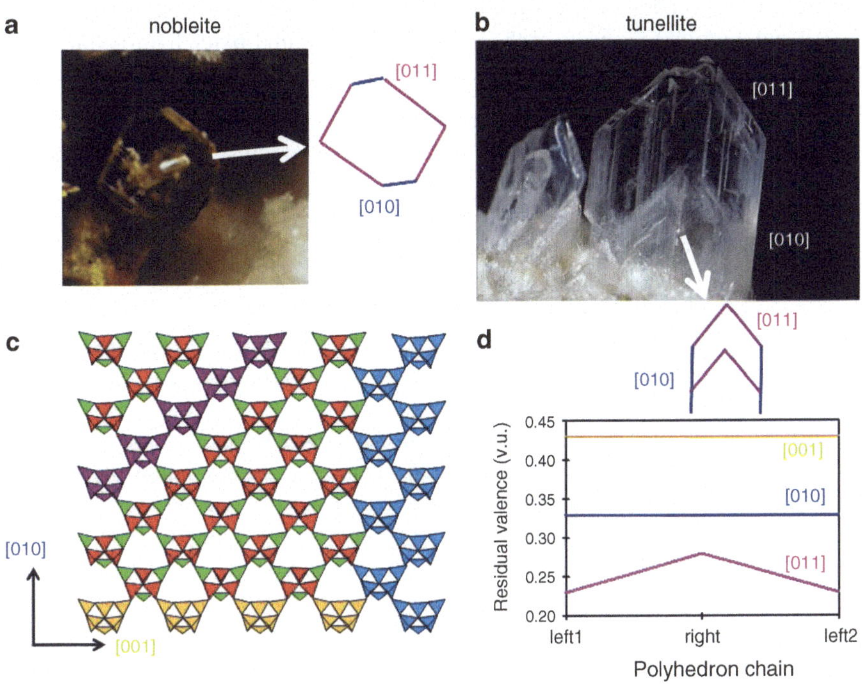

Fig. 17 (**a**, **b**) Photographs of nobleite and tunellite crystals (from [50]); (**c**) sheet of polymerized polyhedra in the structures of nobleite and tunellite; (**d**) sketch of the basal surface of the crystals of nobleite and tunellite and calculated residual valence of the polyhedron chains parallel to the [0 0 1], [0 1 0], and [0 1 1] edges

prominent basal F-faces parallel to the sheets of polymerized polyhedra. The morphology of the F-face itself is defined by edges which are parallel to polyhedron chains within the sheet of polyhedra. Figure 17d shows the normalized residual valence of the left- and right-terminations of the chains parallel to [1 0 0], [1 1 0] and [0 1 0]. Following the arguments given above (edges involving polyhedron chains with low normalized residual valence will grow slowly, whereas edges involving polyhedron chains with high normalized residual valence will grow rapidly), we predict the following dominance of edges for the (1 0 0) faces of nobleite and tunellite: [0 1 1] > [0 1 0] > [0 0 1]. Figure 17 shows sketches of the morphology of the F-face for crystals of nobleite and tunellite, indicating that their morphologies are in accord with our predictions.

8 Summary

Here, we have integrated many individual aspects of the growth and dissolution of minerals into a coherent description of these processes based on bond-valence theory and apply it to aspects of the morphology of uranyl-oxide hydroxyl hydrate and borate minerals. Below, we summarize the main aspects of this work:

1. In borate and uranyl-oxide hydroxyl-hydrate minerals, structural diversity occurs by polymerization of a small number of clusters (or fundamental building blocks).
2. These minerals crystallize from aqueous or hydrothermal solutions, and the FBBs occur as aqueous species in solution.
3. It seems reasonable that crystallization of these minerals occurs by condensation of the clusters in solution.
4. Periodic bond-chains are strongly related to the occurrence of faces on a crystal (and, in turn, the major growth directions of a crystal) [21–23]. We refer to periodic bond-chains as *polyhedron chains*.
5. Anions at the surface (i.e., exposed to an ambient aqueous solution) are called *terminations*, and the incident residual valence at a termination controls its reactivity (i.e., is the driving force for reaction with the aqueous solution).
6. The residual valence of a polyhedron chain controls the growth or dissolution rate at the crystal face associated with that chain and may be calculated as the net residual valence of the terminations along the repeat length of the polyhedron chain.
7. It is well known that growth and dissolution of individual faces on a crystal are controlled primarily by the occurrence of edges, terraces, and kink sites along those edges and terraces. The terminations associated with these features have much higher residual valence than the terminations on the surface expression of the polyhedron chains, making them much more susceptible to (de-) protonation reactions involving species in aqueous solution (i.e., crystallization or dissolution).
8. The bond-valence of an anion termination on a terminating polyhedron chain correlates with the intrinsic acidity constant, pK_a, and with the free energy, ΔG_{AT}, of the corresponding acid–base reaction.

9. Interaction between the ligands and the adjacent aqueous solution produces *activated sites*, and bonds between ions at activated sites and aqueous species catalyze dissolution or crystal growth at an edge.
10. Crystal growth and dissolution processes on an edge are catalyzed by the activated sites and increase with their number and the strength of the bonds between the corresponding anion terminations and the aqueous species.
11. From a bond-valence perspective, at the point of zero charge of a surface, there is a minimum in the number of strong Lewis acids and Lewis bases (i.e., highly charged terminations) on the surface, which results in low bond-valence transfer between surface acceptors and donators, and aqueous species. Thus the lowest interaction between a face and the ambient aqueous solution occurs where the pH of the solution is equal to the point of zero charge of that face, and hence the crystal has very low growth and dissolution perpendicular to that face.
12. We may calculate the pK_a and Lewis basicity for different anion-terminations, and from this calculate the aggregate residual valence along polyhedron chains. Edges involving polyhedron chains with low normalized residual valence will grow slowly, whereas edges involving polyhedron chains with high normalized residual valence will grow rapidly.
13. The relative morphology of prominent basal faces of crystals will be controlled by the relative magnitudes of the residual valence of polyhedron chains parallel to specific edges. Faces should elongate in the direction of chains with low residual valence and should not be defined by edges parallel to chains with high residual valence.

Acknowledgments FCH was supported by a Canada Research Chair in Crystallography and Mineralogy, and FCH and MS were supported by Discovery Grants from the Natural Sciences and Engineering Research Council of Canada.

References

1. Brown ID (2002) The chemical bond in inorganic chemistry: the bond-valence model. Oxford University Press, Oxford
2. Schindler M, Hawthorne FC (2001) A bond-valence approach to the structure, chemistry and paragenesis of hydroxy-hydrated oxysalt minerals: II. crystal structure and chemical composition of borate minerals. Can Mineral 5:1243–1256
3. Hawthorne FC, Schindler M (2008) Understanding the weakly bonded constituents in oxysalt minerals. Z Kristallogr 223:41–68
4. Hawthorne FC (2012) A bond-topological approach to theoretical mineralogy: crystal structure, chemical composition and chemical reactions. Phys Chem Mineral 39:841–874
5. Hawthorne FC (1979) The crystal structure of morinite. Can Mineral 17:93–102
6. Hawthorne FC (1983) Graphical enumeration of polyhedral clusters. Acta Crystallogr A39:724–736
7. Hawthorne FC (1985) Towards a structural classification of minerals: the $^{vi}M^{iv}T_2O_n$ minerals. Am Mineral 70:455–473

8. Ingri N (1963) Equilibrium studies of polyanions containing B^{III}, Si^{IV}, Ge^{IV} and V^{V}. Svensk Kem Tidskr 75:3–34
9. Janda R, Heller G (1979) Ramanspektroskopische. Untersuchungen an festen und in Wasser gelösten Polyboraten. Z Naturforsch 34b:585–590
10. Salentine CG (1983) High-field boron-11 NMR of alkali borates. Aqueous polyborate equilibria. Inorg Chem 22:3920–3924
11. Müller D, Grimmer AR, Timper U, Heller G, Shakibaie-Moghadam M (1993) ^{11}B-MAS-NMR untersuchungen zur anionenstruktur von boraten. Z Anorg Allg Chem 619:1262–1268
12. Maya L (1976) Identification of polyborate and fluoropolyborate ions in solution by Raman spectroscopy. Inorg Chem 15:2179–2184
13. Janda R, Heller G (1979) B–NMR-spektroskopische. Untersuchungen an wäßrigen Polyboratlösungen. Z Naturforsch 34b:1078–1083
14. Christ CL, Truesdell AH, Erd RC (1967) Borate mineral assemblages in the system Na_2O–CaO–MgO–B_2O_3–H_2O. Geochim Cosmochim Acta 31:313–337
15. Schindler M, Hawthorne FC (2001) A bond-valence approach to the structure, chemistry and paragenesis of hydroxy-hydrated oxysalt minerals: III. paragenesis of borate minerals. Can Mineral 5:1257–1274
16. Grenthe I, Fuger J, Konings RJM, Lemire RJ, Muller AB, Nguyen-Trung C, Wanner H (2004) Chemical thermodynamics of Uranium. Nuclear Energy Agency, OECD Nuclear Energy Agency, Data Bank Issy-les-Moulineaux, France
17. Tsushima S, Rossberg A, Ikeda A, Müller K, Scheinost AC (2007) Stoichiometry and structure of uranyl(VI) hydroxo dimer and trimer complexes in aqueous solution. Inorg Chem 46:10819–10826
18. Hawthorne FC, Burns PC, Grice JD (1996) The crystal chemistry of boron. Rev Mineral 33:41–115
19. Langmuir D (1978) Uranium solution-mineral equilibria at low temperatures with applications to sedimentary ore deposits. Geochim Cosmochim Acta 42:547–569
20. Schindler M, Putnis A (2004) Crystal growth of schoepite on the (104) surface of calcite. Can Mineral 42:1667–1681
21. Hartman P, Perdok WG (1955) On the relations between structure and morphology of crystals I. Acta Crystallogr 8:49–52
22. Hartman P, Perdok WG (1955) On the relations between structure and morphology of crystals II. Acta Crystallogr 8:521–524
23. Hartman P, Perdok WG (1955) On the relations between structure and morphology of crystals II. Acta Crystallogr 8:525–529
24. Hawthorne FC, Krivovichev SV, Burns PC (2000) The crystal chemistry of sulfate minerals. Rev Mineral Geochem 40:1–112
25. Burns PC (1999) The crystal chemistry of uranium. Rev Mineral 38:23–90
26. Krivovichev SV, Filatov SK (1999) Structural principles for minerals and inorganic compounds containing anion-centrered tetrahedra. Am Mineral 84:1099–1106
27. Hiemstra T, Venema O, Van Riemsdijk WH (1996) Intrinsic proton affinity of reactive surface groups of metal (hydr)oxides: the bond valence principle. J Colloid Interface Sci 184:680–692
28. Bickmore BR, Rosso KM, Nagy KL, Cygan RT, Tadanier CJ (2003) Ab initio determination of edge surface structures for dioctahedral 2:1 phyllosilicates: implications for acid–base reactivity. Clays Clay Mineral 51:359–371
29. Bickmore BR, Tadanier CJ, Rosso KM, Monn WD, Eggett DL (2004) Bond-valence methods for pKa prediction: critical reanalysis and a new approach. Geochim Cosmochim Acta 68:2025–2042
30. Schindler M, Hawthorne FC (2001) A bond-valence approach to the structure, chemistry and paragenesis of hydroxy-hydrated oxysalt minerals: I. theory. Can Mineral 5:1225–1242
31. Schindler M, Mutter A, Hawthorne FC, Putnis A (2004) Prediction of crystal morphology of complex uranyl-sheet minerals. I. Theory. Can Mineral 42:1629–1649

32. Faure G (1998) Principles and applications of geochemistry: a comprehensive textbook for geology students. Prentice Hall, Upper Saddle River
33. Stumm W (1992) Chemistry of the solid-water interface. Wiley, New York
34. Schindler M, Hawthorne FC (2004) A bond-valence approach to the uranyl-oxide hydroxy-hydrate minerals: chemical composition and occurrence. Can Mineral 42:1601–1627
35. Schindler M, Mutter A, Hawthorne FC, Putnis A (2004) Prediction of crystal morphology of complex uranyl-sheet minerals. II. Observation. Can Mineral 42:1651–1666
36. Finch RJ, Cooper MA, Hawthorne FC, Ewing RC (1996) The crystal structure of schoepite, $[(UO_2)_8O_2(OH)_{12}](H_2O)_{12}$. Can Mineral 34:1071–1088
37. Finch RJ, Hawthorne FC, Ewing RC (1998) Structural relations among schoepite, metaschoepite and "dehydrated schoepite". Can Mineral 36:831–845
38. Piret P (1985) Structure cristalline de la fourmariérite, $Pb(UO_2)_4O_3(OH)_4{\cdot}4H_2O$. Bull Minéral 108:659–665
39. Rufe E, Hochella M Jr (1999) Quantitative assesment of reactive surface area of phlogopite dissolution during acid dissolution. Sci 285:874–876
40. Burns PC, Ewing RC, Hawthorne FC (1997) The crystal chemistry of hexavalent uranium: polyhedron geometries, bond-valence parameters and polymerization of polyhedra. Can Mineral 35:1551–1570
41. Hawthorne FC (1992) The role of OH and H_2O in oxide and oxysalt minerals. Z Kristallogr 201:183–206
42. Hawthorne FC (1994) Structural aspects of oxides and oxysalt crystals. Acta Crystallogr B50:481–510
43. Burns PC, Grice JD, Hawthorne FC (1995) Borate minerals. I. Polyhedral clusters and fundamental building blocks. Can Mineral 33:1131–1151
44. Grice JD, Burns PC, Hawthorne FC (1999) Borate minerals II. A hierarchy of structures based on the borate fundamental building block. Can Mineral 37:731–762
45. Konnert JA, Clark JR, Christ CL (1970) Crystal structure of fabianite, $CaB_3O_5(OH)$, and a comparison with the structure of its synthetic dimorph. Z Kristallogr 132:241–252
46. Moll H, Reich T, Szabó Z (2000) The hydrolysis of dioxouranium (VI) investigated using EXAFS and ^{17}O-NMR. Radiochim Acta 88:411–415
47. http://www.trinityminerals.com/sm/uranium.shtml
48. Karanovic L, Rosic A, Poleti D (2004) Crystal structure of nobleite, $Ca[B_6O_9(OH)_2]{*}3H_2O$, from Jarandol (Serbia). Eur J Mineral 16:825–833
49. Burns PC, Hawthorne FC (1994) Hydrogen bonding in tunellite. Can Mineral 32:895–902
50. Gerhold G, Kampf AR, Bruland K, Ettensohn D, Behnke D (2006) The photo-atlas of minerals for Windows. Los Angeles county museum of natural history, Gem Mineral Council, Los Angeles

Struct Bond (2014) 158: 191–204
DOI: 10.1007/430_2012_84

Published online: 20 April 2013

Structure and Acidity in Aqueous Solutions and Oxide–Water Interfaces

Barry R. Bickmore

Abstract There have been a number of attempts to relate structural descriptors based on the bond-valence theory to the Brønsted acidity of (hydr)oxyacid monomers and oxide surface functional groups, via simple quantitative structure–activity relationships (QSARs). These models show some promise, but since they have been calibrated solely on monomers, it is difficult to know whether oxide surface functional groups are within their domain of applicability. In fact, there are strong reasons, including direct ab initio computation of equilibrium constants for surface functional groups, for doubting whether acidity QSARs based on the bond-valence theory are yet capable of accurately predicting acidity at the level of individual surface functional groups, despite some apparent successes. For progress to continue, we must further develop the relationship between bond valence and structural energy, so that we will be better able to construct widely applicable models

Keywords Acidity · Bond-valence theory · Oxides · Surface complexation models · Surface functional groups

Contents

B.R. Bickmore (✉)
Department of Geological Sciences, Brigham Young University, Provo, UT 84602, USA
e-mail: barry_bickmore@byu.edu

Abbreviations

BVT	Bond-valence theory
G	Global instability index
MUSIC	Multisite complexation
QSAR	Quantitative structure–activity relationship
SBE	Solvation, bond strength, and electrostatic model
SCM	Surface complexation model

1 Introduction

Although the bond-valence theory (BVT) is primarily meant to rationalize and predict molecular structures in solids, chemists naturally try to extend structural models to rationalize and predict reactivity. If a model helps us understand why particular equilibrium structures are preferred, for instance, perhaps quantifying the principles underlying the model can help us predict energetic differences between structural states, which are the bases for both thermodynamic and kinetic theory. The BVT is an excellent vehicle for exploring structure–energy relationships, because it is in some respects quantitatively predictive, and boils down complex, multi-body interactions into a single parameter, the bond-valence sum.

In this chapter, I review several attempts to relate bond valence to equilibrium constants for the acid dissociation of (hydr)oxo-monomers and oxide surface functional groups. Rather than exhaustively reviewing the literature on this subject, I have opted to attempt a concise description of the state of the field. For a number of reasons, reaction energetics at individual surface functional groups is particularly difficult to assess, so models capable of estimating equilibrium constants for these reactions are badly needed.

While there have been some successes in this area, however, BVT principles must be applied carefully to these systems. Bond-valence parameters are calibrated on precisely known structures of crystalline solids, for instance, so the application of the BVT to liquids and solid–liquid interfaces, for which less precise structural information is available, may not be straightforward. Furthermore, BVT-based reactivity models usually involve an implicit assumption that, at least for the purpose of predicting particular types of reaction energies, bond lengths are the dominant aspect of the structure. This may, or may not, be the case, depending on the reactions of interest.

It is important, therefore, to address two questions as we discuss various models. (1) How might the BVT-based structural descriptors employed relate to the potential energy surface? (2) What is the model's domain of applicability? That is, when should such models work, and when should they fail?

2 Bond Valence, Energy, and QSARs

The best-developed link between bond valence and energy involves the valence sum rule (Eq. 2 in chapter "Bond Valence Theory"), which is the backbone of the BVT. The valence sum rule predicts that the summed valence of bonds incident to ion i (ΣS_i) should counterbalance its atomic valence (V_i), and in fact, calculated valence sums usually deviate significantly from the ideal only in cases where the structure is strained [1]. It might be possible, therefore, to link deviation from the ideal valence sum (ΔV) with some predictable energy cost. In fact, a number of studies have shown that ΔV or ΔV^2 can be related to a significant part of the structural potential energy. Salinas-Sanchez et al. [2], for example, defined the Global Instability Index (G) for crystals in terms of ΔV, normalized and averaged over all atoms in a crystal formula unit (see Eq. 23 in chapter "Bond Valence Theory"), and showed that it was a good predictor of phase stability. Perez-Mato et al. [3] showed that the G is proportional to energy maps of certain distortion modes in stuffed tridymite-type structures, and in chapter "Bonding at Oxide Surfaces" of this volume, Poeppelmeier and Enterkin show how the G can be modified to correctly predict the relative stability of surface reconstructions. Rappe and coworkers [4, 5] developed a very successful force field for certain oxide materials, in which one of the potential terms is proportional to ΔV. Adams and coworkers [6–9] have developed potentials based on ΔV^2, which they have successfully used with Reverse Monte Carlo methods to investigate diffusion pathways in glasses. (See also chapter "Practical Considerations in Determining Bond Valence Parameters" in this volume.)

These attempts to explicitly relate ΔV to energy are relatively recent, however, and a more common approach has been to create QSARs by finding correlations between reaction energies and structural descriptors based on bond-valence considerations. Although QSARs of many types are common [10], and can under certain circumstances predict reaction energies to within as little as 0.5 kcal/mol [11], they typically suffer from two problems: (1) Whereas reaction energies reflect the difference between two structural states, QSARs are often keyed to "static" aspects of molecular structure – e.g., the number of particular types of functional groups, the electronegativities of particular atoms, dipole moments, polarizability, molecular shape, and molecular volume [11]. (2) The relationships between these static structural descriptors and reaction energies can be complex, and therefore simple QSAR correlations may only be valid under a restricted set of circumstances, in which other important structural factors are held approximately constant. Bond valence has so far been quantitatively related solely to bond length, and it is likely that other structural factors, such as bond directionality and non-bonded interactions, also affect reaction energies.

3 BVT-Based Brønsted Acidity QSARs

Brønsted acid dissociation reactions have the form $HA \leftrightarrow H^+ + A^-$, where HA is the acid and A^- is the conjugate base. Rather than reporting the raw equilibrium constant (K_a), most workers find it convenient to use the negative logarithm (pK_a), because pK_a is equivalent to the pH at which the acid undergoes 50% dissociation ($a_{HA} = a_{A-}$).

BVT-based acidity models have so far all been QSARs that relate pK_a values to structural descriptors such as ΔV and Lewis base strength (S_B). Equation 1 is the definition of S_B for a single anion.

$$S_B = \frac{|V|}{N_O} \tag{1}$$

Here, N_O is the average observed coordination number in a large number of oxide crystal structures [1], and the same formula can be used to calculate the Lewis acid strength (S_A) of cations (cf. Eq. 7 in chapter "Bond Valence Theory" of this volume). S_A and S_B take into account both the valence and size of atoms, providing an expectation value for individual bond valences. According to the valence matching rule, the most stable structures are formed when the S_A and S_B values of the cations and anions are similar (see Eq. 9 in chapter "Bond Valence Theory" of this volume).

These structural descriptors were first applied to simple acids, for which pK_a values can easily be determined experimentally. However, the practical reason for this is to predict pK_a values for individual functional groups on large molecules and solid–liquid interfaces, which are much more difficult to determine.

3.1 Simple Acids

Brown [1, 12] related the pK_a values of a number of simple acids to S_B of their conjugate bases via Eq. 2 (cf. Fig. 13 in chapter "Bond Valence Theory").

$$pK_a = 14.3 \ln\left(\frac{S_B}{0.135}\right) \tag{2}$$

For the purpose of predicting oxyacid pK_a values, we take S_B as the expectation value for OH...O bonds incident to the O^{2-} ions in the base from surrounding H_2O molecules. That is, we divide the total expected OH...O bond valence by the expected number of OH...O bonds. For example, in $H_3SiO_4^-$, the conjugate base of silicic acid (H_4SiO_4), there is one Si–O bond with an average of 1 v.u. to each of the four O^{2-} ions, and we assume all H–O bonds to be 0.8 v.u. The N_O value for O^{2-} is 4, so that the three OH ligands have two unoccupied bonding sites, while the O^{2-} ligand has three, for a total of nine. However, Brown [1] recommended reducing this to six, because the Si–OH groups do not have enough valence to accept more than a single, weak OH...O bond. The bond valences incident to the OH ligands sum to 1.8 v.u., so that its "unsaturated" (i.e., leftover) valence (V_u) is 0.2 v.u., whereas $V_u = 1$ v.u. for the O^{2-}. The total value of V_u for the molecule is

1.6 v.u., which is divided among the six expected H-bond acceptor sites to obtain $S_b = 0.267$ v.u. Applying Eq. 2, we predict $pK_a = 9.75$, which is close to the measured value of 9.84 [13] (see Table 5 in chapter "Bond Valence Theory" for more examples).

The advantages of this simple relationship are that it can address multiple acid dissociation reactions of the same molecule and be applied to a wide variety of acids. Brown's model proves difficult to apply to individual functional groups on large molecules and surfaces, however, because S_B values are averaged over all the O^{2-} atoms. In addition, the example of silicic acid shows that the rules for counting expected H-bond acceptor sites are somewhat ambiguous. The model may still prove useful for estimating average acidity for these larger systems, but this has not yet been attempted.

Hiemstra et al. [14] attempted to calibrate an acidity QSAR that would be transferable to individual surface functional groups by focusing on individual functional groups in (hydr)oxy-acid monomers. (Hereafter, this will be referred to as the multisite complexation, or MUSIC, method.) In the MUSIC method, pK_a values are related to ΔV of a single O^{2-} ion in the conjugate base via Eq. 3, taking into account Me–O bonds, H–O bonds, and OH. . .O bonds from the surrounding water molecules.

$$pK_a^{int} = -19.8(\Delta V) \tag{3}$$

Here, pK_a^{int} is the "intrinsic" pK_a value, which is corrected for the electrostatic work of removing H^+ from the base. (An analogous electrostatic correction is done for surfaces.) To calculate ΔV for the O^{2-} ion in $H_3SiO_4^-$, one starts with the N_O value of four for O^{2-} ions in oxides to define the number of expected bonding sites. One of the four expected bonds is the Si–O bond, which is assumed to have the average value of 1.0 v.u. We expect three more bonds to the O^{2-}, which we assume are OH. . .O bonds of 0.2 v.u. from the surrounding water molecules. This brings the total bond valence incident to the O^{2-} to 1.6 v.u., so that $\Delta V = -0.4$ v.u. Equation 3 predicts $pK_a^{int} = 7.9$ for $H_4SiO_4 \leftrightarrow H^+ + H_3SiO_4^-$, close to the true value of ~8.5 [13]. In cases where the acid is a hydrated cation, the valence of the H–O bond in the hydroxyl group on the base is assumed to be 0.8 v.u. In summary, one tallies the Pauling (average) bond strength of the Me–O bond, 0.8 v.u. for any H–O bonds, and 0.2 v.u. for each addition expected bond, to account for OH. . .O bonds.

Before we discuss this approach further, it is worth contrasting it to Brown's method. Brown related acidity to the average expected valence of OH. . .O bonds to the base as a whole, and that expected valence was determined by assuming the valence sum rule holds. Hiemstra and coworkers assumed bond strengths for OH. . .O bonds to a single O atom on the base, and related acidity to the deviation of the valence sum on that O from the ideal 2 v.u. Thus, a fundamental feature of the MUSIC method is that it assumes the valence sum rule is *not* obeyed.

Several studies by Bickmore and coworkers [13, 15–17] addressed this and other issues with the MUSIC method. They performed a bond-valence analysis of the output trajectories from ab initio molecular dynamics simulations of liquid water and explicitly hydrated oxo-species, including silicic acid, phosphoric acid, carbonic acid, and their conjugate bases, to show that the valence sum rule is obeyed on a time-averaged basis in such systems. This result does not necessarily imply that the MUSIC method will not work consistently – empirically fitting QSARs may correct systematic errors, after all. But it at least gives notice that the model might be getting the right answers for the wrong reasons, which calls into question its ability to predict outside the calibration set.

Bickmore and coworkers [13] did, in fact, demonstrate that the MUSIC method would not always work outside its calibration set. Equation 3 was calibrated on the pK_a^{int} values for (1) the first acid dissociation of a number of oxyacids (e.g., H_4SiO_4, H_3PO_4, H_2CO_3), and (2) dissociation reactions of hydroxyacids resulting in a neutrally charged base (e.g., $Fe(OH)_2^{+1} \bullet 4H_2O \leftrightarrow H^+ + Fe(OH)_3 \bullet 3H_2O$). The point of doing an electrostatic correction to obtain "intrinsic" pK_a values, however, is to remove the effect of long-range Coulomb interactions, so that only the energy of making and breaking bonds is left to be modeled using bond-valence terms. Thus, it should be possible to predict pK_a^{int} values for multiple dissociations of the same acid. But since ΔV in Eq. 3 is estimated assuming average bond valences for the Me–O bonds, this is impossible. For instance, if all Si–O bonds are taken to be 1.0 v.u., then there should be no difference in pK_a^{int} values for the reactions $H_4SiO_4 \leftrightarrow H^+ + H_3SiO_4^-$ and $H_3SiO_4^- \leftrightarrow H^+ + H_2SiO_4^{2-}$, because ΔV estimates for the Si–O groups on both bases are the same. In fact, the values for these two reactions vary widely (~8.5 and ~10.6, respectively,) and it cannot be the case that the Me–O bond length stays the same throughout such reactions. Curiously, while the pK_a^{int} values through series of acid dissociations vary widely for oxyacids, they are very similar for the hydroxyacids. Therefore, the use of average Me–O bond valences is actually justified for hydroxyacids.

To deal with this problem, one needs to know how the Me–O bond lengths change through multiple acid dissociations, at least in the oxyacids. Detailed structures of aqueous molecules generally cannot be obtained by experiment, unfortunately. Bickmore et al. [13, 15, 16] attempted to address this problem by using density functional theory (DFT) to calculate the Me–O bond lengths of interest for both gas-phase and explicitly hydrated molecules. The purpose of this was to account for progressive bond relaxation as H^+ ions are removed from an acid molecule. They found that they could predict the pK_a^{int} values of successive acid dissociations for a number of (hydr)oxo-monomers within about ± 1 log unit, using the following equation.

$$pK_a^{int} = 60.5 S_B + \alpha l_b - 18.1 \tag{4}$$

In Eq. 4, S_B for the O^{2-} of interest is derived by subtracting from 2 the calculated Me–O bond valence, and then subtracting another 0.8 v.u. for any H–O bonds

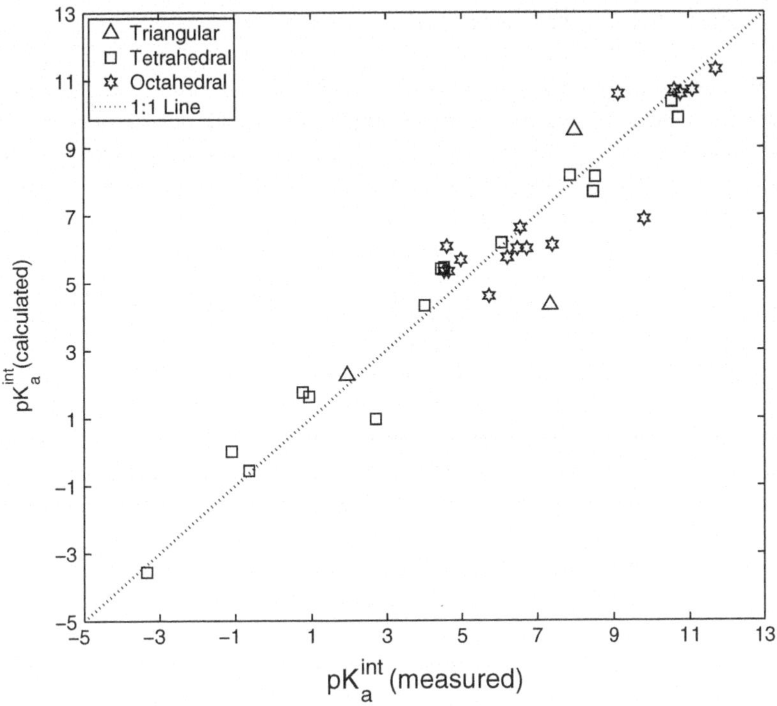

Fig. 1 Predicted (Eq. 4) vs. measured pK_a^{int} values for multiple acid dissociations of a number of (hydr)oxyacids (*triangular*, *tetrahedral*, and *octahedral*) [16]

present. We then divide the result by the number of unclaimed bonding sites. OH. . .O bonds from the solvent are *not* accounted for, just as in Brown's method [1, 12], so that S_B is an expectation value for the OH. . .O bonds. The variable I_b is the fraction ionic character of the Me–O bonds, derived from the difference in Pauling electronegativities between Me and O [18], and α is the regression coefficient obtained for the dependence on I_b. It turns out that the value of α is dependent on molecular shape – 5.3 for hexaquo cations (e.g., Fe^{2+}•$6H_2O$), 20.6 for tetrahedral oxyacids (e.g., H_4SiO_4), and 51.3 for triangular oxyacids (e.g., H_2CO_3). In addition, average (Pauling) bond valences worked better than any derived from calculated bond lengths for the hydroxyacids.

Figure 1 shows the predicted (Eq. 4) vs. measured pK_a^{int} values for a set of tetrahedral oxyacids, triangular oxyacids, and octahedral hydroxyacids. To date, this method is the most chemically accurate BVT-based acidity QSAR for (hydr) oxyacid solution monomers, but in the next section I will show that it is still unclear how successfully any of the methods discussed here can be applied to oxide–water interfaces, although some progress has undeniably been made.

3.2 Surface Functional Groups

The problem with modeling molecular-scale surface acidity is that it is usually impossible to be sure a model is, conceptually, even approximately right. Oxide surfaces can include a number of chemically distinct functional groups, but data on their acidity is often restricted to potentiometric titrations, in which the response of the entire surface is averaged, or data involving the averaged pH response of a single, crystallographically distinct surface, such as measurements of single-surface points of zero charge. Even advanced spectroscopic studies of crystallographically distinct surfaces provide data averaged over the entire surface although there have recently been some interesting attempts to infer the acidity of individual functional groups from such data [19, 20]. The pH response of distinct types of functional groups on aqueous molecules can sometimes be distinguished in a potentiometric titration curve, but surface acidity is significantly affected by the development of surface charge, which is continuously altered as acid–base reactions take place. This smears out the response of individual functional group types, making them indistinguishable from a macroscopic point of view.

Reactions at surface functional groups are typically modeled via surface complexation models (SCMs), which are simply equilibrium chemical models modified to correct for surface electrostatic effects. SCMs model acid–base reactions at surface functional groups via "intrinsic" equilibrium constants and ionic solution species concentrations corrected to account for the electric field around the interface. Thus, the "effective" equilibrium constants account for both chemical and electrostatic effects, and continuously change as surface charging progresses.

There are a number of ways to model the interfacial electric field in SCMs, involving various combinations of a Poisson–Boltzmann distribution and/or one or more parallel-plate capacitors [21]. Originally, modelers treated essentially all the building blocks of an SCM – electrostatic model, site types, site densities, and equilibrium constants – as fitting parameters. It was soon discovered, however, that one could usually fit potentiometric titration data with multiple SCMs that posit widely diverging molecular-scale pictures of the interface [22].

The first attempt to use the BVT to address this deficiency in SCMs was the MUSIC method of Hiemstra and coworkers [14], the point of which is to limit the number of adjustable parameters in SCMs by constraining some of the most important variables. Site types and densities are estimated via crystallographic models and microscopic measurements of the surface area, and surface pK_a^{int} values are estimated via Eq. 3. If we can fit an SCM to titration data with these parameters constrained to empirically or theoretically determined values, we have more confidence that the molecular-scale picture presented by the model is close to the truth.

Dozens of studies have used the MUSIC method to partially parameterize SCMs for various oxide surfaces, so by that measure the method has been very successful. But if the point of the method is to avoid a situation where models successfully explain data based on an inaccurate portrayal of the interface at the molecular scale, we should be careful to examine whether this problem has truly been solved. That is, does the method work because it provides an accurate molecular-scale picture of the interface, or because there are still enough degrees of freedom in the adjustable model parameters to fit the macroscopic data, no matter whether the microscopic picture is essentially accurate?

There are a number of reasons to question whether the MUSIC method can be consistently successful at estimating surface pK_a^{int} values for oxides, but the most important has to do with how the method is ported to surfaces. We have seen that Eq. 3 was calibrated on the structures of solution monomers, for which averaged (Pauling) Me–O bond valences were assumed. But when Hiemstra et al. [14] applied the method to surfaces, they recommended using unrelaxed Me–O bond lengths from bulk crystal structures. In some cases Hiemstra and coworkers followed this procedure, as with goethite (FeOOH), and in others they used Pauling bond valences, as with gibbsite ($Al(OH)_3$). In fact, the bonds at surfaces often do relax significantly, relative to the bulk solid, but even if we apply Eq. 3 only in cases where surface relaxation is negligible, we have to wonder whether we can expect to use it to consistently obtain accurate pK_a^{int} estimates, when it was calibrated using Pauling Me–O valences.

Another issue with the way the MUSIC method is applied to surfaces has to do with the assignment of O^{2-} coordination numbers. When applied to solution monomers, the method assumes a coordination number of four for an O^{2-} ion in a functional group, so any of the four bonding sites not occupied by Me–O and H–O bonds are assumed to be occupied by OH. . .O bonds of 0.2 v.u. When applied to surfaces, however, the method instead assumes that O^{2-} ions bonded to one Me (η) have a coordination number of three, those bonded to two Me (μ_2) have a coordination number of three or four, and those bonded to three Me (μ_3) have a coordination number of four. The reason given for this is to account for steric hindrance of H-bonding from the adjacent water molecules, but the modeler decides whether μ_2 groups have a coordination number of three or four. This results in a difference of 0.2 v.u. in ΔV, which translates into a difference of ~4 log units in the calculated pK_a^{int} (Eq. 3). Whereas traditional SCMs have pK_a^{int} values obtained by a computerized optimization algorithm, here modelers must significantly manipulate some of the pK_a^{int} values by hand. In practice, this kind of adjustment has sometimes been applied even more extensively. E.g., Hiemstra et al. [14] had to assume a coordination number of 4 for the silanol (>Si–OH) groups on SiO_2 surfaces to obtain an appropriate point of zero charge (pH_{PZC}).

Machesky and coworkers have attempted to address some of the aforementioned problems with their seminal work on TiO_2 (rutile) and SnO_2 (cassiterite) [23–25], which are both dominated by their (110) surfaces. Their approach has been to use Eq. 3 to estimate pK_a^{int}, but to obtain the bond valences via DFT geometry optimizations and both classical and DFT molecular dynamics simulations. They also constrained their SCMs and molecular simulations with potentiometric titration data, single-surface pH_{PZC} determinations, and synchrotron-based spectroscopic data. They were able to successfully reproduce the pH_{PZC} values within ~1.4 log units, which is reasonable, considering all the potential sources of error in the calculated bond lengths.

The ability of the method used by Machesky and coworkers to more or less accurately predict the pH_{PZC} values of these oxides is impressive, and it seems likely that they have gotten at least part of the picture right. But which part? The estimated pH_{PZC} values are derived from the calculated pK_a^{int} values, averaged over the entire surfaces, so it may be that this version of the MUSIC method correctly predicts average behavior, but not the acidity of individual functional groups. In fact, Sprik and coworkers [26, 27] performed ab initio MD simulations of the hydrated rutile (110) surface to directly calculate pK_a^{int} values for the individual functional groups. Whereas their calculated pH_{PZC} value was slightly more accurate than that calculated by Machesky and coworkers, they found that their individual pK_a^{int} values were quite different.

Furthermore, while Machesky and coworkers have addressed the issues raised above regarding the use of Pauling Me–O bond valences and arbitrary numbers and strengths of OH...O bonds, they have not addressed the issue of acidity model calibration. That is, how can we expect to obtain accurate pK_a^{int} estimates by inserting precisely calculated bond lengths into a QSAR calibrated on idealized solution monomer structures with averaged bond lengths?

Once again, Bickmore et al. [13, 15, 16] addressed this issue by calibrating their model (Eq. 4) on calculated solution monomer structures, but at this point it is still unclear how to apply Eq. 4 to surface functional groups, because the dependence of pK_a^{int} on the fraction ionic character of the Me–O bonds (α) was shown to depend on the basic molecular shape. Is there a single α value one can use for all surface functional groups, or does it depend strongly on the specific environment? The answers to these questions depend on the physical meaning of the structural descriptors in the model. Why, for instance, is S_B of the conjugate base such a strong predictor of acidity? Bickmore et al. [16] used ab initio molecular dynamics simulations of several explicitly hydrated oxyacids and their conjugate bases to show that S_B is highly correlated with the valence of the strongest H-bond reaching the O^{2-} from the solvent. They reasoned that stronger H-bonds would make it easier for H^+ ions to jump from solvent molecules to O^{2-} ions on the base. This is undoubtedly true, but as mentioned above, equilibrium constants must somehow be related to a change in energy between the products and reactants, whereas all BVT-based acidity QSARs are related exclusively to the structure of the base.

Casey, Rustad, and coworkers [28–32] brought up yet another issue that must be considered. If all BVT-based acidity QSARs are calibrated on solution monomers, how can we be sure they will correctly predict the acidity of, for instance, bridging (μ_2 or μ_3) oxygens on oxide surfaces? Their research has focused on large oxo-molecules that include such groups, and for which the acidities of the individual functional groups can often be determined with a high degree of confidence via potentiometry, UV–vis spectroscopy, and NMR spectroscopy. Structural analysis and molecular modeling have shown that neither these acidities nor their relative order, can be derived from simple bond-valence considerations, although Casey, Rustad, and coworkers have never analyzed the valence of OH...O bonds incident to the functional groups, as both Bickmore's and Machesky's groups did. (In one instance, however, Rustad [32] did count OH...O bonds for use in MUSIC calculations.)

Nevertheless, nobody has successfully addressed the criticisms of Casey, Rustad, and coworkers, and their work serves to underscore two glaring facts about this field. First, we need to be doing more work on systems for which we can tell when we have the right answer. Second, we need to know *why* our models work to be confident that they can provide accurate insights into the details of less well understood systems.

4 Outlook

In this chapter, I have described several attempts to predict the Brønsted acidity of oxo-groups on molecules and oxide surfaces based on bond-valence considerations. If I had to choose a metaphor to describe the state of the field, I would say that there is a lot of smoke, but we haven't yet found the fire. Clearly, there have been enough successes to warrant the conclusion that the BVT can be a powerful tool for relating molecular structure to acidity. However, serious doubts have been cast on the ability of any of the BVT-based acidity QSARs so far proposed to consistently make accurate predictions of the acidity of individual functional groups in diverse structural settings.

There are reasons for hope. Sverjensky and Sahai, for instance, published a series of papers [33–37] in which they developed the solvation, bond strength, and electrostatic (SBE) model. The SBE model fairly accurately predicts the surface acidity of a number of simple oxides based on model terms derived from easily obtained characteristics of each solid, including the Pauling bond valence to bond length ratio and the dielectric constant. While this model does not provide molecular-scale information, it has been so successful at explaining the average acid–base reactivity of oxide surfaces that it seems quite plausible that a similar approach might work well at the scale of individual functional groups.

It has become evident that if this is to happen, we need to better understand the relationship between bond valence and energy. A number of attempts to develop this link were mentioned in Sect. 2 of this chapter, but there is still no way of weighing the relative contributions of bond-valence and other structural factors, such as bond angles and non-bonded interactions, to the potential energy surface. Some preliminary work is being done in that area [38], but it is not well developed.

At the beginning of this chapter, I said that we had to pay special attention to two questions when discussing BVT-based acidity models. (1) How might the BVT-based structural descriptors employed relate to the potential energy surface? (2) What is the model's domain of applicability? To date, there are no clear answers to either.

References

1. Brown ID (2002) The chemical bond in inorganic chemistry: the bond valence model. Oxford University Press, New York
2. Salinas-Sanchez A, Garcia-Muñoz JL, Rodriguez-Carvajal J, Saez-Puche R, Martinez JL (1992) Structural characterization of R_2BaCuO_5 (R = Y, Lu, Yb, Tm, Er, Ho, Dy, Gd, Eu and Sm) oxides by X-ray and neutron diffraction. J Solid State Chem 100:201–211
3. Perez-Mato JM, Withers RL, Larsson A-K, Orobengoa D, Liu Y (2009) Distortion modes and related ferroic properties of the stuffed tridymite-type compounds $SrAl_2O_4$ and $BaAl_2O_4$. Phys Rev B79:064111
4. Cooper VR, Grinberg I, Rappe AM (2003) Extending first principles modeling with crystal chemistry: a bond-valence based classical potential. In: Davies PK, Singh DJ (eds) Fundamental physics of ferroelectrics. American Institute of Physics, Melville, New York
5. Shin Y-H, Cooper VR, Grinberg I, Rappe AM (2005) Development of a bond-valence molecular-dynamics model for complex oxides. Phys Rev B 71:054104
6. Adams S (2001) Relationship between bond valence and bond softness of alkali halides and chalcogenides. Acta Crystallogr B 57:278–287
7. Adams S, Moretzki O, Canadell E (2004) Global instability index optimizations for the localization of mobile protons. Solid State Ionics 168:281–290
8. Adams S, Rao RP (2009) Transport pathways for mobile ions in disordered solids from the analysis of energy-scaled bond-valence mismatch landscapes. Phys Chem Chem Phys 11:3210–3216
9. Adams S, Swenson J (2002) Bond valence analysis of transport pathways in RMC models of fast ion conducting glasses. Phys Chem Chem Phys 4:3179–3184
10. Hinchliffe A (2003) Molecular modelling for beginners. Wiley, Chichester
11. Rappé AK, Casewit CJ (1997) Molecular mechanics across chemistry. University Science Books, Sausalito
12. Brown ID (1981) The bond-valence method: an empirical approach to chemical structure and bonding. In: O'Keefe M, Navrotsky A (eds) Structure and bonding in crystals. Academic, New York
13. Bickmore BR, Tadanier CJ, Rosso KM, Monn WD, Eggett DL (2004) Bond-valence methods for pK_a prediction: critical reanalysis and a new approach. Geochim Cosmochim Acta 68:2025–2042
14. Hiemstra T, Venema P, Van Riemsdijk WH (1996) Intrinsic proton affinity of reactive surface groups of metal (hydr)oxides: the bond valence principle. J Colloid Interface Sci 184:680–692
15. Bickmore BR, Rosso KM, Mitchell SC (2006) Is there hope for multisite complexation modeling? In: Lützenkirchen J (ed) Surface complexation modelling. Elsevier, Amsterdam
16. Bickmore BR, Rosso KM, Tadanier CJ, Bylaska EJ, Doud D (2006) Bond-valence methods for pK_a prediction. II. Bond-valence, electrostatic, molecular geometry, and solvation effects. Geochim Cosmochim Acta 70:4057–4071
17. Bickmore BR, Rosso KM, Brown ID, Kerisit S (2009) Bond-valence constraints on liquid water structure. J Phys Chem A 113:1847–1857
18. Pauling L (1960) The nature of the chemical bond. Cornell University Press, Ithaca
19. Catalano JG, Fenter P, Park C (2007) Interfacial water structure on the (012) surface of hematite: ordering and reactivity in comparison with corundum. Geochim Cosmochim Acta 71:5313–5324
20. Sung J, Shen YR, Waychunas GA (2012) The interfacial structure of water/protonated α-Al_2O_3 (11–20) as a function of pH. J Phys Condens Matter 24:124101
21. Kosmulski M (2001) Chemical properties of material surfaces. Marcel Dekker, Basel
22. Westall JC, Hohl H (1980) A comparison of electrostatic models for the oxide/solution interface. Adv Colloid Interface Sci 12:265–294

23. Machesky ML, Predota M, Wesolowski DJ, Vlcek L, Cummings PT, Rosenqvist J, Ridley MK, Kubicki JD, Bandura AV, Kumar N, Sofo JO (2008) Surface protonation at the rutile (110) interface: explicit incorporation of solvation structure within the refined MUSIC model framework. Langmuir 24:12331–12339
24. Vlcek L, Zhang Z, Machesky ML, Fenter P, Rosenqvist J, Wesolowski DJ, Anovitz L, Predota M, Cummings PT (2007) Electric double layer at metal oxide surfaces: static properties of the cassiterite-water interface. Langmuir 23:4925–4937
25. Rosenqvist J, Machesky ML, Vlcek L, Cummings PT, Wesolowski DJ (2009) Charging properties of cassiterite (α-SnO_2) surfaces in NaCl and RbCl ionic media. Langmuir 25:10852–10862
26. Cheng J, Sprik M (2010) Acidity of the aqueous rutile TiO2(110) surface from density functional theory based molecular dynamics. J Chem Theory Comput 6:880–889
27. Sulpizi M, Sprik M (2010) Acidity constants from DFT-based molecular dynamics simulations. J Phys Condens Matter 22:284116
28. Casey WH, Rustad JR (2007) Reaction dynamics, molecular clusters, and aqueous geochemistry. Annu Rev Earth Planet Sci 35:21–46
29. Casey WH, Rustad JR, Banerjee D, Furrer G (2005) Large molecules as models for small particles in aqueous geochemistry research. J Nanopart Res 7:377–387
30. Casey WH, Rustad JR, Spiccia L (2009) Minerals as molecules – use of aqueous oxide and hydroxide clusters to understand geochemical reactions. Chem Eur J 15:4496–4515
31. Casey WH, Swaddle TW (2003) Why small? The use of small inorganic clusters to understand mineral surface and dissolution reactions in geochemistry. Rev Geophys. doi:10.1029/2002RG000118
32. Rustad JR (2005) Molecular dynamics simulation of the titration of polyoxocations in aqueous solution. Geochim Cosmochim Acta 69:4397–4410
33. Sahai N, Sverjensky DA (1997) Sovation and electrostatic model for specific electrolyte adsorption. Geochim Cosmochim Acta 61:2827–2848
34. Sahai N, Sverjensky DA (1997) Evaluation of internally consistent parameters for the triple-layer model by the systematic analysis of oxide surface titration data. Geochim Cosmochim Acta 61:2801–2826
35. Sverjensky DA (1994) Zero-point-of-charge prediction from crystal chemistry and solvation theory. Geochim Cosmochim Acta 58:3123–3129
36. Sverjensky DA (2005) Prediction of surface charge on oxides in salt solutions: revisions for 1:1 (M+L-) electrolytes. Geochim Cosmochim Acta 69:225–257
37. Sverjensky DA, Sahai N (1996) Theoretical prediction of single-site surface-protonation equilibrium constants for oxides and silicates in water. Geochim Cosmochim Acta 60:3773–3797
38. Bickmore BR, Wander MCF, Edwards J, Maurer J, Shepherd K, Meyer E, Johansen WJ, Frank RA, Andros C, Davis M (2012) Electronic structure effects in the vectorial bond-valence model. Am Miner (in press)

Struct Bond (2014) 158: 205–232
DOI: 10.1007/430_2013_98

Published online: 8 March 2013

Bonding at Oxide Surfaces

James A. Enterkin and Kenneth R. Poeppelmeier

Abstract Concepts in chemical bonding when combined with physics-based energetic considerations can lead to a more complete understanding of the structure, stability, and reactivity of oxide surfaces. While this symbiosis has long been understood for bulk structures, chemical bonding considerations have historically been used less frequently for surfaces. In this chapter, we analyze the chemical bonding of published surface structures of $SrTiO_3$ and MgO using bond valence sum analysis. Bond valence theory compares favorably with complex quantum mechanical calculations in assessing surface structures and explains the experimentally observed surface structures in a readily comprehensible manner. Bond valence theory also helps explain discrepancies between DFT predicted surface stability and experimentally observed surface structures, accurately predicts the adsorption of foreign species onto surfaces, and can be used to predict changes in surface structures.

Keywords Adsorbates · Bond valence sum · Reconstruction · Surface structure · Surfaces

Contents

J.A. Enterkin (✉)
Chemical Sciences and Engineering Division, Argonne National Laboratory, Argonne, IL, USA
e-mail: james.a.enterkin@gmail.com

K.R. Poeppelmeier
Chemical Sciences and Engineering Division, Argonne National Laboratory, Argonne, IL, USA

Department of Chemistry, Northwestern University, Evanston, IL, USA

1 Introduction

Surfaces have traditionally been more difficult to solve and understand than bulk structures. This is summed up in the quote, commonly attributed to Wolfgang Pauli, "God made the solid state, but he left the surface to the devil." Oxide surfaces have proven a particular challenge, often viewed as inherently different from the bulk. Theories about the driving forces behind surface structure formation include the minimization of "dangling bonds" [1] or reduction of Coulomb forces [2]. Many believe that polar surfaces must be different from the bulk since they require "charge compensation" (see, for instance, [3–5] and references therein). Recent results reveal that surface structures share more in common with the bulk than previously believed [6].

Approaches to understanding bulk structures can be roughly divided into physical and chemical methods. The former generally consists of minimizing the potential energy of a structure. The latter consists of understanding the localized chemical bonds through methods such as bond valence. In bulk structures, these methods are complimentary: each provides useful and important information necessary for a more complete understanding of bulk structures.

The physics approach has dominated attempts to understand surface structures. Despite the many demonstrated uses in bulk structures, Brown's 2009 review [7] listed only one case where bond valence has been applied to surface structures. Ruberto and coworkers' examination of the κ-Al_2O_3 (001)/(00$\bar{1}$) surface [8, 9] considers bond valence sums in a discussion of polarity compensation. In contrast and as seen in the previous chapter, bond valences are often applied to surfaces at aqueous interfaces. The difference is, in part, historical; that is chemists study aqueous interfaces while physicists examine clean surfaces in vacuum conditions. Bond valence considerations can be a useful companion to the physics-based investigations, potentially leading to predictions of what surface structures may form and what reactions may occur.

There are, of course, differences between bonding at the surface and in the bulk. Each surface atom forms fewer and shorter bonds than a bulk atom, and they can adopt coordination geometries not observed in the bulk. Bond valence theory should apply to the surface just as to the bulk, however. Bond valence theory predicts that shorter bond lengths are necessary when the number of bonds is fewer in order to conserve the overall bond valence.

Since 2009, several papers have been published which consider the bond valence of surfaces. Many have involved the L. Marks and Poeppelmeier groups at Northwestern University and deal with the surfaces of strontium titanate ($SrTiO_3$). These include a discovery of a homologous series of surface structures on $SrTiO_3$ (110) [6], studies of the (2 × 2) [10], c(4 × 2) [11], and ($\sqrt{13} \times \sqrt{13}$)$R$33.7° [12] surface reconstructions[1] on the $SrTiO_3$ (100) surface, an examination of water adsorbed on the $SrTiO_3$ (100) surface [13, 14], and a review of several previously proposed surface structures on $SrTiO_3$, MgO, and NiO from a bond valence perspective [15].

While all of bond valence theory should apply, these papers have focused mainly on bond valence sums. Bond valence theory has been used to confirm surface models calculated via DFT, or conversely to show their unsuitability [15]. Bond valence theory has also been used to explain why certain structures are observed to be more stable, for example in the homologous series of surface structures on $SrTiO_3$ (110) [6]. Additionally, bond valence theory has been used to understand and predict how surfaces will rearrange, and where foreign species will adsorb to surfaces [13–15]. Finally, bond valence theory has explained why predictions do not match observed surface structures [13–15].

The most commonly considered surface adsorbates have been hydrogen and hydroxide. Hydrogen bonding creates a slight difficulty, requiring different parameters for R_0 and perhaps even b for hydrogen bonds of different lengths due to the asymmetry of X–H–X bonds, which is best modeled by different values for the short and long portions of the hydrogen bond (see Sect. 8.1.1 in Chap. 2 of this volume [16]). It is unclear which of the various literature values for R_0 is best. For the works dealing with hydrogen on surfaces, an R_0 of 0.957 Å has been used, the length of an O–H bond in gaseous H_2O. This was chosen because the bond distance of gas phase H_2O, like the surfaces being considered, has no significant H–X interactions. For simplicity, $b = 0.37$ was maintained. While determining an R_0 value from a single parameter is far from an optimal solution (see Chap. 3 by Adams [17] for more on determining bond valence parameters), it has proved sufficient for the small number of hydroxylated surface structures so far considered.

To aid in analysis of these surface structures, a new metric was defined: surface instability index (SII) [15]. The SII is calculated similar to the global instability index (G, Eq. 2.21 of Brown [16]), except that only the atoms in the surface structure and the first bulk layer are included. Enterkin and coworkers demonstrated

---

[1] Surface reconstructions are herein referred to by Wood's notation: (N × M) $R\varphi$. N and M refer to the length of the lattice vectors for the surface reconstruction, where the vectors are N and M times the length of the bulk lattice vectors in the plane of the surface. $R\varphi$ indicates that the surface lattice vectors N and M are rotated by φ degrees with respect to the bulk lattice vectors in the same plane. When $R\varphi$ is omitted, the angle φ is 0°. Other symmetry indicators can also be included, for example the $SrTiO_3$ (100) c(4 × 2) reconstruction has lattice vectors 4 and 2 times the (010) and (001) bulk lattice vectors, in the same direction (0° rotation), while the "c" indicates that it is a centered unit cell.

that inclusion of the top bulk layer along with the surface atoms was most representative of the actual stability of the surface structure [15]. They calculated instability indices for the atoms in the surface structure only, for the surface plus the top bulk layer, and for the surface plus top two bulk layers. Excluding the top bulk layer neglected the instability associated with the strain imposed upon the bulk by the surface. Including more bulk layers caused SII to converge towards the bulk G value and become less representative of the surface. A bulk instability index was also calculated for the central most stoichiometric unit in the model and was shown to be useful, for example in checking whether a sufficiently thick slab had been used in DFT surface structure calculations.

In this chapter, we first discuss how the bond valence method can work in a complementary manner with DFT surface calculations, similar to how they are known to complement bulk DFT calculations. We then review several known and proposed surface structures on the perovskite $SrTiO_3$ and the rock salt magnesium oxide (MgO) from a bond valence perspective. We continue to examine a few cases where, similar to solid–liquid interfaces, adsorbates from the atmosphere may be interacting with oxide surfaces. In these discussions, we will show how bond valence can explain and even predict surface reconstructions and interactions of the surface with adsorbates. Finally, we will discuss other issues present at surfaces where bond valence analysis could be of value to ongoing work.

2 Bond Valences and DFT Calculations

Application of bond valence theory, and metrics such as the bond valence sum, has been complicated in the case of surfaces because there are relatively few cases for which the atomic coordinates are precisely known. Most of those few cases, where exact coordinates are known, come from DFT calculations. It is thus important to consider the relationship between DFT and bond valence.

For bulk structures, bond valence theory provides much information that is complementary to DFT calculations. For instance, it has been used to correct DFT bond lengths [18], and G follows the same trends as DFT calculated energy for bulk structural instabilities [19].

2.1 Bond Valence and DFT Structural Convergence

The bond valence model also works well for surface analysis, as indicated by the complementary nature of bond valence sums and DFT structural convergence. Figure 1 shows data during a DFT structural minimization of a hydroxylated MgO (111) surface structure [20]. The G and DFT calculated energy have nearly

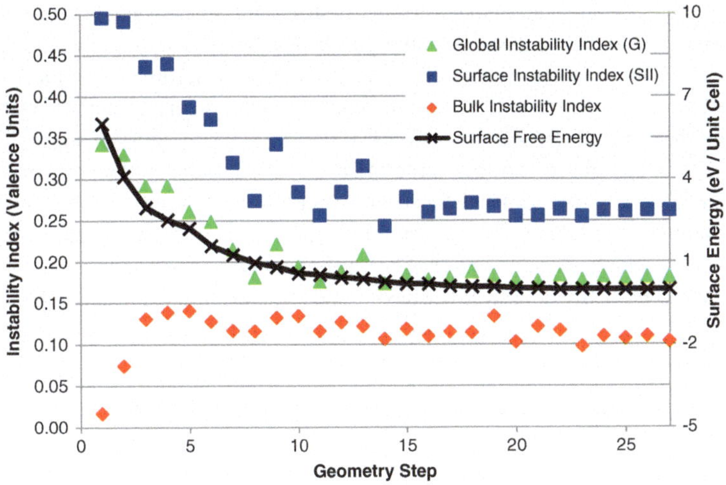

Fig. 1 Global, surface, and bulk instability indices and energy plotted as a function of geometry optimization step for a hydroxylated MgO (111) [15]. Surface from [20]. Energy is relative to the final energy. Figure adapted with permission from [15]

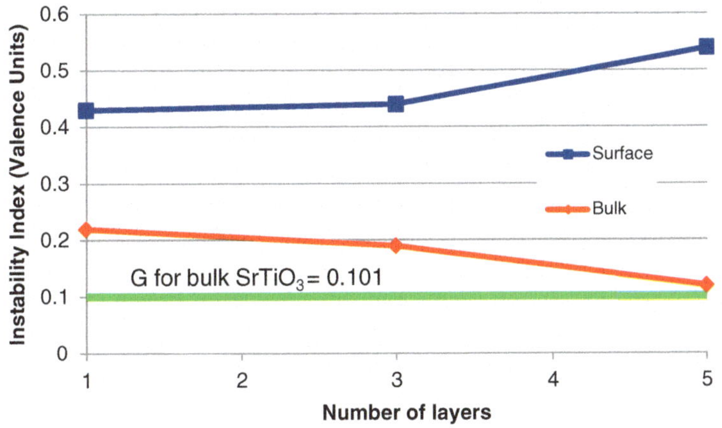

Fig. 2 Change in SII and bulk instability index (instability index for the innermost stoichiometric layer in the slab) with number of layers in calculation slab for the $SrTiO_3$ (110) halfO_2 surface. Figure adapted with permission from [15]

identical trends, demonstrating that most of the energy reduction can be attributed to optimization of local bonding. The improved bonding occurs primarily at the surface, as indicated by the SII. The instability index of the innermost stoichiometric layer in the slab increases as expected; long-range strains from the surface rearrangements perturb the bonding in the center of the slab.

Bond valence sums can also determine if sufficient layers have been used in the DFT surface calculations. If a thick enough slab has been used, then the center of the slab will have bond valence sums the same as the bulk structure. If the bond valence sums of the central layers differ from those of the bulk material, then the slab is too thin. Enterkin et al. [15] recently demonstrated this, using slabs of three different thicknesses in calculations of the same $SrTiO_3$ (110) TiO faceted structure (Fig. 2). While the bond valence sums of the surface species were improved for models with smaller numbers of layers, the bulk bond valence sums differed significantly from those in bulk $SrTiO_3$. A necessary, but not sufficient, condition for an accurate model is that the atoms at the central layer have bond valence sums similar to the bulk.

2.2 *Analysis of DFT Calculated Structural Models*

Bond valence can also be used to check calculated structures. Perhaps a simple approach is to calculate the bond valence sums and SII for the structure. For bulk structures, G values over 0.20 normally indicate problems with the structure. Surfaces have been shown to have bonding similar to the bulk, and therefore it is reasonable to assume the SIIs over 0.20 would indicate problems with a surface structure. However, fewer surface structures have been analyzed by bond valence methods; therefore, it is not yet certain that the same cutoff works equally well for surfaces. It is possible that the bonding may be sufficiently different and a different cutoff for SII will be found more appropriate in the future.

A stable atomic coordination will also have electrons filling space in all directions. This phenomena is well known and has been considered at least since the introduction of valence shell electron pair repulsion theory [21]. In bond valence theory, this has been formalized as the valence vector sum rule: "In a stable coordination sphere the sum of the bond valence vectors around an ion is zero" [22] (See Sect. 6.2 in Chap. 2 of this volume by Brown [16]). For metal cations at oxide surfaces, this requires bonds to O anions equally in all directions. As lone electron pairs can also fulfill this requirement for O anions, they can be stable with a nonzero valence vector sum, i.e. with bonds distributed unequally through space (See Sect. 7.1 in Chap. 2 of this volume by Brown [16]). Quantitative work on surfaces has so far been limited to bond valence sums. Since bond valence sums ignore these geometric concerns, metal terminated structures will be less stable than indicated by a bond valence sums analysis. This is seen in the differences between the SII and the DFT calculated energies for cation terminated surfaces. Valence vector sums should be as useful for surface structures as for bulk structures. So far, however, the valence vector sum rule has been applied only qualitatively to surface structures, and so we will limit our discussion in this chapter to applying it qualitatively.

The $SrTiO_3$ (111) surface provides a good example of how bond valence theory can aid in the analysis of DFT calculated and predicted surface structures. Several surface reconstructions have been observed, but none have been solved, and the

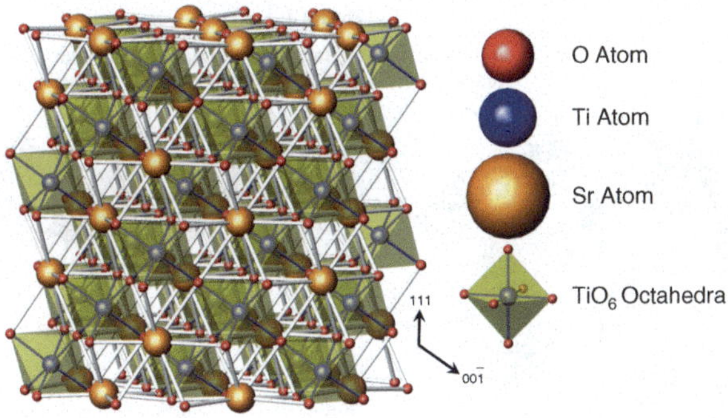

Fig. 3 $SrTiO_3$ in the (111) direction (*vertical axis*) is composed of alternating layers of SrO_3^{4-} and Ti^{4+}. Figure shown terminated by SrO_3^{4-} layers on *top* and *bottom*, which would require a surface structure of nominal +2 charge per 1 × 1 column to maintain valence neutrality of the surface structure

structures remain unknown (see [23] and references therein). Several surface structural models have been proposed based upon DFT calculations. Recently, bond valence was used to analyze six DFT calculated models and determine which were reasonable [15].

In the (111) direction, $SrTiO_3$ is composed of alternating layers of SrO_3^{4-} and Ti^{4+} (Fig. 3). This leads to a polar surface. To stabilize such a surface, the terminal layer must have half the nominal charge of the bulk layers. In this case, a 2+ surface atop an SrO_3^{4-} layer or a 2− surface atop a Ti^{4+} layer. From a chemical perspective, the challenge is to find an appropriate valence neutral surface with reasonable oxidation. Enterkin and coworkers [15] analyzed six of the structures from Marks and coworkers [24, 25] (termed models 3–8 by Marks and coworkers, see Fig. 4) which had stoichiometry that allowed Sr, Ti, and O to have valences charges of 2+, 4+, and 2−, respectively.

Model 3 is SrO rich at the surface, consisting of an SrO_3 termination with 1/3 of the oxygen sites vacant. The bond valence sums are somewhat close to the atomic valences, leading to an SII of 0.21, indicating the possibility of problems. The surface Sr has coordination sphere with one half completely empty, and thus it fails the valence vector sum test and is less stable than indicated by the SII.

Model 4 has a Ti layer termination, with a single O atom directly above each terminal Ti, resulting in TiO_4 tetrahedra that corner share with bulk Ti octahedra. The terminal O atom is dangling and under-bonded by nearly half a valence unit (bond valence sum = −1.56), while the Sr in the top bulk layer is slightly under-bonded (bond valence sum = 1.82). This leads to a high SII of 0.29. The Sr in the top bulk layer has a half-filled coordination sphere, and therefore this model is less stable than indicated by the SII.

Model 5 surface consists of an SrO_3 termination with a half-filled Ti layer on top. All bond valence sums are within 0.30 of the atomic valence, leading to an SII of 0.17. However, it also fails the valence vector sum test.

Model 6 has the same stoichiometry as model 5, but only half the Ti in the top layer occupy bulk-like positions. The rest are located directly above a TiO_6 octahedron

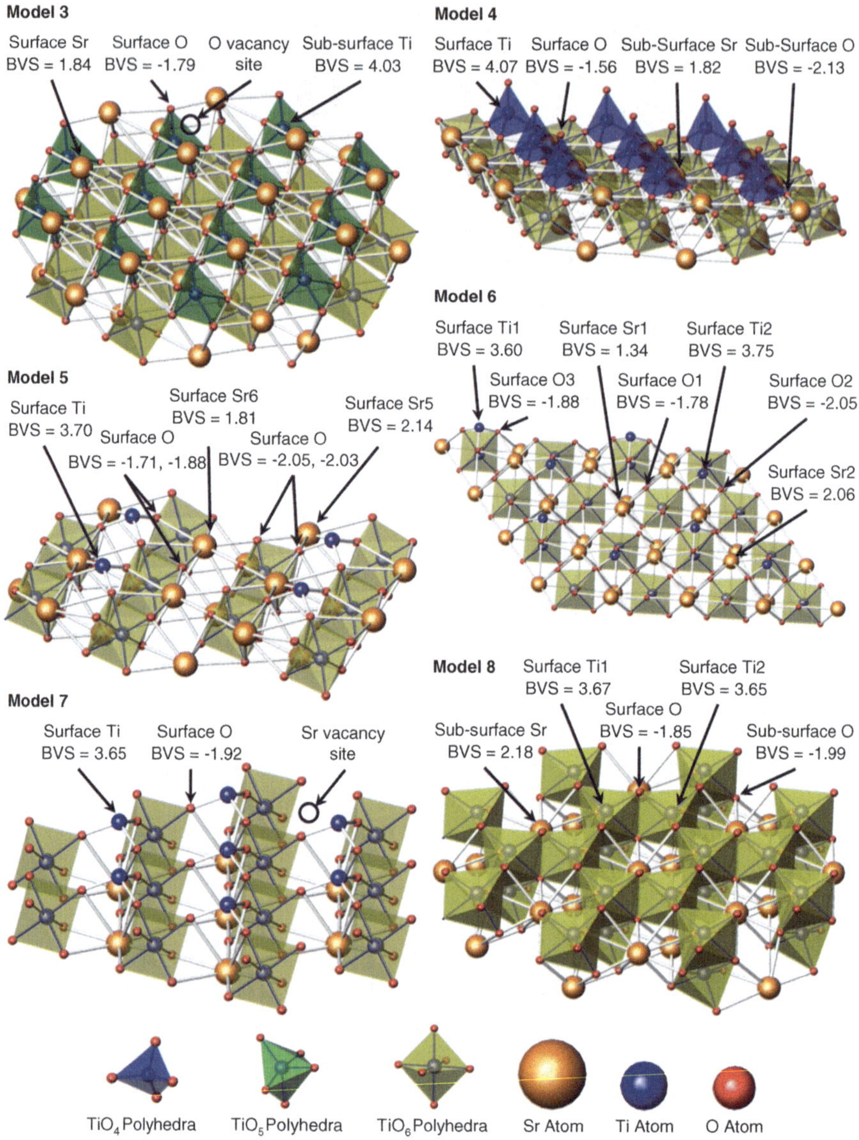

Fig. 4 *Top* to *bottom*: $SrTiO_3$ (111) surface structure models 3 through 8 considered in [24]

from the layer below. The surface atoms are more drastically under-bonded, which leads to a higher SII of 0.22. Yet again, the structure does not satisfy the valence vector sum rule and is less stable than indicated by the SII.

Models 7 and 8 have equal amounts of excess TiO_2 at the surface. Model 7 is terminated in a bulk-like Ti layer atop an O_3 layer (SrO_3 layer with vacant Sr sites), resulting in a surface TiO_2 stoichiometry. The surface Ti form TiO_3 trigonal

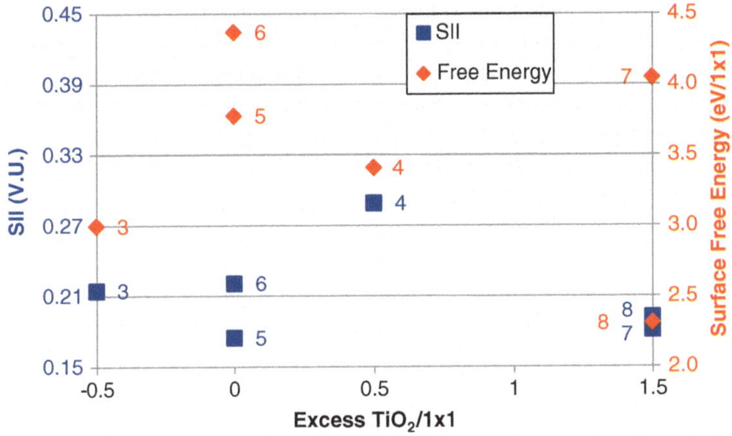

Fig. 5 Comparison of surface free energy [24] and SII for $SrTiO_3$ (111) surface models. Differences between SII and surface free energy can be accounted for when considering the valence vector sums. Figure adapted with permission from [15]

pyramids which corner share with bulk Ti octahedra. All bond valence sums are near the atomic valences, except for the outer most surface Ti, which is significantly under-bonded (bond valence sum = 3.65). The Ti also fails to satisfy the valence vector sum rule. Thus, although the SII is reasonable (0.18) the structural model is likely not a stable one.

Model 8 was an example where bond valence analysis showed that the slab used in the original calculations was not thick enough. Enterkin and coworkers therefore repeated the DFT optimization of model 8 with a thicker slab, after which the bond valence sums for the central layers matched those of bulk $SrTiO_3$. Model 8 has a Ti_2O_3 layer atop an SrO_3 termination, again resulting in a surface with TiO_2 surface stoichiometry. In model 8, however, the surface Ti form edge sharing octahedra, half of which corner share with three bulk Ti octahedra while the other half face share with one bulk Ti octahedra. The surface is somewhat under-bonded, which leads to a borderline SII of 0.20. Of the models considered, model 8 is the only one which qualitatively satisfies the valence vector sum rule.

For models 3 through 8 (Fig. 4), the SII fall in the order model 5 < model 7 < model 8 < model 3 < model 6 < model 4. The first five of these are all close together (0.17 < SII < 0.22) while model 4 is significantly higher (SII = 0.29). Except for model 8, all of these have at least one metal atom at the surface with an incomplete coordination sphere: for models 3 and 4 an Sr atom, for model 7 a Ti atom, and for models 5 and 6 an Sr atom and half a Ti atom per (1 × 1) unit cell. These do not satisfy the valence vector sum rule (see Sect. 6.2 in Chap. 2 of this volume by Brown [16]) and are less stable than indicated by the SII (Fig. 5). Of the five models with similar SIIs, model 8 has the most reasonable bonding. This agrees with the DFT calculations [24].

The bond valence analysis of the small reconstructions from Marks et al. [24] is consistent with the DFT energies. Since DFT surface energies can only be evaluated relative to other known surface energies, the DFT calculations only demonstrate which models were more or less stable relative to the other models. Bond valence sums offer a way to state from an absolute perspective which are appropriately coordinated and therefore might exist on the observed surfaces. In most cases, the SII are in general larger than for the cases where experimentally determined structures were analyzed. The cases with acceptable SII fail to satisfy the valence vector sum rule. This indicates that the motifs are unlikely to exist in the experimentally observed structures. The exception is model 8, which would be a reasonable configuration, especially if combined with another motif that helped decrease the average bond length, similar to how the chains are interspersed with rings in the homologous series of structures on $SrTiO_3$ (110) (see Sect. 3, below) [6, 15].

2.3 Bond Valence to Decrease DFT Calculation Cost

Enterkin et al. [15] also propose that bond valence can be used to reduce the cost of DFT calculations. Minimizing the G prior to performing a DFT structural relaxation can significantly decrease the cost of the DFT calculation. This can be done by hand, changing the position of a few atoms in an initial structural model so they have reasonable bond valence sum values. In principle a bond valence sum-based structural optimization could be carried out in a matter of seconds on a standard laptop computer even for large structures. Such a program may also provide a particularly useful alternative to force-field type calculations of surfaces and defects, as good force-fields for surfaces and defects are often lacking.

2.4 Bond Valence as a Check on DFT Functionals

Finally, bond valence sums may provide a check on the accuracy of a DFT functional. The $(\sqrt{2} \times \sqrt{2})$-R45° reconstruction on the (100) surface of $SrTiO_3$ provides an example of where bond valence analysis indicates that the inaccuracies of a DFT functional may explain the difference between observation and calculations. (For more on the $SrTiO_3$ (100) surface, see Sect. 5, below.) On the $SrTiO_3$ (100) surface, three reconstructions with a double TiO_2 layer have been observed: (2×1), (2×2), and $c(4 \times 2)$. The $(\sqrt{2} \times \sqrt{2})$-R45° reconstruction has the same stoichiometry and is calculated to be lower in energy than any of the three observed surfaces but has never been observed experimentally. A bond valence sum analysis of the $(\sqrt{2} \times \sqrt{2})$-R45° surface structure noted that although the bond valence sums of surface atoms were good, it has over-bonded bulk Sr and O atoms (all Sr have bond valence sums ≥ 2.32), which leads to a high G of 0.21 [15]. The high bulk Sr atomic coordination occurs because the surface structure couples to an antiferroelectric bulk distortion which persists to large depths in a DFT calculation. The DFT method used in those calculations (generalized gradient approximation) is known to overestimate the bulk antiferroelectric distortion [26–28]. This leads to a similarly large Sr bond valence sum in equivalently calculated bulk $SrTiO_3$. While the case of the $(\sqrt{2} \times \sqrt{2})$-R45° surface remains open, the bond valence analysis revealed that the DFT functional used may account for the difference between observed and calculated surface structures.

This could be taken one step further, to evaluating functionals in general, or at least how good they are for a specific system. Since current DFT methods have problems in accurately calculating solid materials in general and solid surfaces in particular, much work has been undertaken to develop more accurate functionals for solids and solid surfaces [29–31]. Since bond valence parameters are empirically derived from crystal structures and are universal, they should provide a good check on these new functionals. A more accurate functional should yield structures with bond valence sum values that are closer to those derived from crystal structures. Similarly, such a functional would be expected to give a lower SII for surface structures experimentally known to be stable, at least as compared to those which are known from experiment to be unstable.

3 Observed Reconstructions on Oxide Surfaces

When looking at surface structures, the question of why specific structures are observed often arises. A total energy approach tells us that the observed structures form because they have the lowest free energy. Even when the observed structures have the lowest free energy, the question often remains: *why* do those surfaces have the lowest free energy?

Bond valence can explain why a specific surface is the most stable. This was recently the case for the $SrTiO_3$ (110) surface. Bond valence analysis explained why particular members of a homologous series of $(n \times 1)$ structures were observed while others were not [6, 15]. The observed members of the homologous series were calculated to be lower in energy than other proposed members of the series or other model surfaces. A bond valence analysis revealed *why* the observed surfaces were the most stable.

In the (110) direction, $SrTiO_3$ is composed of alternating layers of $SrTiO^{4+}$ and $O_2{}^{4-}$ (Fig. 6). This leads to a polar surface, similar to the (111) surface of $SrTiO_3$: a nominal charge of ± 2 per surface unit cell will maintain valence neutrality. Perhaps the simplest way of solving the "polar catastrophe" is to terminate the (110) surface in a half-occupied $O_2{}^{4-}$ layer (halfO_2, Fig. 6) [32–40], or variations thereupon [32–36]. Other commonly considered models for the (110) surface are the TiO faceted model and the Sr faceted model. The TiO faceted model has TiO^{2+} rows in bulk-like positions atop an $O_2{}^{4-}$ termination and is equivalent to the $(\infty \times 1)$ member of the homologous series (see below). The Sr facet model consists of an Sr^{2+}-adatom in a bulk-like position atop at $O_2{}^{4-}$ termination [32–40].

The homologous series consists of chains of corner sharing TiO_4 tetrahedra, which are interspersed by rings of similar corner sharing TiO_4 tetrahedra after every

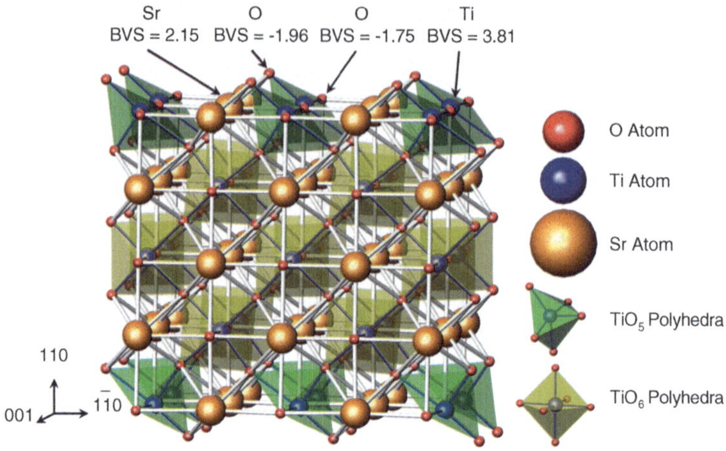

Fig. 6 $SrTiO_3$ in the (110) (*vertical*) direction composed of alternating layers of $SrTiO^{4+}$ and O_2^{4-}. The commonly considered half-O_2 surface model, with the terminal O_2 layer 50% occupied, is shown on *top* and *bottom* (110) facets of the slab

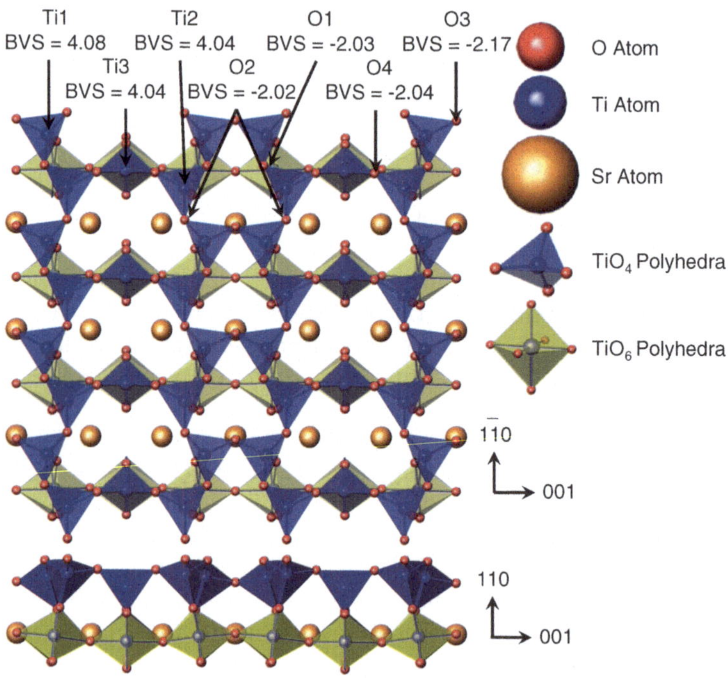

Fig. 7 The $n = 3$ member of the $(n \times 1)$ homologous series of surface structures on $SrTiO_3$ (110) viewed from above (*top*) and parallel to surface (*bottom*). Figure adapted with permission from [6]

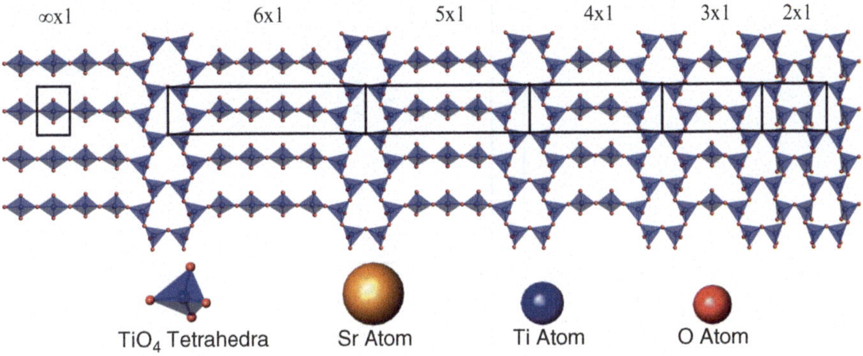

Fig. 8 Homologous series of surface structures on $SrTiO_3$ (110). Figure adapted with permission from [6]

n chain tetrahedra (see Fig. 7 for the $n = 3$ structure, Fig. 8 for entire series) [6]. While titanium normally has a coordination number greater than four in the bulk, there are inherently fewer bonds at the surface. By decreasing the bond length, reasonable bond valence sums can be maintained. By rearranging to a tetrahedral geometry, the valence vector sum rule is satisfied as well. The $n = 3$ and $n = 4$ structures have the best bond valence sums, and the absolute values of the bond valence sums decrease as n increases (Table 1). This is not surprising, as the excess TiO_2 at the surface also decreases with increasing n: the lower the value of n, the more Ti and O atoms are packed into the same area. In general, the coordination in the rings is too high, while that in the chains is too low. Thus interspersing chains with rings at the proper interval leads to structures with optimal bond valence sums. Not only does this agree with DFT calculations of the surfaces (Table 1, Fig. 9), but it also explains why the observed structures are the most stable.

The different members of the homologous series often coexist as intergrowths. Only the (3 × 1) has been observed as a sole phase. The (3 × 1) and (4 × 1), (4 × 1) and (5 × 1), and (5 × 1) and (6 × 1) have been observed as intergrown pairs. The (3 × 1) and (4 × 1) have the best bond valence sums, followed by the (5 × 1), and all have been observed as a major phase on the surface. The (6 × 1) is somewhat underbonded and has only been observed as intergrowths with a (5 × 1) structure. Higher n structures are even more underbonded and have not been observed. On the other end of the series, the overbonded (2 × 1) structure has not been observed. The observed structures have the lowest free energy because they have the optimal bonding.

The (∞ × 1) structure is equivalent to the often calculated TiO faceted model and is higher in free energy than either the lower n members of the homologous series, or the halfO_2 structure. The halfO_2 structure is somewhat reasonable (SII = 0.21). A similar model, with the terminal oxygen bridging between two Sr atoms, has bond valence sums that deviate significantly from the expected values (SII = 0.41), indicating any structure similar to this must be considered highly doubtful, in

Table 1 Comparison of bond valence across homologous series on $SrTiO_3$ (110)

Reconstruction	$(\infty \times 1)$	(6×1)	(5×1)	(4×1)	(3×1)	(2×1)
TiO_2 units per nm^2	4.637	6.183	6.492	6.956	7.728	9.274
Average Ti bond valence sum	3.19	3.87	3.92	3.99	4.05	4.10
Average O bond valence sum	−1.21	−1.87	−1.97	−1.99	−2.07	−2.16
SII	0.57	0.17	0.10	0.07	0.07	0.13
Experimental observation	Not observed	Intergrowths in (5×1)	Major phase	Major phase	Sole phase	Not observed

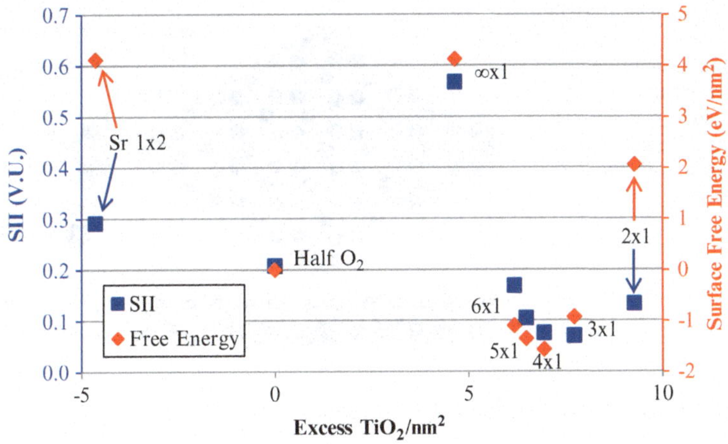

Fig. 9 Comparison of surface free energy [6] and SII for $SrTiO_3$ (110) surface structures. Figure adapted with permission from [15]

agreement with Heifets and coworkers' calculations that this termination is significantly higher in energy than the halfO_2-A [32].

The Sr faceted model has terminal Sr atoms with good bond valence sums [15]. The bond valence sum, however, does not reveal the full instability of the terminal Sr atom. The majority of the coordination sphere for the apical Sr atom is empty. While the valence vector sum has not been calculated, it has been qualitatively noted that this leads to a less stable structure than indicated by the SII [22]. Bond valence theory therefore predicts that this and related structures are not stable, and they have not been observed experimentally.

It is likely that such application of bond valence theory could prove useful in general to understand surfaces. For example, Lazzeri and Selloni [41] described a homologous series of surface structures of $(1 \times n)$ structures ($n = 3$–$6, \infty$) on TiO_2 anatase (001) (although they do not term it a homologous series). Those structures differ in how frequently a row of TiO_2 units is added to the bulk termination (the (1×1) being the $n = \infty$ case). While they do not mention bond valence sum, they report that the bond lengths increased as n increased, which would result in decreased coordination and lower bond valence sum. This is the same as observed for the $SrTiO_3$ (110) homologous series. Interspersing the bulk-like termination anatase (001) termination with TiO_2 rows increased the bonding, just as interspersing the chains with rings increased bonding in the $SrTiO_3$ (110) homologous series. While Lazzeri and Selloni conclude in their work that the number of undercoordinated surface titanium atoms cannot explain the series of reconstructions, they considered only number of bonds, and not bond strength. This is one example (among many) where a bond valence analysis would likely prove helpful in understanding a surface.

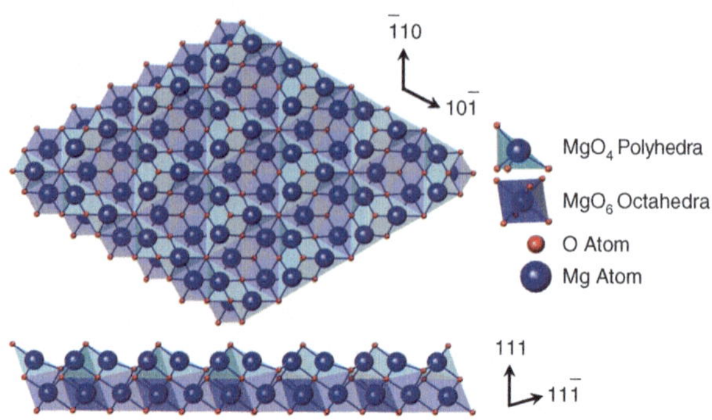

Fig. 10 O terminated MgO octapolar surface viewed from top (*top*) and side (*bottom*). An Mg terminated octapolar surface is formed by replacing each O with an Mg and each Mg with an O

4 Adsorbates on Surface Structures

Bond valence theory offers great potential in explaining the adsorption of foreign species upon surfaces. Under-bonded sites, for example, are far more likely than optimally bonded or over-bonded sites to adsorb foreign species, thereby forming more bonds and increasing their bond valence sums. An understanding of adsorption behavior can lead to a deeper understanding of other properties, such as surface reactivity. As seen in Chap. 7 of this volume [42], bond valence theory has proven useful at this task at aqueous interfaces. Although it has rarely been used for surfaces at a vacuum or gaseous interface, it should be just as useful. Recently the hydroxylated (111) surfaces of the rocksalts MgO [20] and NiO [43] were studied in ultra high vacuum conditions. These structures were then reexamined from a bond valence perspective [15], which shed additional light on the behavior of adsorbed hydroxyl groups on these surfaces. Similar results were obtained on both the MgO and NiO (111) surfaces. The MgO (111) surface will be reviewed here as an example of how bond valence aids in the understanding of surface adsorbates.

In the (111) direction, MgO is composed of alternating magnesium and oxygen layers. This creates a polar surface, similar to the $SrTiO_3$ (111) and (110) surfaces. The most commonly considered surfaces are those proposed by Wolf, which he termed octapoles [2]. They have a p(2 × 2) periodicity with surface atoms in bulk-like positions but are missing ¾ of the atoms in the top layer and ¼ of the atoms in the second layer, essentially creating (100) nanofacets (Fig. 10, Table 2). Because the cation and anion layers have identical geometries and the cations and anions have equal but opposite charges, either magnesium or oxygen atoms could compose either layer, as long as the layers alternate between magnesium and oxygen. Thus there are two types of octapolar surfaces, depending on which type of atom is in the top layer. While such structures are predicted to be stable and have been the subject of much theoretical work [20, 44, 45], they have never been definitively observed experimentally.

Whereas the O terminated octapole is calculated to be slightly lower in energy via DFT, the SII of the Mg terminated octapole is lower (0.13 vs. 0.16). The Mg

Table 2 Bond valence sums for MgO octapolar structure models

	Mg-oct			O-oct			MgH-oct			OH-oct		
SII	0.13			0.16			0.16			0.09		
Layer	Atom	Mult.	BVS	Atom	Mult.	BVS	Atom	Mult.	BVS	Atom	Mult.	BVS
Surface	Mg_1	1	1.83	O_1	1	−1.70	Mg_1	2	1.84	H_2	2	1.02
										H_1	2	1.02
										O_{14}	2	−1.87
										O_1	2	−2.15
Sub-surface	O_1	3	−1.96	Mg_1	3	1.91	H_1	2	1.04	Mg_1	4	1.96
							H_2	2	1.00			
							O_2	2	−2.17	Mg_2	2	2.07
							O_1	4	−1.82			
							O_{15}	2	−2.43			

Table used with permission from [15]

terminated octapole, however, has a terminal cation with most of its coordination sphere empty, in violation of the valence vector sum rule. Thus, the bond valence analysis agrees with the DFT energetics of the octapolar surfaces. Hydroxylated versions of the Mg and O terminated octapoles were also considered by Ciston and coworkers [20]. The MgH octapole is like the dry Mg octapole, with an OH group in the vacant O position in the second layer, and one other O atom in that layer also hydroxylated. In the OH octapole, the terminal O is hydroxylated, and additional OH group bridges two Mg atoms from the second layer. The hydroxylated octapoles were calculated to be low in energy, but with the O termination the lower of the two. This agrees well with the bond valence analysis, where the SII is lower for the OH termination (SII $= 0.09$ vs. 0.16). Like the dry Mg octapole, the MgH octapole fails to satisfy the valence vector sum rule and is even less stable than predicted by bond valence sum alone.

While a surface structure with (2×2) periodicity has been observed, Ciston and coworkers concluded that the observed diffraction intensities do not match with an octapolar structure [20]. Instead, the diffraction data indicates a (2×2)-α type structure. In this structure, the atoms in the terminal layer can occupy any or all of three different possible sites (Fig. 12) and the different occupations are virtually indistinguishable crystallographically. Assuming that Mg and O can only have oxidation states of ± 2 and no adsorbates are present, two of the three sites must be occupied in order to maintain valence neutrality. Like the octapolar structures, the (2×2)-α- can have either a magnesium bulk termination with the surface sites filled by oxygen ((2×2)-α-O), or an oxygen bulk termination with the surface sites filled by magnesium ((2×2)-α-Mg). The (2×2)-α-Mg structure considered by Ciston has all three sites occupied by Mg atoms, would not be valence neutral, and does not appear to exist. A (2×2)-α-Mg structure with two sites occupied would be valence neutral, but the terminal Mg atoms would not satisfy the valence vector sum rule.

The (2×2)-α-O structures considered by Ciston and coworkers are stoichiometric and have two sites occupied by O atoms and the third site vacant, thereby maintaining overall valence neutrality [20]. The three models differ only in which site is left vacant: site 1 is vacant in model 1, site 2 in model 2, and site 3 in model 3 (Fig. 12). The bond valence sums agree with the DFT energies: the SIIs fall in the same order as the DFT energies. In models 1 and 2, the bond valence sum for the O atom in sites 1 and 2 is significantly closer to the atomic valence of 2− than the O atom in site 3. It fits that model 3, where site 3 is vacant, should have the best overall SII (0.25, compared to 0.46 and 0.49 for models 1 and 2, respectively). In model 3, the under-coordination is less severe and more spread throughout the surface structure. By similar logic, the O atom in site 2 is slightly better coordinated than the O atom in site 1. The overall ordering of the coordination of the three oxygen sites is site 2 > site 1 >>site 3. While these models match the diffraction data, they have high surface free energy and SII.

Ciston and coworkers found evidence of hydroxyl groups on the surface via X-ray photoelectron spectroscopy. The hydroxylated versions of the (2×2)-α structures were found to be superior to the clean (2×2)-α structures in terms of

Table 3 Comparison of bond valence sums for MgO (2 × 2)-α surface sites

		2 × 2-α-O1		2 × 2-α-O2		2 × 2-α-O3		2 × 2-α-OH1		2 × 2-α-OH2		2 × 2-α-OH3	
SII		0.46		0.49		0.25		0.22		0.20		0.37	
DFT calculated free energy		28.32		31.96		22.58		13.80		13.72		17.26	
		Atom	BVS	Atom	BVS	Atom	BVS	Atom	BVS	Atom	BVS	Atom	BVS
Surface	Site 1			O_1	−1.83	O_1	−1.51	O_8	−1.59	H_1	0.99	H_2	1.00
										O_8	−2.12	O_8	−2.23
	Site 2	O_1	−2.02			O_2	−1.67	H_1	1.00	O_1	−1.77	H_1	1.00
								O_1	−2.28			O_1	−2.42
	Site 3	O_2	−0.92	O_2	−0.89			H_2	1.03	H_2	1.03	O_2	−0.90
								O_2	−1.93	O_2	−1.94		

Table adapted with permission from [15]

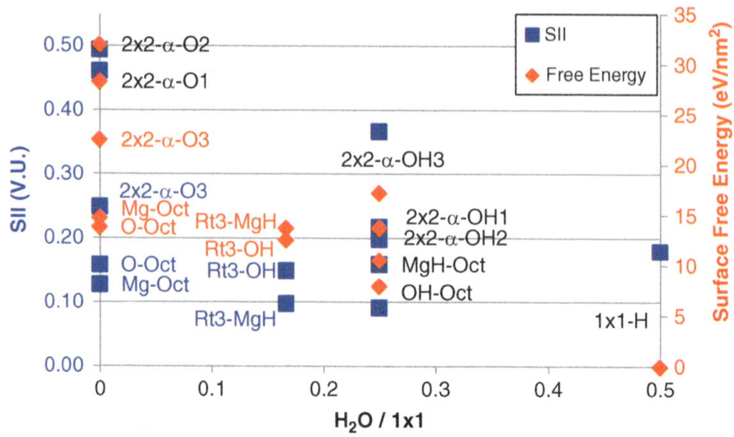

Fig. 11 Comparison of surface free energy [20] and SII [15] for MgO (111) surface structures. Figure adapted with permission from [15]

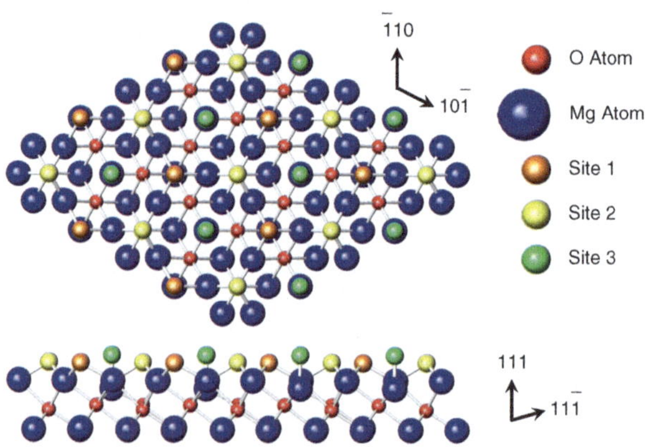

Fig. 12 MgO (111) (2 × 2)-α surface structures viewed from top (*top*) and side (*bottom*)

free energy and SII. If a water molecule were to dissociate upon a dry (2 × 2)-α surface, then all three sites would become occupied: two by OH groups, one by an O atom. In (2 × 2)-α-OH1 site 1 is occupied by the O atom, in (2 × 2)-α-OH2 site 2, and in (2 × 2)-α-OH3 site 3; the other two sites in each case being occupied by hydroxyl groups (Table 3, Figs. 11 and 12). From the O site stability revealed in the bond valence sum analysis of the dry (2 × 2)-α-O structures, one can predict a stability for the hydroxylated structures of model 2 > model 1 >>model 3. This is exactly the trend of the DFT calculated energies for these three structural models. The bond valence sums of the hydroxylated (2 × 2)-α-OH models further confirm this order of stability, both for structures and for individual sites.

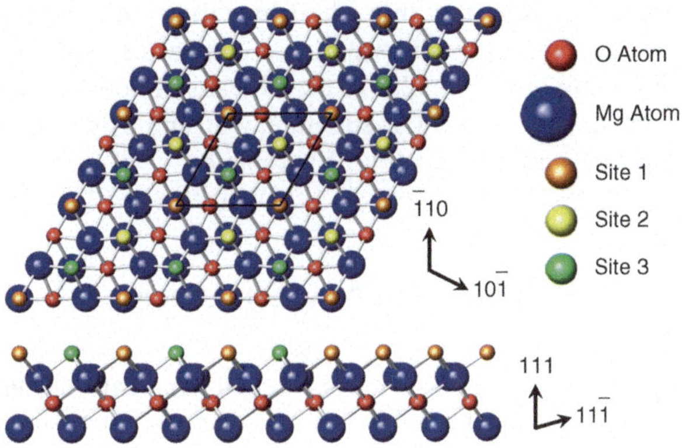

Fig. 13 MgO (111) surface viewed from top (*top*) and side (*bottom*). The (√3 × √3)-R30° unit cell is outlined; in the (√3 × √3)-R30° structure, sites 2 and 3 are occupied by a one hydroxyl group and one oxygen atom per unit cell, while site 1 is vacant. In the 1 × 1-H structure all three sites are occupied by hydroxyl groups

Table 4 Bond valence sums for MgO 1 × 1-H, and √3 × √3-R30°-OH structural models

1 × 1-H			√3 × √3-R30°-OH		
SII = 0.18			SII = 0.15		
Atom	Mult.	BVS	Atom	Mult.	BVS
H	1	1.03	H	1	1.00
O	1	−2.30	O (OH)	1	−2.35
			O	1	−1.73
Mg	1	2.22	Mg	3	1.99

Table adapted with permission from [15]

The dry (2 × 2)-α-O structures have high enough SIIs that they are not likely to exist [15], which agrees with Ciston and coworkers' experimental findings of structures transitioning away from the (2 × 2)-α-O structure as hydroxide was removed through annealing [20]. If a dry (2 × 2)-α-O structure were to exist, it would readily adsorb foreign species: hydroxyl groups would bond most strongly to site 3, with the second hydrogen slightly more likely at site 1 than at site 2. Removal of an H_2O group, conversely, would most likely remove the O from site 3. The bond valence analysis indicates both where water molecules will adsorb and dissociate and where they are likely to re-associate and desorb.

This is consistent with the observations in Ciston et al. [20], who reported that the surface transitioned between the (2 × 2)-α and (√3 × √3)-R30° reconstruction. Samples were initially prepared with a hydroxylated (1 × 1) structure. The (1 × 1)-H structure is a simple (1 × 1) oxygen termination, with a hydrogen bound to each O atom. Annealing this at higher temperatures resulted in a (√3 × √3)-R30° termination, with one third of the hydroxyl content. This was attributed to the loss of some water from the surface. The (√3 × √3)-R30°-OH structure has one O atom and one OH group per (√3 × √3) unit cell atop a magnesium termination, both in sites equivalent to site 2 in the (2 × 2)-α structures. (An Mg terminated (√3 × √3)-R30°-MgH structure was also considered, but it was determined to be kinetically inaccessible [20]. It also fails to satisfy the valence vector sum rule [15]). Upon further annealing at higher temperatures, the (√3 × √3)-R30°-OH structure transitioned to a (2 × 2)-α-OH structure. Ciston et al. [20] concluded that this transition occurred through dehydration and oxygen rearrangement, followed by rehydration as the sample cooled in air. This is consistent with an oxygen atom moving to site 2, followed by adsorption and dissociation of water at site 3, leaving a (2 × 2)-α-OH structure, with site 3 hydroxylated, as expected from bond valence sum analysis (Fig. 13, Table 4).

5 Adsorbates and New Structures

As was shown in the previous section, bond valence analysis can be used to understand how adsorbates bind to structures and how structures reorder. Bond valence analysis can go one step further and *predict* whether adsorbates will bind to a surface and where they will bind. This was done recently for the $SrTiO_3$ (100) surface. In doing so, the bond valence analysis also explained discrepancies between calculation and experiment.

Several $SrTiO_3$ (100) surfaces have been crystallographically solved from surface diffraction data, including the (2 × 1) [46–52], (2 × 2) [50, 52–55], and c(4 × 2) [48, 50, 51, 56–58] surface structures, each with a TiO_2 double layer. A (√2 × √2)-R45° [50, 52] TiO_2 double layer surface structure has been calculated to be lower in energy than any of the three solved surfaces of the same stoichiometry, yet it has never been observed. As discussed above in Sect. 2.4, this may be due to the failings of the functional used in calculations, which leads to high bond valence sums for the bulk strontium atoms (>2.32). The (2 × 2) and c(4 × 2) structures are calculated to be similar to each other in energy and to have significantly lower free energy than the (2 × 1) structure. This has raised questions as to why the (2 × 1) structure is experimentally observed.

Bond valence analysis confirms that the (2 × 2) and c(4 × 2) structures should be preferable to the (2 × 1) structure, at least as the initial models of the structure exist. Both are composed of TiO_5 polyhedra which edge and corner share in different patterns. The (2 × 2) structure [50, 52–55] shows nearly ideal bond valence sums, although the top bulk layers are only slightly over-bonded, leading to an acceptable SII of 0.14. The c(4 × 2) structure [48, 50, 51, 56–58] is also reasonable (SII = 0.17). The surface layer has good bond valence sums, except for an O atom in the middle of four TiO_5 polyhedra, which is significantly over-bonded (bond valence sum = −2.56). Overall, the bond valence sums are similar to the (2 × 2) structure, which agrees well with DFT calculations where the two structures are similar in calculated surface energy (within 0.06 eV).

The (2 × 1) structure [46–52] is surprisingly under-bonded at the surface for a structure which has been crystallographically solved (Fig. 14). The top two surface O atoms, including the "dangling oxygen" are both under-bonded (both with bond valence sum = −1.74) as is the Ti atom not bonded to the "dangling oxygen" (bond valence sum = 3.65), leading to a high SII of 0.22. This is in agreement with DFT calculations, which find the (2 × 1) structure higher in surface free energy than the (2 × 2) or c(4 × 2) by >3.5 eV/nm^2. The undercoordination suggests that this structure may be unstable as is, and would be more stable if it were able to form

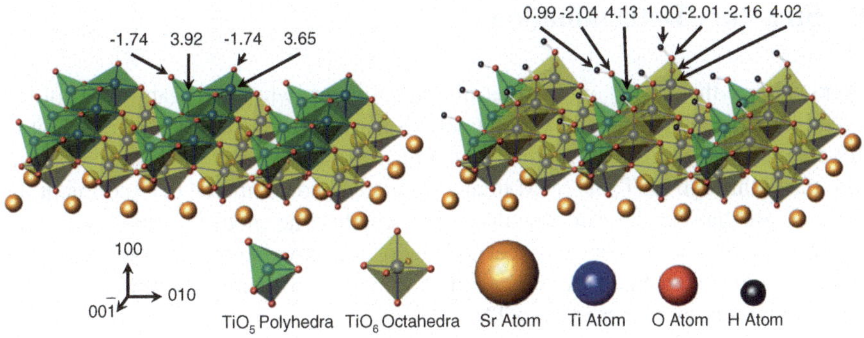

Fig. 14 SrTiO3 (100) (2 × 1) surface structures. Left: dry structural model. Right: wet structure. Bond valence sums are indicated for each unique surface atom

additional bonds to increase coordination. The bond valence analysis also allowed the prediction of where the adsorbates would bond: it was hypothesized that a hydrogen atom would bond to the under-bonded "dangling oxygen," while a hydroxide group would bond to the under-bonded surface titanium atom [59]. This dissociated water molecule would maintain the overall valence neutrality.

Unfortunately, 2D diffraction data could not distinguish between the dry and wet (2 × 1) surfaces, as light scattering hydrogen atoms could not be found, and the additional oxygen atom was directly above a much stronger scattering titanium atom. X-ray photoelectron spectroscopy confirmed that hydroxide groups were present on the (2 × 1) surface and were not easily removed even upon annealing at 750°C [14]. For comparison, hydroxide groups on the c(4 × 2) structure were removed by annealing at 300°C. The DFT optimized wet (2 × 1) structure showed a marked improvement in bond valence sums, with an SII decrease from 0.22 to 0.12 (Fig. 14). Further, the DFT energetic showed that when all surfaces were hydroxylated, the energy of the (2 × 1) was comparable to the c(4 × 2) and (2 × 2) structures [14]. As final confirmation, the simulated STM images of the wet (2 × 1) surface matched experimental STM images [13]. Thus it was the presence of dissociated water, predicted via a bond valence analysis, which reconciled the differences between DFT and experiment.

Another example of applying bond valence theory to predict new structures comes from the $(\sqrt{13} \times \sqrt{13})R33.7°$ surface reconstruction on $SrTiO_3$ (100) [12]. This structure was crystallographically solved, and bond valence sums were used to confirm the structure. The structure is composed of edge and corner sharing TiO_5 polyhedra. Kienzle and coworkers then extrapolate from the solved structure and create models for two other structures, a (3 × 3) and a $(\sqrt{5} \times \sqrt{5})R26.6°$ by tiling the valence-neutral units found in the $(\sqrt{13} \times \sqrt{13})R33.7°$ reconstruction in different patterns. They conclude that many structures, both ordered and glass-like, are possible on the $SrTiO_3$ (100) surface, as long as bond valence requirements are locally satisfied.

6 Structure Determination

Bond valence theory can aid in determining structures when they cannot be solved experimentally. For example, when a periodicity is known from STM images, it is not trivial to determine an atomic structure. Normally models will be proposed, calculated through DFT, the STM image simulated and compared to experiment. In the end, all that can be said definitively is whether the model is consistent with experiment or not. Recently Becerra-Toledo and coworkers incorporated bond valence sums into this process, greatly aiding in structure determination of a series of structures observed via STM [11].

The authors of this study propose models for "diline" and "triline" features that appear in STM images of the $SrTiO_3$ surface. These features appear either upon annealing a c(4 × 2) reconstructed surface or concurrent with the c(4 × 2) reconstruction. The model structures for the "diline" and "triline" are therefore based on the c(4 × 2) reconstruction. In proposing the "diline" model structure, bond valence sums were used as a constraint. In the "triline" structure, reduced "Ti^{2+}" was observed in the X-ray photoelectron spectra. The authors model for the "triline" structure contained some significantly reduced titanium atoms, with bond valence sum of 2.87 when using the Ti^{3+} - O^{2-} R_0 value. This was compared to the bond valence sum for Ti in a rocksalt "TiO structure" taken from the literature, using the same R_0 value.

7 Multiple Valence States

Except for the Becerra-Toledo study of the c(4 × 2) related structures [11], bond valence has so far been applied at the surface solely on structures where each surface atom is in its optimal oxidation state. Bond valence theory is well equipped to deal with oxidation and reduction. Bond valence sums have been especially useful in bulk structures where an element occurs in multiple oxidation states within the same structure, as it provides an excellent way, sometimes the only way, to assign oxidation state to each occurrence of the element [7] (see Chap. 2 of this volume [16]). Bond valence theory should prove equally useful for similar applications on surfaces.

One set of structures where such an analysis would be beneficial are the Sr adatom models proposed for several $SrTiO_3$ ($n \times n$) reconstructions [60–62]. These models match experimental STM images quite well. However, having an Sr adatom means that the structure must somewhere be reduced by 2 electrons per Sr adatom. A bond valence analysis would quickly reveal the location and effects of this reduction. This could then be compared with experimental data, such as XPS, to see if the reduction in the model structure matches with some experimental observation, as was done by Becarra-Toledo and coworkers for the $SrTiO_3$ "trilines" [11]. Including other considerations from bond valence theory, such as the valence vector sum rule, would help reveal whether and how such a metal adatom could be stable on a surface. This is but one system among many where bond valence theory could aid in the analysis of nonideal valence states at surfaces.

8 Conclusions

Chemical bonding goes far in explaining surface structures, with a bond valence sum calculation capable of illuminating much of this stability. A bond valence analysis has the power to predict where water is likely to adsorb and dissociate on a surface. Under-bonded structures are likely to adsorb foreign species. A structure without any under-bonded species on the surface might be more robust and less likely to adsorb any foreign species. Conversely, something on an over-bonded surface structure might dissociate in order to lower the bond valence sums. Bond valence theory, in predicting adsorption of foreign species, has led to the reconciliation of calculated and observed surfaces. Bond valence theory has also been used to assess calculated structural models absolutely, whereas DFT alone could only determine if structures were better or worse than other models, such as in the case of the $SrTiO_3$ (111) surface.

Chemical bonding models of surfaces generally agree with physics-based calculations and with experiment, just as they do for bulk structures. In each case examined so far where the SII and DFT calculated energy disagree, other chemical bonding theories (usually the valence vector sum rule) could explain this discrepancy. Additionally, the bond valence model has been shown to easily describe concepts, such as polar surfaces, which have been most difficult to deal with from a purely physics-based approach. In some cases bond valence sums may help for systems which are difficult for DFT. In the most difficult cases, it may prove necessary to use chemical and physics-based methods in conjunction to achieve the most complete understanding of a surface. For instance, the fact that in DFT the $(\sqrt{2} \times \sqrt{2})$-R45° (001) $SrTiO_3$ surface has the lowest energy, but it does not have the lowest SII implies that their might be a problem with this surface where DFT is underestimating the oxygen–oxygen nonbonded repulsions.

Clearly, a chemical bonding approach can enhance nearly any surface study. The complementary nature of the chemistry and physics-based models can lead to a greatly enhanced understanding of surface structure, chemistry, and reactivity. As with bulk materials, the best way to move forward is to consider the chemistry and physics simultaneously, with each providing insight that is difficult to reach from the alternate approach.

Acknowledgment This work is supported in part by the U.S. Department of Energy, Basic Energy Sciences, Office of Science, under contract # DE-AC02-06CH11357, and in part by the Institute for Catalysis in Energy Processing, a collaborative research effort between the Northwestern University Center for Catalysis and Surface Science and Argonne National Laboratory, funded through the US Department of Energy, Office of Basic Energy Science, award number DE-FG02-03-ER15457.

References

1. Pashley MD (1989) Electron counting model and its application to island structures on molecular-beam epitaxy grown GaAs(001) and ZnSe(001). Phys Rev B Condens Matter 40(15):10481–10487
2. Wolf D (1992) Reconstruction of NaCl surfaces from a dipolar solution to the Madelung problem. Phys Rev Lett 68(22):3315–3318
3. Goniakowski J, Finocchi F, Noguera C (2008) Polarity of oxide surfaces and nanostructures. Rep Prog Phys 71(1):016501
4. Deak DS (2007) Strontium titanate surfaces. Mater Sci Technol 23(2):127–136
5. Noguera C (2000) Polar oxide surfaces. J Phys Condens Matter 12(31):R367–R410
6. Enterkin JA, Subramanian AK, Russell BC, Castell MR, Poeppelmeier KR, Marks LD (2010) A homologous series of structures on the surface of $SrTiO_3$ (110). Nat Mater 9(3):245–248
7. Brown ID (2009) Recent developments in the methods and applications of the bond valence model. Chem Rev 109(12):6858–6919
8. Ruberto C, Yourdshahyan Y, Lundqvist BI (2002) Stability of a flexible polar ionic crystal surface: metastable alumina and one-dimensional surface metallicity. Phys Rev Lett 88(22): 226101
9. Ruberto C, Yourdshahyan Y, Lundqvist BI (2003) Surface properties of metastable alumina: a comparative study of kappa- and alpha-Al_2O_3. Phys Rev B Condens Matter 67(19):195412
10. Lin Y, Becerra-Toledo AE, Silly F, Poeppelmeier KR, Castell MR, Marks LD (2011) The (2 × 2) reconstructions on the $SrTiO_3$ (001) surface: a combined scanning tunneling microscopy and density functional theory study. Surf Sci 605(17–18):L51–L55
11. Becerra-Toledo AE, Marshall MSJ, Castell MR, Marks LD (2012) c(4 × 2) and related structural units on the $SrTiO_3$ (001) surface: scanning tunneling microscopy, density functional theory, and atomic structure. J Chem Phys 136(21):214701
12. Kienzle DM, Becerra-Toledo AE, Marks LD (2011) Vacant-site octahedral tiltings on $SrTiO_3$ (001), the ($\sqrt{13} \times \sqrt{13}$) $R33.7°$ survace, and related structures. Phys Rev Lett 106(17):176102
13. Becerra-Toledo AE, Castell MR, Marks LD (2012) Water adsorption on $SrTiO_3$ (001): I. experimental and simulated STM. Surf Sci 606(7–8):762–765
14. Becerra-Toledo AE, Enterkin JA, Kienzle D, Marks LD (2012) Water adsorption on $SrTiO_3$ (001): II. water, water, everywhere. Surf Sci 606(9–10):791–802
15. Enterkin JA, Becerra-Toledo AE, Poeppelmeier KR, Marks LD (2012) A chemical approach to understanding oxide surfaces. Surf Sci 606(3–4):344–355
16. Brown D (2013) Bond valence theory. Struct Bond. doi:10.1007/430_2012_89
17. Adams S (2013) Practical considerations in determining bond valence parameters. Struct Bond. doi:10.1007/430_2013_96
18. Bickmore BR, Rosso KM, Nagy KL, Cygan RT, Tadanier CJ (2003) Ab initio determination of edge surface structures for dioctahedral 2: 1 phyllosilicates: implications for acid–base reactivity. Clays Clay Miner 51(4):359–371
19. Etxebarria I, Perez-Mato JM, Garcia A, Blaha P, Schwarz K, Rodriguez-Carvajal J (2005) Comparison of empirical bond-valence and first-principles energy calculations for a complex structural instability. Phys Rev B Condens Matter 72(17):174108
20. Ciston J, Subramanian A, Marks LD (2009) Water-driven structural evolution of the polar MgO (111) surface: an integrated experimental and theoretical approach. Phys Rev B Condens Matter 79(8):085421
21. Gillespie RJ, Nyholm RS (1957) Inorganic stereochemistry. Q Rev Chem Soc 11(4):339–380
22. Harvey MA, Baggio S, Baggio R (2006) A new simplifying approach to molecular geometry description: the vectorial bond-valence model. Acta Crystallogr Sect B-Struct Sci 62: 1038–1042
23. Chiaramonti AN, Lanier CH, Marks LD, Stair PC (2008) Time, temperature, and oxygen partial pressure-dependent surface reconstructions on $SrTiO_3$ (111): a systematic study of oxygen-rich conditions. Surf Sci 602(18):3018–3025

24. Marks LD, Chiaramonti AN, Tran F, Blaha P (2009) The small unit cell reconstructions of $SrTiO_3$ (111). Surf Sci 603(14):2179–2187
25. Marks LD, Chiaramonti AN, Tran F, Blaha P (2009) CIF depository for the small unit cell reconstructions of $SrTiO_3$ (111). http://www.wien2k.at/Depository/SurfSci_STO111_Marks/
26. Wu ZG, Cohen RE, Singh DJ (2004) Comparing the weighted density approximation with the LDA and GGA for ground-state properties of ferroelectric perovskites. Phys Rev B Condens Matter 70(10):104112
27. Tinte S, Stachiotti MG, Rodriguez CO, Novikov DL, Christensen NE (1998) Applications of the generalized gradient approximation to ferroelectric perovskites. Phys Rev B Condens Matter 58(18):11959–11963
28. Heifets E, Kotomin E, Trepakov VA (2006) Calculations for antiferrodistortive phase of $SrTiO_3$ perovskite: hybrid density functional study. J Phys Condens Matter 18(20):4845–4851
29. Perdew JP, Ruzsinszky A, Csonka GI, Vydrov OA, Scuseria GE, Constantin LA, Zhou XL, Burke K (2008) Restoring the density-gradient expansion for exchange in solids and surfaces. Phys Rev Lett 100(13):136406
30. Perdew JP, Ruzsinszky A, Csonka GI, Constantin LA, Sun JW (2009) Workhorse semilocal density functional for condensed matter physics and quantum chemistry. Phys Rev Lett 103(2):026403
31. Stroppa A, Kresse G (2008) The shortcomings of semi-local and hybrid functionals: what we can learn from surface science studies. New J Phys 10(6):063020
32. Heifets E, Goddard WA, Kotomin EA, Eglitis RI, Borstel G (2004) Ab initio calculations of the $SrTiO_3$ (110) polar surface. Phys Rev B Condens Matter 69(3):035408
33. Heifets E, Kotomin EA (2000) Atomistic simulation of $SrTiO_3$ and $BaTiO_3$ (110) surface relaxations. Thin Solid Films 358(1–2):1–5
34. Heifets E, Kotomin EA, Borstel G (1999) Atomistic calculations of (110) surface relaxation for perovskite titanates. Surf Rev Lett 6(6):1215–1219
35. Heifets E, Kotomin EA, Maier J (2000) Semi-empirical simulations of surface relaxation for perovskite titanates. Surf Sci 462(1–3):19–35
36. Kotomin EA, Heifets E, Dorfman S, Fuks D, Gordon A, Maier J (2004) Comparative study of polar perovskite surfaces. Surf Sci 566:231–235
37. Bottin F, Finocchi F, Noguera C (2003) Stability and electronic structure of the (1×1) $SrTiO_3$ (110) polar surfaces by first-principles calculations. Phys Rev B Condens Matter 68(3):035418
38. Bottin F, Finocchi F, Noguera C (2003) Ab-initio study of the polar $SrTiO_3$ (110)(1×1) surfaces. Surf Sci 532:468–471
39. Bottin F, Finocchi F, Noguera C (2005) Facetting and $(n \times 1)$ reconstructions of $SrTiO_3$ (110) surfaces. Surf Sci 574(1):65–76
40. Eglitis RI, Vanderbilt D (2008) First-principles calculations of atomic and electronic structure of $SrTiO_3$ (001) and (011) surfaces. Phys Rev B Condens Matter 77(19):195408
41. Lazzeri M, Selloni A (2001) Stress-driven reconstruction of an oxide surface: the anatase TiO_2(001)-(1×4) surface. Phys Rev Lett 87(26):266105
42. Bickmore B (2013) Structure and acidity in aqueous solutions and oxide–water interfaces. Struct Bond. doi:10.1007/430_2012_84
43. Ciston J, Subramanian A, Kienzle D, Marks LD (2010) Why the case for clean surfaces does not hold water: structure and morphology of hydroxylated nickel oxide (111). Surf Sci 604 (2):155–164
44. Finocchi F, Barbier A, Jupille J, Noguera C (2004) Stability of rocksalt (111) polar surfaces: beyond the octopole. Phys Rev Lett 92(13):136101
45. Zhang WB, Tang BY (2008) Stability of MgO(111) polar surface: effect of the environment. J Phys Chem C 112(9):3327–3333
46. Erdman N, Poeppelmeier KR, Asta M, Warschkow O, Ellis DE, Marks LD (2002) The structure and chemistry of the TiO_2-rich surface of $SrTiO_3$ (001). Nature 419(6902):55–58
47. Johnston K, Castell MR, Paxton AT, Finnis MW (2004) $SrTiO_3$ (001)(2×1) reconstructions: first-principles calculations of surface energy and atomic structure compared with scanning tunneling microscopy images. Phys Rev B Condens Matter 70(8):085415

48. Castell MR (2002) Scanning tunneling microscopy of reconstructions on the $SrTiO_3$ (001) surface. Surf Sci 505(1–3):1–13
49. Naito M, Sato H (1994) Reflection high-energy electron-diffraction study on the $SrTiO_3$ surface-structure. Physica C (Amsterdam, Neth) 229(1–2):1–11
50. Warschkow O, Asta M, Erdman N, Poeppelmeier KR, Ellis DE, Marks LD (2004) TiO_2-rich reconstructions of $SrTiO_3$ (001): a theoretical study of structural patterns. Surf Sci 573 (3):446–456
51. Erdman N, Marks LD (2003) $SrTiO_3$ (001) surface structures under oxidizing conditions. Surf Sci 526(1–2):107–114
52. Iles N, Finocchi F, Khodja KD (2010) A systematic study of ideal and double layer reconstructions of ABO_3 (001) surfaces (A = Sr, Ba; B = Ti, Zr) from first principles. J Phys Condens Matter 22(30):305001
53. Herger R, Willmott PR, Bunk O, Schleputz CM, Patterson BD, Delley B (2007) Surface of strontium titanate. Phys Rev Lett 98(7):076102
54. Herger R, Willmott PR, Bunk O, Schleputz CM, Patterson BD, Delley B, Shneerson VL, Lyman PF, Saldin DK (2007) Surface structure of $SrTiO_3$ (001). Phys Rev B Condens Matter 76(19):195435
55. Jiang QD, Zegenhagen J (1995) $SrTiO_3$ (001) surfaces and growth of ultra-thin $GdBa_2Cu_3O_{7-X}$ films studied by LEED/AES and UHV-STM. Surf Sci 338(1–3):L882–L888
56. Erdman N, Warschkow O, Asta M, Poeppelmeier KR, Ellis DE, Marks LD (2003) Surface structures of $SrTiO_3$ (001): a TiO_2-rich reconstruction with a c(4 × 2) unit cell. J Am Chem Soc 125(33):10050–10056
57. Castell MR (2002) Nanostructures on the $SrTiO_3$ (001) surface studied by STM. Surf Sci 516(1–2):33–42
58. Jiang QD, Zegenhagen J (1999) c(6 × 2)and c(4 × 2)reconstruction of $SrTiO_3$ (001). Surf Sci 425(2–3):343–354
59. Enterkin JA (2010) A chemical approach to understanding oxide surface structure and reactivity. Northwestern University, Evanston
60. Kubo T, Nozoye H (2001) Surface structure of $SrTiO_3$ (100)-($\sqrt{5} \times \sqrt{5}$)-R26.6°. Phys Rev Lett 86(9):1801–1804
61. Liborio LM, Sanchez CG, Paxton AT, Finnis MW (2005) Stability of Sr adatom model structures for $SrTiO_3$ (001) surface reconstructions. J Phys Condens Matter 17(23):L223–L230
62. Kubo T, Nozoye H (2003) Surface structure of $SrTiO_3$ (100). Surf Sci 542(3):177–191

Struct Bond (2014) 158: 233–250
DOI: 10.1007/430_2012_83

Published online: 6 April 2013

Bond Valences in Education

I. David Brown

Abstract The bond valence theory's simplicity, its rigor, and predictive power make it ideally suited for introducing the concept of the chemical bond in introductory courses. This chapter suggests how the theory might be presented in a classroom. It starts with a critique of the bond models currently used and then shows how different aspects of the theory might be introduced, beginning with the simple exercise of plotting the lines of field for an array of charges like those found in the ionic model. The generation of stable bonds using the valence matching rule leads naturally to a discussion of bonding geometry and the influence of lone pairs. Deriving the classic ball-and-stick model from bond valence theory gives an opportunity to discuss how one might define a covalent bond if this were thought to be useful. The chapter ends showing how the theory can be applied to aspects of chemical reactivity.

Keywords Bond geometry · Bond valences · Covalent bonds · Hydrogen bonds · Ionic model · Lewis acids and bases · Lone pairs · Teaching · Valence matching

Contents

I.D. Brown (✉)
Brockhouse Institute for Materials Research, McMaster University, Hamilton, ON, Canada LS8 4M1
e-mail: idbrown@mcmaster.ca

1 Why Do We Need Another Model of Chemical Bonding?

Chemists introduced the modern concept of the atom around 1800 and that of the bond around 1860. With these two ideas, they were able to develop a topological model that systematized the bewildering variety of organic compounds which they were then discovering. But the evidence for the reality of atoms and bonds was entirely circumstantial. The lack of evidence for the physical existence of either atoms or bonds caused physicists to be cautiously skeptical; they refused to accept the reality of either atoms or bonds. Not until 1900 were they convinced, and then only when they found ways to measure the properties of individual atoms and were able to develop quantum mechanics to show how electrons and nuclei combined to form atoms.

The reality of the atom was then unquestioned, but the bond remained a hypothetical entity; there was nothing in quantum mechanics that led to a definition that bore any resemblance to the chemists' ball-and-stick model of the bond. The success of the bond model is undeniable; even at the present day it is universally used to describe chemical structure, particularly in organic chemistry, but the failure to find a theoretical basis for the model has been an embarrassment. This failure was unavoidable as long as people sought it in quantum mechanics, since there is an essential incompatibility between the chemical bond and the quantum views of the atom.

In order to understand this incompatibility we need to examine the often overlooked assumption of the bond model that all chemical structures can be explained in terms of an interaction between nearest neighbor atoms, as determined by the valences of the two atoms. The valence in the original ball-and-stick model is today represented by two concepts, the number of bonding electrons supplied by the atom (the atomic valence) and the number of localized bonds that the atom forms (the coordination number). For good reason, neither of these concepts can be derived from the quantum model.

The quantum model was developed using the physics approach of finding the lowest energy arrangement of electrons around a given array of nuclei. This energy can be calculated by solving the Schrödinger equation, but the Coulomb potential used in this calculation requires that it explicitly includes the interactions between each charge and every other charge in the system, a condition that is clearly incompatible with a localized bond picture. Further, in the quantum description of an atom, or more generally of a molecule, the electrons cannot be individually identified. One cannot identify particular portions of the total electron density as being the contribution of the valence electrons. All that can be measured is the total electron density; the contributions of the individual electrons remain unknowable.

The collection of electrons behaves as a single entity, making it impossible to count how many valence electrons are present, much less where they are located. If the electrons cannot be located, they cannot be counted, and the bonds cannot be counted if they are not localized. Neither the valence of the atom nor the number of bonds the atom forms can be determined using quantum mechanics.

This does not mean that the bond model is wrong; it has proved itself remarkably useful and will continue to do so. It does mean that we can never find the physical basis for the bond model by appealing to quantum mechanics. If a theory of the chemical bond is to be developed, it must be based on a different view of the atom, one that recognizes, on the one hand, the functional difference between the valence and core electrons and, on the other, that a force field is used that gives a description of the atomic interactions in terms of nearest neighbors alone.

Why, one might ask, is it necessary to have a theoretical derivation of the chemical bond when the model already works so well? There are several reasons. Firstly, any theory starts with a number of assumptions which define the scope of the theory. For example, the assumptions made in Sect. 5 of [1] show why the ionic model is able to describe the properties of covalent as well as ionic bonds, even though it would seem self-evident that the ionic model should be restricted to ionic compounds. Secondly, a theory allows for the derivation of theorems that may not have already been discovered empirically. For example, by treating the ionic model as a capacitive electrical circuit, bond valence theory is able to predict the lengths of the bonds in a simple and straightforward way as is shown in Sect. 5 of [1]. Thirdly, the classification of bonds is much simpler because the concepts of the model are all mutually compatible. For example, bond valence theory makes no distinction between ionic and covalent bonds, and the arrangement of bonds around an atom is determined by the properties of its ligands rather than by the supposed configuration of orbitals that have no real existence (Sect. 6.2 of [1]). Fourthly, bond valence theory is based on a simple but physically correct atomic picture, which can be analyzed to understand the physical origin of bonding anomalies such as dative bonds (Sect. 2 of [1]). For all of these reasons it is an excellent theory for introducing the concept of the chemical bond.

Since a localized bond model requires information neither about the distribution of electron density nor about the energy of the molecule, quantum mechanics is not needed. Bond valence theory determines the number of bonding electrons (atomic valence) by counting how many electrons have a small ionization energy, and it develops the bond picture using the electrostatic field rather than the electrostatic potential. In the ionic model representation of bond valence theory, a bond exists between any two atoms linked by lines of electric field and the number of these lines linking two atoms (the electrostatic flux) is a measure of the number of electrons used to form the bond. This is the foundation on which the model is built.

The quantum mechanical and bond valence theories can thus be seen as complementary descriptions, both being exact representations of the Coulomb force, one expressed through the electric potential, the other through electric field. The quantum model gives the energy of a molecule or crystal and describes its excited states. It also gives the distribution of the electron density, but it does not give the valence of the atom or a description of the structure in terms of localized bonds. Bond valence theory, on the other hand, uses the number of valence electrons given by the ionization energies to describe the bond network as well as the bonding geometry of the molecule or crystal, but it says nothing about the location of the electrons or the energy of the system. Quantum theory gives a physicist's view of the atom, and bond valence theory gives a chemist's view. It describes how a collection of atoms will arrange themselves in space, and the chemical properties that derive from the resulting structure.

Bond valence theory is developed in Chap. 2 of this volume [1]. This chapter shows how it might be used to introduce the concept of the chemical bond in an introductory chemistry course. Details of the theory are not repeated here, though pointers to the relevant sections of [1] are given. The purpose of the rest of this chapter is to suggest ways in which the bond valence theory might be used to present chemical bonding in a classroom setting.

2 Introducing the Bond Valence Theory

The force that holds atoms and molecules together is the electrostatic Coulomb force, traditionally represented by the Coulomb potential, but the electric field has properties that seem to be custom designed for the description of chemical bonds. The Faraday lines of field give a quantitative graphical view; the number of field lines being equal to the charge that generates them. Since it is a characteristic of atoms and chemical compounds that positive and negative charges are uniformly distributed over short distances, the lines of field that leave from a positive charge always terminate on a nearby negative charge. The electric field is therefore an ideal way to illustrate a localized bonding interaction.

The ionic model is a straightforward way to introduce chemical bonding. Students can be given a picture of an array of positive and negative charges such as that shown in Fig. 1. If they are asked to draw the lines of field, they will likely end up with a picture like that shown in Fig. 2.

Figure 2 immediately leads to the definition of a bond: any two atoms that are linked by lines of field are connected by a bond. Further, the number of lines (the electrostatic flux) is a direct measure of the strength of the bond. If the charges of the ions are set equal to their atomic valence (the number of electrons the atom uses for bonding), the electrostatic flux that forms the bond is called the bond valence. From this simple construction comes the most important rule of the bond valence theory: the valence sum rule (Eq. 2 in [1]), which states that the valence of an atom is equal to the sum of the valences of the bonds it forms. This is Gauss' law of electrostatics.

It should also be apparent that if two atoms are brought closer to each other, the number of lines of field linking them will increase. This leads us to expect a correlation between the valence (electrostatic flux) and the length of the bond (Eq. 4 in [1]). Although the length of a bond is not likely to be of great interest in

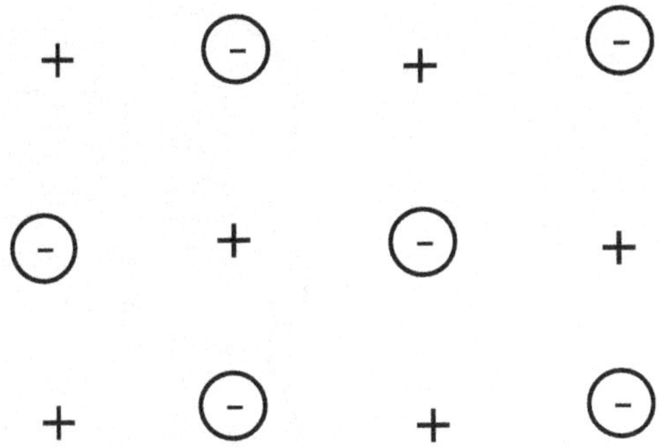

Fig. 1 An array of positive and negative charges as found in the ionic model

Fig. 2 The charge array of Fig. 1 with lines of field shown

an introductory course, this correlation is the essential link between the bond valence theory and experiment. Using this correlation, we can convert observed bond lengths into bond valences to see how an atom distributes its valence between the bonds it forms. It allows us to test whether the predictions of the model are correct, but it can also be used to test whether a given structure is correct by checking how well the sums of the experimentally determined bond valences agree with the atomic valences.

The ionic model provides a simple and painless introduction to bond valence theory. It introduces not only the concept of a bond but also the notion that an atom shares its valence among the bonds that it forms in a way that can be easily determined by experiment. In a more advanced course students who have done some elementary circuit theory in physics could be introduced to the network equations (Eq. 14 in [1]) derived by showing that the bond network is equivalent to a capacitive electrical circuit. The bond valence is the quantity that in other models is approximated by bond order or bond number, but only the bond valence rigorously obeys the valence sum rule.

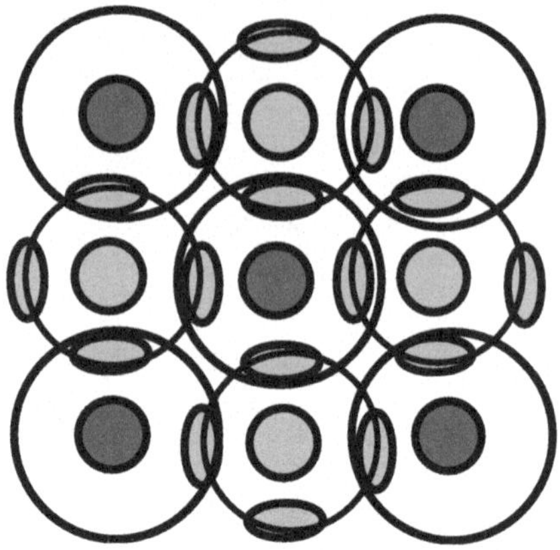

Fig 3 The filled valence shell of the anion (*light core*) is split into four bonding segments which overlap with the valence shell of the cations (*dark core*)

As shown in Sect. 5 of [1] the ionic model is by no means restricted to structures containing only ionic bonds; it applies to the majority of structures with localized bonds including many organic compounds. It is limited only by its inability to describe the bonds formed between two cations or between two anions. If such bonds are present, a more sophisticated model is needed, one in which there is no distinction between the anion and the cation. The core-and-valence-shell picture of the atom is described in Sect. 2 of [1].

One can convert the ionic model to the core-and-valence-shell picture using Fig. 3 where the lightly shaded circles are the anion cores and the darkly shaded ones the cation cores. The valence shells are indicated by the larger unshaded circles. The electrons in the overlap region, which in the ionic model are assumed to be associated with the anions, are now divided into the bond segments shown as shaded ovals in Fig. 3. As the cation and anion cores each contributes an equal numbers of valence electrons to a given segment, the electrostatic flux linking each overlap segment to the anion core is the same as the flux linking it to the cation core, and both are equal to the bond valence. All that remains is to recognize that each bond segment lies at the intersection of the valence shells of the two atoms it links. There is now no distinction between the cation and anion since both are treated the same way. This allows the core-and-valence-shell model to be used to describe any localized bond, cation–cation and anion–anion bonds as well as covalent and ionic bonds.

The core-and-valence-shell picture shown in Fig. 2 in [1] provides a simple but realistic picture of an atom that can be used to discuss how a bond behaves when two atoms come together. Their valence shells overlap and the electrons in the overlap region now experience the positive charges of both nuclei, pulling them towards the overlap region, holding the two atoms together and forming the bond as shown in Fig. 4 in [1]. In this respect bond valence theory is similar to the Lewis model, but without the restriction that a bond requires an integral number of electron pairs. Removing this restriction removes the need to introduce the unnecessary concept of resonance. The picture is also compatible with the electron density picture of the atom, though simplified by focusing on what is important for the bond model, namely the number of valence electrons, and leaving the quantum description to be added later.

Two properties of the atom are important in bonding: its valence, i.e., the number of electrons it uses for bonding and its size. The size is difficult to define precisely since the atom does not have a sharp surface and bonds tend to be longer or shorter according to their valence, but the size is important in determining how many bonds an atom can form. The reverse idea can also be used. The number of bonds an atom forms can be used as a measure of its size. Coordination numbers for a particular atom are determined by experiment, and they show some variation, depending on the composition of the compound and nature of the ligands. Some elements such as sulfur(VI) are found with only one coordination number to oxygen (four), but others such as sodium are known with all the coordination numbers between three and twelve. In this case one can take the average (6.2), and since oxygen is the most common ligand, the average can be restricted to oxygen ligands. It may be different for other ligands, but the trends will be the same. From these average coordination numbers one can calculate the average valence of the bonds the atom forms. This is known as the bonding strength of the atom (Eq. 7 in [1]) since it can be used as an estimate of the valence of the bonds the atom is expected to form. The result is the list of bonding strengths shown in Table 1 in [1]. This table divides the bonding strengths into those of cations and those of anions because most compounds can be described using the ionic model.

3 Bonding Strength, Electronegativity, and Valence Matching

The valence matching rule (Eq. 9 in [1]) follows directly from the definition of bonding strength defined in the previous section. It expresses the idea that for a bond to form between two atoms their bonding strengths should be the same, or at least similar. It is normally found that stable bonds will not be formed between atoms with bonding strengths that differ by more than a factor of two, unless one of the atoms has a valence shell with lone pairs, a case treated more fully in Sect. 5. Sect. 11.2 of [1] shows how bonding strengths can be used to construct a network of bonds, pointing out how the network is dominated by the atoms with larger bonding strength because they form stronger bonds. When a salt dissolves in water, for example, it is the weaker bonds that break first; the more strongly bonded complexes remain intact, usually in the form of complex ions. When Na_2SO_4 reacts with water, the weak Na–O bonds (0.17 valence units (vu)) are broken but the strong S–O bonds (1.5 vu) of the SO_4^{2-} groups remain intact. A similar effect occurs on melting.

This may not be the best place to introduce the role of lone pairs (Sect. 5), but it is necessary at this point to introduce the bonding strengths of anions which are determined by the octet rule (Eq. 11 in [1]). The important property of anions is that their negative valence is always less than their highest cation valence as can be seen in Table 1 in [1].

An unexpected result is that the bonding strengths of atoms in their highest valence state, S_E, reproduce the electronegativity scale (Eq. 8 in [1]), providing a direct link between chemical structure and electronegativity. If the core-and-valence-shell picture has already been discussed, one can also point out that the valence of an atom is equal to the charge on the core, and the coordination number scales as the surface area of the atom, i.e., it varies as the square of the radius. Consequently, the bonding strength of an atom in its highest valence state is proportional to the electric field of the core at its valence shell, the place where all the chemistry takes place.

Students can be offered exercises in deciding which atoms in a given chemical formula are cations and which are anions based on the rule that all the anions have higher electronegativities, S_E, than any of the cations, and that the sum of the valences of all the ions must be zero since the compound is electrically neutral. They could use the valence matching rule to show which atoms will be bonded. For example, what is the simplest compound containing Ca, Al, and O? Here clearly oxygen must be the anion ($S_E = 2.0$ vu) with Ca ($S_E = 0.27$ vu) and Al ($S_E = 0.57$ vu) as the cations. Since the valence of O is −2 vu, and Al is +3 vu, there must be at least two Al atoms giving a formula of $CaAl_2O_4$. Aluminum has the highest bonding strength ($S_{Al} = 0.57$ vu) with the bonding strength of oxygen next ($S_O = -0.50$ vu). Al–O bonds will form first with a valence of 0.5 vu corresponding to a coordination number around Al of six and around O of three since there are twice as many oxygen atoms as aluminum. The oxygen atoms have a residual valence of −0.5 vu leftover after the Al–O bonds are formed, which is enough to form two bonds each of 0.25 to calcium ($S_{Ca} = 0.27$ vu), leaving calcium with eight bonds. This gives a correct description of the bonding in one of the several forms which $CaAl_2O_4$ is known to adopt [2].

4 Bonding Geometry

With an understanding of bonds and bonding strength, this may be a good place to discuss bonding geometry, firstly by pointing out that the closer two atoms approach each other, the larger the flux linking them. This can be introduced by asking the students to draw the lines of field for an array of unequally spaced cations (A) and anions (B) such as that shown in Fig. 4. The correlation between bond valence and bond length is described in Chap. 3 of this volume and for many bond types it can be calculated using Eq. 4 in [1].

If the students have already been introduced to circuit theory in their physics courses, they may be familiar with the use of the Kirchhoff equations to solve electrical networks. By showing that a chemical bond in the ionic model is really an electric capacitor, the students can be shown that the ionic model is equivalent to a capacitive electric circuit that can be solved to predict the bond valences, hence the bond lengths. The only information needed for this exercise is the bond network, viz: how the atoms are linked (Sect. 6.1 of [1]). Such a presentation shows how results from one field (physics) can often be used in a different field (chemistry). Electrical circuits that have the same properties as a non-electrical system (in this

A B A B A B

Fig. 4 A chain of unequally spaced cations (A) and anions (B) having valences of +1 and −1 vu, respectively. The exercise is to draw the lines of field

Table 1 Frequently encountered coordination numbers to oxygen for atoms in their highest oxidation state

Group	1	2	13	14	15	16	17
Period 2	Li	Be	B	C	N		
	4,6	4	3,4	3	3		
Period 3	Na	Mg	Al	Si	P	S	Cl
	4–7	6	4,5,6	4	4	4	4

case a chemical bond network) are called equivalent circuits, because similar mathematical ideas are present. In both cases the node equation is a physical restriction that must be obeyed (the sum of the bond valences must equal the atomic valence) while at the same time the loop equation ensures that the valence of each atom is distributed among the bonds as equally as possible. The network equations illustrate why the bonds formed by an atom are not all the same length, but depend on the coordination numbers of all the atoms in the compound.

Knowing how to calculate bond lengths may not be important in an introductory class, but knowing how the bonds are arranged around an atom is. The popular explanation in terms of orbitals is misleading since assigning electrons to particular orbitals gives a false view of how electrons behave in an atom. If the orbital model were correct, all the elements in periods 2 and 3 of the Periodic Table would form four bonds pointing towards the corners of a tetrahedron, but not all of these elements adopt this arrangement as shown in Table 1. Nearly half of the elements that are tetrahedrally coordinated are also found with other coordination numbers. The orbital model explains three- and four-coordination by assuming the orbitals hybridize with triangular and tetrahedral symmetry, respectively, but one third of the elements in this table are found in five, six, or seven coordination which the orbital model cannot even explain, let alone predict. Higher coordination numbers are dismissed as being the result of "ionic bonding" with the implication that ionic bonds are somehow unpredictable, but even the orbital model is unable to predict whether an atom will adopt two, three, or four coordination; it only explains how the bonding can be described once the coordination number is known.

Even a cursory glance at the numbers in Table 1 shows that the real determinant of the coordination number is the size of the atom with the largest atom being sodium and smallest nitrogen. Some atoms, particularly the alkali metals, are found with more than one coordination number. This allows the atom to adopt a wider range of bonding strengths in order to match the bonding strength of the counterion. According to the principle of maximum symmetry (Eq. 1 in [1]) the bonds are assumed to adopt the most symmetric possible arrangement around the atom, ideally an arrangement in which all the bonds are equivalent by symmetry (Sect. 6.2 of [1]). For three coordination this is a triangle, for four it is a tetrahedron, and for six it is an octahedron. Five and seven coordination are rarely found as they have lower symmetry. There are other factors that affect the bond geometry. For example, the symmetry may be lowered by steric constraints or by electronic anisotropies such as the presence of stereoactive lone pairs described in Sect. 7 of [1] and Sect. 5.

The bonding of oxygen around zinc illustrates the way in which the same atom can adopt different coordination numbers. Zinc adopts a high symmetry coordination with four (tetrahedral) and six (octahedral) coordination, but not with five, where at least two symmetry-independent bonds are needed in a three-dimensional structure. Although zinc has an average coordination number of 5.0, only 15% of zinc atoms are found with five coordination while 58% are four coordinate and 35% are six coordinate. Zinc has a bonding strength of $2/5.0 = 0.40$ vu. When it is octahedrally coordinated it forms bonds with a valence of $2/6 = 0.33$ vu but when it is tetrahedrally coordinated it forms bonds with a valence of $2/4 = 0.50$ vu. Since oxygen has a bonding strength of -0.50 vu, ZnO favors the higher bonding strength and is tetrahedrally coordinated. On the other hand, carbonate has a bonding strength of only 0.22 vu, causing zinc in $ZnCO_3$ [3] to adopt octahedral coordination where the average bond valence of 0.33 vu is a better match.

5 Lone Pairs

The influence of lone pairs on geometry is a topic that cannot be avoided. This is a good place to introduce the subject since the discussion of the arrangement of bonds around an atom offers a natural entrée into a classroom discussion of the role of lone pairs. The good news is that the valence shell electron pair repulsion (VSEPR) model [4] is compatible with bond valence theory. VSEPR gives a good description of the geometry of atoms containing lone pairs that are fully stereoactive, but it does not address the question of why some lone pairs are inactive or only partially stereoactive. The only difference between the VSEPR model and bond valence theory lies in replacing the discrete electron pair explanation of the VSEPR model by a flexible distribution of valence electrons and lone pairs.

Lone pairs are found in some of the heavier cations, e.g., Tl^{+}, Pb^{2+} and Bi^{3+}, but by far the most important lone pairs are those found in anions, which always contain lone pairs because their anion valence is always less than their maximum cation valence (Sect. 5 of [1]). Oxygen is the most commonly found anion and its stereochemistry is dominated by its two lone pairs as described in Sect. 7.1 of [1]. The presence of lone pairs gives anions the flexibility to form bonds that are stronger than those normally allowed by the valence matching rule (Eq. 9 in [1]), but it can only do this if the bonding and lone pair electrons rearrange themselves within the valence shell, giving rise to the asymmetric coordination described by the VSEPR model. The influence of the lone pair as a function of the bonding strength of the cation is illustrated in Table 2 in [1] and Fig. 5. When the cation bonding strength is less than the anion bonding strength ($S_O = -0.5$ vu) all the bonds are equivalent and the lone pair is inactive. When the bonding strength of the

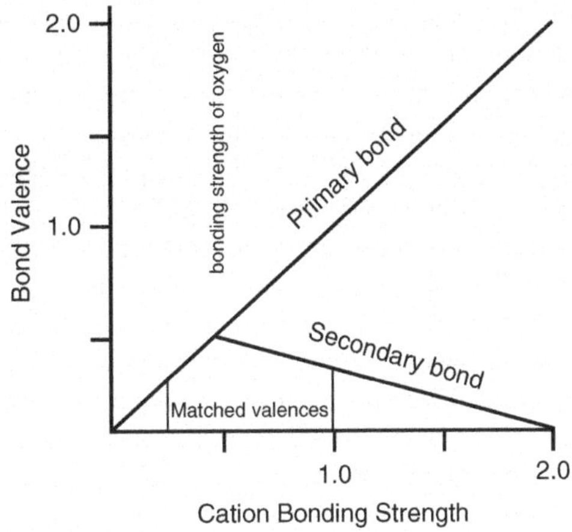

Fig. 5 The valences of the primary and secondary bonds in oxygen as a function of the bonding strength of the principal cation

cation is larger than that of the anion, the bonds split into short (primary) bonds and long (secondary) bonds. The primary bonds link to the strongly bonding cation, forming bonds whose valence is equal to the bonding strength of the cation and whose geometry is given by the rules of the VSEPR model. The secondary bonds are longer and link to cations with a lower bonding strength. As the cation bonding strength increases, so does the valence of the primary bonds causing the secondary bonds to become weaker, eventually leading to the case where the secondary bonds vanish as shown in Fig. 5. This latter is the situation described by the VSEPR model which ignores the presence of secondary bonds. Bond valence theory includes the VSEPR model but supplements its description of the arrangement of the ligands by predicting when the lone pairs will be stereoactive. It also predicts the valences and lengths of the primary and secondary bonds. The angles between the bonds are determined, not by the arrangement of electron pairs, but by the solid angles occupied by the electrostatic flux linking the core to the bonding electrons and lone pairs (Table 3 in [1]).

6 Ball-and-Stick (Covalent) Model

Although the ionic model is often thought to apply only to compounds composed of ionic bonds, Sect. 5 of [1] shows that its application is almost universal; only homoionic and delocalized bonds are excluded. The ionic model describes covalent and ionic bonds without distinction because the electrostatic flux depends only on the number of valence electrons and not on whether the bonding electrons lie closer to the anion or cation, or somewhere in between. It has been customary to label strong bonds as covalent, but attempts to define a covalent bond quickly run into problems.

The ideal covalent bond is most closely associated with the original ball-and-stick model in which the bond is assumed to be formed by a pair of electrons. Such bonds all have valences of 1.0 vu. This model can be derived from the ionic model or more generally from the core-and-valence-shell picture, by considering only those atoms whose valence is equal to their coordination number, a condition originally assumed in the ball-and-stick model of organic chemistry. In a compound where all the atoms obey this restriction, the bonds have a valence of exactly 1.0 vu (see Sect. 8 of [1]). This is the situation found for carbon and hydrogen in hydrocarbons where all the bonds are electron pair bonds and all are equivalent. The ball-and-stick model implicitly assumes that there is some intrinsic stability associated with the electron pair bond, perhaps because of the frequency with which these bonds are found in molecules, but attempts to extend the model beyond hydrocarbons involve introducing complications such as resonance hybrids and hydrogen bonding.

Bond valence theory assigns no special stability to electron pair bonds, but instead shows that the topological constraints responsible for the formation of molecules are also the constraints that favor for electron pair bonds. If C–H bonds have a valence of 1.0 vu, the hydrogen atom must terminate the bond network, so that all hydrocarbons are necessarily molecular. Similarly, halogens bonded to atoms with bonding strengths of 1.0 vu or greater can form only one bond; hence, they too will terminate the bond network and result in the formation of molecules. Because it is more difficult to visualize the extended structure of a solid than to visualize the structure of a molecule, the teaching and theory of chemical structure has tended to focus on the bonding in molecules. The unrecognized consequence is that this overemphasizes the importance of electron pair bonds and the consequent assumption that they have a special stability.

When organic molecules contain nitrogen or oxygen atoms, the conditions for the ball-and-stick model are no longer satisfied, and because these atoms have atomic valences that are too large to be saturated by an electron pair bond, they also form secondary bonds. In general, organic compounds containing these atoms do not strictly obey the rules of the ball-and-stick model, but they continue to be correctly described by bond valence theory which predicts, for example, how the formation of hydrogen bonds changes the chemistry of the molecule.

6.1 Ionic Versus Covalent

One of the many attractive features of bond valence theory is that it provides a single description of a bond that spans the whole range of bond types with valences ranging upwards from zero. Because bonds form a continuum, the theory makes no distinction between ionic and covalent bonds; all bonds obey the same set of rules. Yet the use of the terms “ionic” and “covalent” is so widespread that their usefulness cannot be questioned. Some might define covalent bonds as those that share bonding electrons and ionic bonds as those formed by the electrostatic attraction of oppositely charge ions, but this is unsatisfactory because it is impossible to know where the valence electron are. An alternative definition identifies a bond as ionic when the electronegativity difference between the terminal atoms is greater than some threshold. This is more precise, but the choice of threshold is always arbitrary. If it is useful to make the distinction, a convenient rule of thumb is to label the primary bonds formed by an anion with stereoactive lone pairs as covalent; and to label all other bonds as ionic. According to this definition, the S–O bonds in Na_2SO_4 are covalent, but Na–O bonds are ionic. The short (primary) bonds of 0.57 vu in corundum, Al_2O_3, would be called covalent, but the longer (secondary) bonds of valence 0.43 vu would be called ionic (see Table 2 in [1]). The exception would be the bonds in hydrocarbons which have no lone pair anions but are normally considered to be covalent. It is a simple rule which captures the essence of the terms while avoiding the temptation to assign degrees of ionicity or covalency to a particular bond. The bond valence is sufficient to identify where on the continuum a particular bond lies, and one should avoid the temptation to believe that the rules that apply to ionic bonds are different from those that apply to covalent bonds.

7 Lewis Acid and Base Strengths of Complex Ions

The presence of strongly bonding cations causes oxygen to form strong primary bonds, leaving the unused oxygen valence to form secondary bonds. This creates complex anions in which, for example, primary S^{6+}–O^{2-} bonds with valences of 1.5 vu create a complex SO_4^{2-} anion within which the bonds can be described as covalent (Sect. 6.1). The oxygen atoms have unused residual valence of −0.5 vu with which they form secondary bonds to a cation with a smaller bonding strength, such as Na^+. As this secondary bonding strength is equal to that of water, the Na–O bond is easily broken in aqueous solution, but the bonding strength of the S^{6+} atom is a poor match for water and the S–O bonds remain unbroken. The complex sulfate anion is not decomposed by the water and remains intact even in solution. The residual (external) valence of the sulfate anion is −2 and assuming each oxygen atom forms a total of four bonds, one primary and three secondary, the sulfate anion has a coordination number of twelve giving it a bonding strength of −0.17 vu. This is also a measure of the Lewis base strength of the sulfate ion.

In the case of the sulfate ion, all the oxygen atoms are equivalent and all have the same Lewis base strength, but in the disulfate ion, $S_2O_7^{2-}$, formed by two tetrahedral sulfate ions sharing a bridging oxygen atom (Fig. 10 in [1]), the bridging and terminal oxygen atoms are chemically distinct. The two bonds formed by the bridging oxygen atom to sulfur cannot have valences greater than 1.0 vu, and if their valences are both 1.0 vu, the atomic valence of oxygen is already saturated. The bonding strength (or Lewis base strength) of this oxygen is close to zero but that of the six terminal oxygen atoms is $-2/18 = -0.11$ vu. The bonding strength

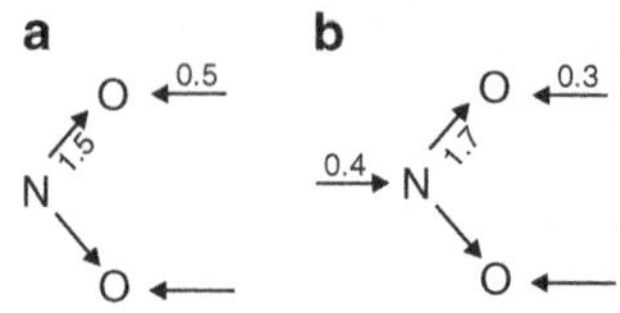

Fig. 6 The structure of (**a**) the free nitrite ion and (**b**) the nitrite ion coordinated to a cation. The valence of the bonds is shown. For simplicity the oxygen atoms are shown as forming only one external bond, but they normally would be expected to form three with valences of (**a**) −0.17 vu and (**b**) −0.10 vu. The bond valence is positive if the arrow points away from the atom and negative if it points towards the atom

of the disulfate ion is only −0.11 vu compared to the bonding strength of −0.17 vu of the sulfate ion.

More interesting is the behavior of the nitrite ion, NO_2^-. Within the complex ion the oxygen atoms are the anions ($V = -2$ vu) and the nitrogen is the cation ($V = +3.0$ vu). The N–O bonds have valences of 1.5 vu, which satisfies the valence of nitrogen but leaves the oxygen atoms each with a residual valence of −0.5 vu, as shown in Fig. 6a. The lone pairs on all the atoms are stereoactive, making the Lewis base strengths of the oxygen atoms (corresponding to their secondary bonds) equal to −0.17 vu, the same as that of the sulfate ion. The nitrogen has an unfilled coordination site opposite its stereoactive lone pairs, but as it is a cation and its valence is fully satisfied by the two N–O bonds, it has a Lewis acid strength of zero. Nevertheless, it can form bonds, typically to transition metals. These are traditionally described as dative bonds in which both electrons are supplied by the lone pair, but this explanation is an oxymoron, since lone pairs are, by definition, non-bonding, hence unable to form a bond. In practice, since nitrogen has a larger electronegativity than any of the transition metals, it can form an M–N bond in which nitrogen acts as the anion and M as the cation. Contrary to the dative bond explanation, it does not use its lone pairs for this purpose, rather it shares the external valence of the nitrite ion, −1.0 vu, which is normally expressed only by the oxygen atoms.

Figure 6b shows that although the nitrogen acts as a cation in relation to its bonds to oxygen, it has a Lewis base strength and acts as an anion with respect to the external bonds. The residual charge of −1 vu on the nitrite anion is distributed in Fig. 6b over all three atoms rather than just the two oxygen atoms as shown in Fig. 6a. Contrary to what one would expect, when the nitrogen atom forms an external bond, the N–O bonds get shorter, not longer. The net valence on N is still +3 and the lone pair is still non-bonding. In calculating the valence sum around nitrogen the valence of the external bond is subtracted from that of the internal bonds as suggested by the directions of the arrows in Fig. 6b. As the bonding electrons in Fig. 6b are more uniformly distributed around the nitrogen than in Fig. 6a, the lone pair is less stereoactive.

Water is an example of a complex which has a Lewis base function (through oxygen) and a Lewis acid function (through hydrogen). For a neutral complex such as water, the sum of the valences of the external bonds around the complex must be zero. The residual positive valence of the Lewis acid functions must equal the residual negative valence of the Lewis base functions. The formation of hydrogen bonds is only possible if the oxygen atom also forms external bonds with the same total valence.

8 Hydrogen, Hydrogen Bonds, and Water

At some point, every course on chemistry has to recognize that water is the most pervasive and important compound in terrestrial chemistry, since our very existence depends on its unique chemistry. Its properties arise from its ability to form hydrogen bonds whose unusual properties derive from the unique nature of the hydrogen atom: the only atom with no core electrons and no possibility of having a lone pair of electrons.

The unique properties of hydrogen and hydrogen bonds are described in Sect. 8.1.1 of [1]. Suffice it to say here that without an electron core, there is nothing to prevent the whole hydrogen atom, electron and nucleus, from moving into the overlap region, and without a lone pair, there is nothing to prevent hydrogen from forming a second bond (see Fig. 11 in [1]). Because the valence shells of the two atoms to which hydrogen is bonded are both full they are unable to overlap with each other, causing the O–H–O bond to be stretched. According to the distortion theorem (Eq. 5 in [1]), the hydrogen atom moves off center towards the atom with the larger Lewis base strength. The valences of the two bonds formed by hydrogen in the O–H. . .O hydrogen bond are 0.8 and 0.2 vu. Because the two O–H bonds in the H_2O molecule have a valence of 0.8 vu, the oxygen atom has a residual valence of −0.4 vu, which if the oxygen atom is four coordinate, gives it a Lewis base strength of −0.2 vu, leaving the water molecule ideally matched to itself.

The magnitude of the 8:2 valence split and the 180° O–H. . .O angle of the normal hydrogen bond can be predicted from the observed closest approaches of oxygen atoms in other compounds [5]. Because the hydrogen bond can be found in a wide range of chemical contexts, some hydrogen bonds are more, and some less, asymmetric than this norm (see Fig. 12 in [1]). Those that are more symmetric are linear and have the terminal atoms closer together. These are known as short or "strong" hydrogen bonds, though in reality because they are strained, they are weaker than normal hydrogen bonds even though the valence of the acceptor (longer) H. . .O bond is larger. The more asymmetric hydrogen bonds are longer and usually bent. The important chemistry of the hydrogen bond arises from the valence of the H. . .O bond still being large enough to have a significant effect on the chemistry, but weak enough to be easily broken. Clearly any hydrogen bond will be broken in aqueous solution as it is well matched to the bonding strength of water, a property that makes life possible.

Because the hydrogen bond is normally asymmetric, hydrogen has two different bonding strengths, 0.8 vu and 0.2 vu (see Table 1a in [1]), with the result that it often links two structural components with different Lewis base strengths. If one focuses only on isolated molecules, hydrogen is described as forming just one bond, but this overlooks the near ubiquitous presence of hydrogen bonding under ambient conditions. The formation of a hydrogen bond reduces the valence of the donor O–H bond from 1.0 to 0.8 vu, thereby activating the molecule's Lewis base functions by −0.2 vu for each hydrogen bond it forms. For example, ammonia, NH_3, would appear to be a simple stable molecule, but the hydrogen atoms will always form hydrogen bonds if any hydrogen bond acceptor is present, and in doing so, they activate the Lewis base function of the nitrogen atom, resulting in nitrogen being able to form a fourth bond with a valence of about 0.6 vu, and incidentally forcing its lone electron pair to lose its stereoactivity. Without the formation of hydrogen bonds, the nitrogen atom would be unable to act as a Lewis base because its 3.0 vu of valence are fully committed to the three N–H bonds. Many of the Lewis base functions of molecules depend on the formation of hydrogen bonds, and only if the hydrogen bonds are taken into account are the chemical properties of the molecule properly described.

Knowing the size of the bonding strengths of water leads directly to a discussion of the properties of solutions (Sect. 9 of [1]). In sodium chloride both ions have bonding strengths of 0.17 vu. They are not only well matched to each other but also well matched to water (Eq. 9 in [1]). Consequently NaCl not only readily dissolves in water, but it also readily crystallizes out when the water is removed. This is why the oceans are salty and inland seas are surrounded by salt flats. SiO_2 in contrast is insoluble because the bonding strengths of both silicon (1.0 vu) and oxygen (−0.5 vu) are less well matched to water than they are to each other. Although the silicon-oxygen match is not perfect, the Si–O bond is stable because oxygen has lone pairs which allows the bond to form as long as the lone pairs on oxygen become stereoactive, explaining why the Si–O–Si bonds in quartz are bent (Chap. 2.7.1). $MgCO_3$ has ions that are well matched to each other (Mg^{2+}: 0.33 vu and CO_3^{2-}: −0.22 vu) but less well matched to water giving it a very low solubility. Other examples of solubility are discussed in Sect. 9.2 of [1].

9 Reactivity

A compound in which the valences are well matched is likely to be stable, whereas one in which the bonding strengths differ by a factor of two is at its limit of stability and likely to react with any compound that will lead to a better valence match. The solubility of simple compounds is governed by the reaction between the compound and water. It should provide students with an opportunity to predict which compounds will dissolve in water, and what kinds of solids will be found if the water is removed. Compounds in which the cation has a much larger bonding strength than the anion crystallize with hydrated cations as this lowers the bonding strength of the cation, while those in which the anion has the larger bonding strength tend to crystallize with hydrogen atoms attached to the anion as this tends to lower the bonding strength of the anion.

For example, consider what structure might result from a combination of Mg^{2+}, Cl^{7+}, and O^{2-}. Chlorine has the largest bonding strength of 1.75 vu and it must make four such bonds to oxygen to give ClO_4^- which has a bonding strength of 1/12 = 0.08 vu (see Table 1c in [1]). This bonding strength is too small to form bonds with magnesium (S_{Mg} = 0.33 vu) so anhydrous $Mg(ClO_4)_2$ is not easy to prepare and if it is prepared it will be hygroscopic, drawing in water to form $Mg(H_2O)_6$ complex ions with a bonding strength of 0.17 vu. When the water is removed the anhydrous compound is not recovered, but rather crystals of the well-matched $Mg(H_2O)_6(ClO_4)_2$ are formed. Other examples can be found in Sect. 10 of [1].

10 Epilogue

The bond valence theory is a chemist's model of chemical structure. It gives a graphical picture of the chemical bond based on traditional electron counting rules. It uses classical physics concepts because, since the atoms in compounds under ambient conditions are in their ground state, quantum descriptions are in most cases not needed. The model is derived from a picture of the atom that is simple enough for an introductory course, but physically accurate enough to allow it to be used as the basis for introducing the full quantum treatment when this is required. The theory leads naturally into the more advanced quantum description that students will meet later, but it avoids the unphysical assumptions inherent in the simplified orbital and Lewis models often found in introductory texts.

It is a theory that involves only simple mathematics – a computer is rarely necessary – yet its derivations lead to theorems that make it far more predictive than most other models of chemical structure. These are not just a set of simplified rules designed to provide an easy introduction to chemical structure, but theorems that have shown themselves to be robust in solving complex research problems such as understanding the chemistry that occurs at surfaces when crystals grow or dissolve (see Chaps 6–9 of this volume), or with the diffusion of electrically conducting ions through amorphous solids described in Chap. 5 of this volume.

Clearly bond valence theory does not contain the whole story. It cannot be used to calculate energies or determine the distribution of the electrons that give rise to bonding. This information requires a quantum treatment using the Coulomb potential, but bond valence theory provides information that quantum theory cannot. It can be used to determine the atomic positions that are required before quantum calculations can be made. If the quantum approach is a physicist's description of the atom, the bond valence theory is the corresponding chemist's description of the bond.

Bond valence theory provides more quantitative predictions of structure and reactivity than other models currently in the chemical curriculum. At the same time it gives a picture of bonding that allows problem structures to be analyzed in physical terms. In short, it provides the long-missing theory behind the popular but empirical model of the chemical bond.

References

1. Brown ID (2013) Chapter 2 of this volume
2. Lazic B, Kahlenberg V, Konzett J, Kaindl R (2006) Solid State Sci 8:589–597 (ICSD 172781)
3. Effenberger H, Mereiter K, Zeeman J (1981) Z Kristallogr 156:233–243 (ICSD 100679)
4. Gillespie RJ, Hargittai I (1991) The VSEPR model of molecular geometry. Allyn and Bacon, Needham Heights
5. Brown ID (2002) The chemical bond in inorganic chemistry: the bond valence model. Oxford University Press, Oxford, pp 77 ff

Appendix A: Glossary

This appendix defines the terms and conventional symbols used in bond valence theory. Words written in italics are defined in this appendix.

Actual coordination number The *coordination number* observed for an atom in a given compound

Anion The atom with the larger *electronegativity* forming a *bond*. Anions are uncharged except in the *ionic model* where the anion carries a negative charge equal to its negative *valence*

Anion bonding strength, S_B The expected value of the *valence of a bond* formed by an *anion*. It can be estimated by dividing the anion's (negative) *valence* by its expected *coordination number*. This is the same as *Lewis base strength*

Anion valence, V_B The *valence of an atom* when acting as an *anion*. V_B is a negative number

Antibonding region A region on the side of an atom opposite to the *bond*. Electrons in this region tend to weaken the bond

Atom type An atom of a particular element with a particular *valence* (oxidation state)

Atomic valence, V The number of electrons in the *valence shell* that an atom uses in bonding. The units of atomic valence are *valence units*, equivalent to one bonding electron. *Cation valences* are taken as positive, *anion valences* as negative

Bipartite graph The graph of a *bond network* in which the atoms are of two kinds (e.g., *cation* or *anion*) with no bonds occurring between atoms of the same kind

Bond The chemical link between two neighboring atoms. In the *ionic model*, two atoms are bonded if and only if they are linked by *electrostatic flux*

Bond network A topological description of the way in which atoms in a system are linked by *bonds*

Bond strain index Root mean square deviation between the observed and *ideal bond valence* averaged over all the *bonds* in the formula unit

Bond valence, S The number of *valence electrons* an atom uses to form a given *bond*. In the *ionic model* it is equal to the *electrostatic flux* linking two ions. The units of *bond valence* are *valence units (vu)*, equivalent to one bonding electron pair. Bond valences are considered to be directed from the *cation* to the *anion*

Bonding electrons The electrons in the *valence shell* of an atom that are used for forming *bonds*

Bonding region The region between two atoms where the *valence shells* overlap. Electrons in this region form the *bond*

Bonding strength An estimate of the *valence of the bonds* formed by an ion. See *anion*- and *cation-bonding strength, Lewis acid*- and *base-strength*

Cation The atom with the smaller *electronegativity* forming a *bond*. Cations are uncharged except in the *ionic model* where the cation carries a positive charge equal to its positive *valence*

Cation bonding strength, S_A The expected value of the *valence of a bond* formed by a *cation*. It can be estimated by dividing the *cation's (positive) valence* by its expected *coordination number*. This is the same as the *Lewis acid strength*

Cation valence, V_A The *valence of an atom* when acting as a *cation*. V_A is a positive number

Chemical bond A localized interaction between two neighboring atoms. See *bond*

Complex A group of two or more atoms linked together by strong (*primary* or *covalent*) bonds. Complexes may be ions carrying a formal charge (*residual valence*) or neutral molecules

Coordination number, N The number of *bonds* formed by, or the number of atoms surrounding, a given atom. See *actual coordination number*

Coordination sphere The environment of an atom including all its *ligands*

Covalent bond A bond may be referred to as a covalent bond if it has a large *bond valence*. A *primary bond* is usually regarded as a covalent bond.

Electron pair bond A *bond* of *valence* 1.0 *vu* involving exactly one pair of electrons

Electronegativity, S_E The property of an element that gives a measure of the electric potential at the surface of the atomic core. The scale used in the bond valence model sets the electronegativity equal to the maximum *(cation) bonding strength* of the atom. This is equal to the number of electrons in the *valence shell* divided by the average *coordination number* adopted by the atom in this valence state when bonded to oxygen ligands

Electrostatic flux The electrostatic flux linking two atoms in the *ionic model* is the *bond valence*

External bond A *bond* formed between an atom in a *complex* and an atom that is not part of the *complex*. In general the *valence* of an external bond is smaller than that of a *bond* internal to the complex

Global instability index, G (also GII) The root mean square deviation of the *bond valence* sums from the *atomic valences* averaged over all atoms in the formula unit

Heteroionic bond A *bond* formed between a *cation* and an *anion*

Homoionic bond A *bond* formed between two *cations* or between two *anions*

Ideal bond valences *Bond valences* predicted for a *valence compound* using the *network equations*

Ionic model A model in which every atom is assigned as either a *cation* or an *anion*, each carrying a charge equal to its *valence*. A *bond* exists in this model between two atoms if they are linked by *electrostatic flux*. This model can be applied to any *valence compound* without regard to whether the bonding is ionic or covalent

Lewis acid strength, S_A The *cation bonding strength*, specifically when this is the *bonding strength* for the *external bonds* formed by an atom of a *complex*. It is equal to the *residual valence* of an atom acting as a Lewis acid divided by the expected number of *external bonds*

Lewis base strength, S_B The *anion bonding strength*, specifically when this is the *bonding strength* for the external bonds formed by an atom of a *complex*. It is equal to the *residual valence* of an atom acting as a Lewis base divided by the expected number of *external bonds*

Ligand An atom bonded to the atom under consideration

Lone pair A pair of *valence shell* electrons not involved in bonding

Network equations The Kirchhoff equations used to solve the capacitive electric circuit equivalent to the *bond network* of a *valence compound*

N_O The average *coordination number* when oxygen is the *ligand*

Overbonded Refers to an atom whose *bond valence sum* exceeds its *atomic valence*

Overlap region The region in which the *valence shells* of two atoms overlap to form a *bond*

Primary bond A bond with a larger than average *valence*, formed by an atom with one or more stereoactive *lone pairs* of electrons

Residual valence The valence available to an atom in a *complex* for forming *external bonds* after the valences of the internal bonds have been satisfied

S_A *Lewis acid strength*, see *cation bonding strength*

S_B *Lewis base strength*, see *anion bonding strength*

S_E *Electronegativity*

Secondary bond A bond with a smaller than average *valence*, formed by an atom with one or more stereoactive *lone pairs* of electrons

Surface instability index The root mean squared deviation of the *bond valence sum* from the *atomic valence* averaged over the atoms forming the surface of a solid

Underbonded Refers to an atom whose *bond valence sum* is less than its *atomic valence*

Valence of an atom, see *Atomic valence*

Valence of a bond, see *Bond valence*

Valence compound A compound having a *bond network* with a *bipartite graph*. All the bonds have an *anion* at one end and a *cation* at the other

Valence shell The outer electron shell of an atom containing the electrons involved in bonding. It contains a number of *bonding electrons* equal to the *valence of the atom* and possibly non-bonding *lone pairs*

Valence unit (vu) 1 valence unit of *atomic valence* corresponds to one electron, 1 vu of *bond valence* corresponds to one pair of electrons since each atom contributes one electron to the *bond*

Valence vector A vector with the magnitude of the *bond valence* directed from the *cation* to the *anion*

V_A *Cation valence* (positive)

V_B *Anion valence* (negative)

V_E *Lone pair*

V_R *Residual valence*

vu *Valence unit*

Appendix B: Programs Using Bond Valences

The appendix lists a number of programs and other sources that can be used for calculating and manipulating bond valences.

BOND_STR
Author: Rodriguez-Carvajal J
Physica B (1993) 192:55–69
On http://extras.springer.com (links to extra material server) this list holds links to the computer programs
FULLPROF, implemented in FULLPROF, calculates bonding geometry and bond valences using CFL or CIF input files.

BONDVAL (Bond Valence Wizard)
Authors: Orlov IP, Popov KA, Urusov VS
J Struct Chem (1998) 39:575–579
On http://extras.springer.com (links to extra material server) this list holds links to the computer programs
Bondval solves the network equations for a given bond network entered from the keyboard

KDist
Author: Knizek K
On http://extras.springer.com (links to extra material server) this list holds links to the computer programs
A freeware program for Windows, implemented in Kalvados, is used with powder diffraction and crystal structure files. It can read structures from CIF and DFT output files and has a function to calculate bond valence sums for the structure. The user can modify or update (by hand) what R_0 values you wish to use. It also has an option to optimize structures based on bond valence sums.

SoftBV
Author: Adams S
On http://extras.springer.com (links to extra material server) this list holds links to the computer programs
An online program for calculating bond valences together with a discussion of bond valence parameters and their dependence of atom softness.

SPuDS
Authors: Lufaso MW, Woodward PM
Acta Cryst (2001) B57:725–738
On http://extras.springer.com (links to extra material server) this list holds links to the computer programs
A program for predicting the structures of perovskite related-compounds. See Chap. 4 in this volume for details.

VALENCE
Author: Brown ID, Hormillosa C, Healy S, Stephen T
J Appl Cryst (1996) 29:489–480
On http://extras.springer.com (links to extra material server) this list holds links to the computer programs
Calculates bond valences for bond lengths entered from the keyboard and vice versa. It has other features that allow bond valences sums and average bond lengths to be calculated. It can also be used to determine bond valence parameters for a series of coordination environments entered on the keyboard. Only available in DOS or C++ source code. A Windows interface is needed and would be welcomed.

VALMAP
Author: Gonzales-Platas J, Gonzales-Silgo C, Ruiz-Peres C
J Appl Cryst (1998) 31:826–827
J Appl Cryst (1999) 32:341–344
On http://extras.springer.com (links to extra material server) this list holds links to the computer programs
A program for plotting maps based on the valence sum of a target atom placed at arbitrary points in the crystal. Various functions are available for calculating the bond valences or other quantity such as valence differences.

VaList
Author: Wills AS
On http://extras.springer.com (links to extra material server) this list holds links to the computer programs
VaList requires an input list of bond lengths in CIF, GSAS, ICSD or FULPROF formats. It calculates bond valences and their sums around individual atoms as well as the global instability index and occupation numbers where a site is occupied by two different atoms.

3DBVSMAPPER
Authors: Sale M, Avdeev M
J Appl Cryst (2012) 45:1054–1056
A program for automatically generating bond-valence landscapes embedded in the Accelrys Materials Studio

BOND VALENCE PARAMETERS (see also SoftBV)
On http://extras.springer.com (links to extra material server) this list holds links to the computer programs
Compiled by I.D. Brown
A compendium of some 1,700 published (and some unpublished) bond valence parameter sets. The user should consult the original papers to confirm the suitability of these parameters for their own purposes.

Index

The manufacturer's authorised representative in the EU is Springer Nature Customer Service Centre GmbH, Europaplatz 3, 69115 Heidelberg, Germany. If you have any concerns regarding our products, please contact ProductSafety@springernature.com

Printed and bound by CPI Group (UK) Ltd, Croydon, CR0 4YY
22/09/2026
02221063-0014